Solar Resources

Solar Heat Technologies: Fundamentals and Applications
Charles A. Bankston, editor-in-chief

1. *History and Overview of Solar Heat Technologies*
Donald A. Beattie, editor

2. *Solar Resources*
Roland L. Hulstrom, editor

3. *Economic Analysis of Solar Thermal Energy Systems*
Ronald E. West and Frank Kreith, editors

4. *Fundamentals of Building Energy Dynamics*
Bruce Hunn, editor

5. *Solar Collectors, Energy Storage, and Materials*
Francis de Winter, editor

6. *Active Solar Systems*
George Löf, editor

7. *Passive Solar Buildings*
J. Douglas Balcomb and Bruce Wilcox, editors

8. *Passive Cooling*
Jeffrey Cook, editor

9. *Solar Building Architecture*
Bruce Anderson, editor

10. *Fundamentals of Concentrating Systems*
Lorin Vant-Hull, editor

11. *Distributed and Central Receiver Systems*
Lorin Vant-Hull, editor

12. *Implementation of Solar Thermal Technology*
Ronal Larson and Ronald E. West, editors

Solar Resources

edited by Roland L. Hulstrom

The MIT Press
Cambridge, Massachusetts
London, England

This book was set in Times Roman by Asco Trade Typesetting Ltd., Hong Kong, and was printed and bound by Halliday Lithograph in the United States of America.

Library of Congress Cataloging-in-Publication Data

Solar resources/edited by Roland L. Hulstrom.

p. cm.—(Solar heat technologies: fundamentals and applications; 2)
Includes index.
ISBN 0-262-08184-9
1. Solar energy—United States. 2. Solar energy—Research—United States.
I. Hulstrom, Roland. II. Series: Solar heat technologies; 2
TJ809.95.S68 vol. 2.
[TJ811.5.U6]
621.47—dc 19 88-37308
CIP

Contents

Series Foreword

Charles A. Bankston

This series of twelve volumes summarizes research, development, and implementation of solar thermal energy conversion technologies carried out under federal sponsorship during the last eleven years of the National Solar Energy Program. During the period from 1975 to 1986 the U.S. Department of Energy's Office of Solar Heat Technologies spent more than $1.1 billion on research, development, demonstration, and technology support projects, and the National Technical Information Center added more than 30,000 titles on solar heat technologies to its holdings. So much work was done in such a short period of time that little attention could be paid to the orderly review, evaluation, and archival reporting of the significant results.

It was in response to the concern that the results of the national program might be lost that this documentation project was conceived. It was initiated in 1982 by Frederick H. Morse, director of the Office of Solar Heat Technologies, Department of Energy, who had served as technical coordinator of the 1972 NSF/NASA study "Solar Energy as a National Resource" that helped start the National Solar Energy Program.

The purpose of the project has been to conduct a thorough, objective technical assessment of the findings of the federal program using leading experts from both the public and private sectors, and to document the most significant advances and findings. The resulting volumes are neither handbooks nor textbooks, but benchmark assessments of the state of technology and compendia of important results. There is a historical flavor to many of the chapters, and volume 1 of the series will offer a comprehensive overview of the programs, but the emphasis throughout is on results rather than history.

The goal of the series is to provide both a starting point for the new researcher and a reference tool for the experienced worker. It should also serve the needs of government and private-sector officials who want to see what programs have already been tried and what impact they have had. And it should be a resource for entrepreneurs whose talents lie in translating research results into practical products.

The scope of the series is broad but not universal. It is limited to solar technologies that convert sunlight to heat in order to provide energy for application in the building, industrial, and power sectors. Thus it explicitly excludes photovoltaic and biological energy conversion and such thermally driven processes as wind, hydro, and ocean thermal power. Even with this limitation, though, the series assembles a daunting amount of information. It represents the collective efforts of more than 200 authors and editors. The volumes are logically divided into those dealing with general topics, such as the availability, collection, storage, and economic analysis of solar energy, and those dealing with applications.

The present volume covers the solar radiation resource. A knowledge of the total available solar radiation is, of course, essential to the design and evaluation of solar thermal energy conversion applications. In addition, the development of radiation concentration devices and some optically selective and radiation-sensitive materials requires detailed information on the spatial and spectral composition of the incident radiation. The U.S. government has supported an assessment of the solar resource since 1975. The assessment has included field measurements and monitoring, model and algorithm development, instrument development and calibration, and the compilation and distribution of insolation data bases. This book presents that work and the results it produced—detailed information on models and algorithms, descriptions of insolation data bases, and the means for obtaining and using such data bases. Volumes 5–11 of this series offer more information on the use of these data in the design and analysis of various solar applications.

Preface

Roland L. Hulstrom

Following the 1973 oil embargo, the U.S. government initiated a program to develop and use solar energy. This led to individual programs devoted to developing various solar radiation energy conversion technologies: photovoltaic and solar-thermal conversion devices. Nearly concurrently, it was recognized that understanding the available insolation resources was required to develop and deploy solar energy devices and systems. It was also recognized that the insolation information available at that time (1973) was not adequate to meet the specific needs of the solar energy community. Federal efforts were initiated and conducted to produce new and more extensive information and data. The primary federal agencies that undertook such efforts were the Department of Energy (DOE) and the National Oceanic and Atmospheric Administration (NOAA). NOAA's efforts included activities performed by the National Weather Service (NWS) and the National Climatic Data Center (NCDC).

This book has two main objectives: to report some of the insolation energy data, information, and products produced by the federal efforts and to describe how they were produced. Products include data bases, models and algorithms, monitoring networks, instrumentation, and scientific techniques. The scope of products and results does not include all those produced by past federal efforts. The book's scope and subject matter are oriented to support the intent and purpose of the other volumes in this series. In some cases, other pertinent material is presented to provide a more complete coverage of a given subject.

Acknowledgments

Roland L. Hulstrom

First, I must thank my authors, who devoted a great deal of their personal time to perfecting their chapters. I must also recognize a group of individuals who provided encouragement and support for the book: Charles A. Bankston, CBY Associates, Inc.; Paul Berdahl, Lawrence Berkeley Laboratory; Eldon Boes, Sandia Laboratory; Eugene Clark, Trinity University; Kinsell Coulson, Mauna Loa GMCC Observatory; John Davies, McMaster University; Edwin Flowers, U.S. Department of Commerce—NOAA/SRF; Bernard Goldberg, Smithsonian Radiation, Biology Laboratory, John Hay, University of British Columbia; John Hickey, The Eppley Lab, Inc.; S. A. Klein, University of Wisconsin—Madison; Fred Koomanoff, DOE; Frank Quinlan, National Climatic Data Center; Ari Rabl, Princeton University; Nagarajo Rao, Oregon State University; Michael Riches, DOE; Michael Rigney, U.S. Department of Commerce—NOAA/NWS; Ronald Stewart, State University of New York—ASRC; Valentine Szwarc, Solar Energy Research Institute; Donald Watt; and John Woo.

Fred Koomanoff, Michael Riches, Mort Prince, and Lloyd Herwig of DOE are recognized for their management and support of the DOE solar radiation energy resource assessment project. Frederick H. Morse, director of the Office of Solar Heat Technologies, DOE, was responsible for including this book as part of the Solar Heat Technologies: Fundamentals and Applications project.

Solar Resources

1 Introduction

Roland L. Hulstrom

Introduction

The earth receives vast amounts of energy from the sun in the form of incident solar radiation (insolation). Solar radiation drives the earth's weather and sustains life. As a result of increasing costs, uncertain availability, and potentially severe environmental impacts of other energy sources, solar energy emerges as a clean and renewable energy source for man's heating, cooling, electricity, and fuel needs.

For many applications, however, the insolation must be converted to a different form, such as heat, electricity, or a fuel (liquid, gas, or solid)—a technology known as solar energy conversion. Several solar energy technologies are currently being developed, including photovoltaics (converting sunlight to electricity), solar thermal (converting sunlight to heat), solar thermal-electric (converting sunlight to heat and then to electricity), and biomass (converting sunlight to biomass and then to fuels). The ultimate goal of developing solar energy conversion technologies is to provide viable ones that can produce suitable, reliable, and cost-competitive energy sources.

Since solar energy conversion devices change insolation into other forms of energy, insolation is *the* energy resource for solar energy conversion. The amount and characteristics of the insolation, along with the device's energy conversion characteristics, directly determine the amount and characteristics of the useful heat or power produced. The cost and technical characteristics of the heat and power produced determine the viability of a solar energy conversion technology in the energy economy. Therefore, understanding the nature of insolation is necessary before the nation can properly develop solar energy technologies and understand their role in overall national energy security and stability.

1.1 The Federal Role

In 1974, the formal federal role in assessing and characterizing solar energy was authorized by the Solar Energy Act, P.L. 93–473 Sec. 5., entitled "Resource Determination and Assessment," which established "... a solar energy resource determination and assessment program with the objective of making regional and national appraisal of all solar energy resources, including data on insolation, wind, sea thermal gradients, and potentials for photosynthetic conversion."

The goals defined by the 93rd Congress as part of this law were to develop

1. better methods for predicting the availability of all energy resources over long periods and in different geographic locations,
2. advanced meteorological, oceanographic, and other instruments, methodology, and procedures necessary to measure the quality and quantity of all solar resources periodically, and
3. activities, arrangements, and procedures for collecting, evaluating, and disseminating information and data relating to solar energy resource assessment.

The U.S. Department of Energy (DOE) has the federal role in researching and developing solar energy conversion devices because it is the lead federal agency for energy.

1.2 The History of Federal Insolation Resource Assessment

New, significant federal efforts in insolation resource assessment resulted from the oil embargo in 1973. Before this event, insolation was generally considered as one of the many climate variables having significance in meteorology, agriculture, and the environment. After the oil embargo, insolation was considered an important national energy resource.

To develop any energy resource requires fully understanding, characterizing, and assessing it; hence, following the 1973 oil embargo, the federal government began to study this resource. Between 1973 and 1975, the National Science Foundation (NSF) was responsible for defining and coordinating the federal insolation resource assessment efforts. As part of these efforts, NSF awarded a grant to the National Oceanic and Atmospheric Administration (NOAA), including the National Weather Service (NWS), to conduct selected activities. NSF and NOAA essentially set out to achieve the solar energy resource assessment goals set forth by the 93rd Congress.

In FY (Fiscal Year) 1976, the newly formed Energy Research and Development Administration (ERDA) assumed the federal role for insolation resource assessment, which was part of a broader program entitled the Environmental and Resource Assessment Program (ERAP). As part of this program, ERDA and NOAA signed an Interagency Agreement (preceded by a Memorandum of Understanding) that established a NOAA role to provide improved understanding of the environment, effective monitoring and prediction, and increased knowledge of the ocean's resources and the nature of the resources. NOAA's tasks were to establish a new national insolation monitoring network, to correct previous (1952–1975) NWS data, to generate estimated insolation using models, to produce a national insolation data base, and to disseminate this information.

In late 1977, ERDA became DOE, which continued funding ERAP through FY 1978. In FY 1979, DOE discontinued ERAP as a formal, line-item budget program, but established a semiformal program, entitled the Insolation Resource Assessment Program (IRAP), to continue these assessment efforts. The DOE management responsibilities were assigned to the Division of Distributed Technology, Photovoltaic Energy Systems Branch. Subsequently, DOE assigned the technical guidance and day-to-day management responsibilities to the Solar Energy Research Institute (SERI). The DOE Photovoltaics Energy Systems Branch obtained IRAP funding from various DOE technology programs.

During early 1979, DOE and NOAA signed a Letter of Understanding stipulating that DOE would (1) fund the operation of the NOAA/NWS insolation monitoring network through 1980, (2) support the NOAA budget request for FY 1981 and beyond before the Office of Management and Budget (OMB) and Congress, (3) fund any special requirements beyond FY 1981, and (4) consult with NOAA on insolation monitoring matters to maximize the benefits to the solar energy technology community. NOAA agreed to request funds in FY 1981 to maintain and operate a network and to respond to proposals that support other federal or nonfederal users of solar resources on processing or presenting (forecasting) solar information.

The semiformal IRAP continued through FY 1980, and the DOE Photovoltaics Energy Systems Branch was able to continue a variety of efforts in insolation resource assessment. One of these was to build a laboratory and staff at SERI, managed in the Resource Assessment and Instrumentation Branch.

In FY 1981, IRAP was essentially discontinued because funding for all solar energy programs decreased sharply. The funding obtained, mostly from the DOE Photovoltaics Program, was allocated to SERI's Resource

Assessment and Instrumentation Branch to maintain a core national laboratory and the national data bases and to perform research in high-priority, selected areas. This situation remained about the same through FY 1985.

In 1984, the House Science and Technology Committee, the American Solar Energy Society (ASES), and SERI's Resource Assessment and Instrumentation Branch concluded that the understanding of and the existing information on insolation resources were not adequate to meet the needs of the solar technologies and future applications. They recommended an expanded and focused federal role. In FY 1986, the Resource Assessment Program was reestablished and assigned to DOE. Subsequently, SERI's Resource Assessment and Instrumentation Branch became the DOE lead center for insolation resource assessment. Note that throughout this entire period (from 1973 to the present), daylighting resource assessment and research efforts were the responsibility of DOE's Solar Buildings Program. Chapter 8 discusses such efforts and their results in more detail.

Since 1973, many groups have stated the need for insolation resource assessment and characterization, including NSF, ERDA, the Electric Power Research Institute (EPRI), ASES, the U.S. House of Representatives Committee on Science and Technology, and SERI. The current DOE-SERI Resource Assessment Program is following their recommendations by addressing the following fundamental needs:

1. a national historical insolation data base,
2. a national benchmark monitoring network, and
3. research data sets and models that fully describe and predict the nature of the insolation resources.

The national historical insolation data base provides a collection of fundamental insolation and meteorological data encompassing all major climates gathered over a long time at several sites in the United States. This data base is used to guide the research and development of solar energy technologies, assess typical field performance of solar energy systems, and assess the regional insolation resources across the United States. The specific objectives of such a data base are that they include

1. the fundamental components, consisting of the hourly direct beam component and the total (i.e., global) and diffuse (i.e., sky) insolation on a horizontal surface,

2. monthly mean bias errors of 5%,

3. a period of record of 30 years, 1961 to 1990, and

4. approximately 300 stations.

The national monitoring network provides highly accurate continuous measurements of the fundamental insolation components and related parameters. Such data will be used to update continuously and build the historical data base and to document any trends in the insolation resources. Such trends may be a result of changes in cloudiness, natural events (e.g., volcano eruptions), or air pollution. The specific requirements of the national monitoring network are to

1. measure the hourly direct beam component and total and diffuse insolation,

2. measure insolation with a better than 5% accuracy,

3. operate continuously,

4. measure fundamental related parameters, including cloud cover and temperature, and

5. have approximately 30–40 benchmark stations that represent the major climates across the United States.

DOE, SERI, NOAA, and NWS cooperate to ensure the operation of the national insolation monitoring network and the dissemination of archival data.

A variety of research insolation data sets and models are being developed to describe fully and predict the nature of the insolation resources. In many cases, the research data sets are necessary to develop the models. These research data sets and models are needed to

1. characterize the small-scale spatial characteristics of insolation and predict site-specific insolation resources,

2. characterize and spectral nature of the solar radiation resource,

3. characterize the short-term (less than 1 hour) temporal nature of the insolation resource,

4. characterize the spatial distribution of solar radiation across the sky dome and near the solar disk,

5. convert fundamental insolation data, as described for the national data

base and monitoring network, to the insolation available to a variety of solar collectors (e.g., building walls and windows, tilted flat-plate collectors, concentrators),

6. forecast the insolation resources available to a solar collector and system over 1-hour to 3-day periods, and

7. simulate insolation characteristics given basic meteorological data.

These capabilities will provide data and models to guide the research and development of solar energy conversion technologies; to optimize and evaluate candidate designs, devices, and systems; to evaluate field performance of candidate devices and systems; to select sites for systems and plants; to assess solar resources on a small scale; and to provide information on operating and managing a solar system, plant, or both after installation.

1.3 Book Overview

The remainder of this book covers insolation data bases and resources in the United States, insolation models and algorithms, insolation monitoring networks, solar radiation instrumentation, spectral terrestrial solar radiation, insolation forecasting, and illuminance models and resources in the United States.

Chapter 2 provides a listing and describes available insolation data sets and products and sources of information pertaining to U.S. insolation resources. It also sets forth how to obtain such information. The description of each data set or product allows the reader to decide whether or not the data set or product would be useful. This listing of data sets and products includes major ones produced snce 1976 and other pertinent data sets. With respect to U.S. insolation resources, examples of available products and sources of more detailed information are provided. Chapters 3, 4, and 5 provide further details regarding the production and nature of the insolation data sets and products.

Chapter 3 presents an overview of how solar radiation transferring through the earth's atmosphere is modeled. Emphasis is given to describing algorithms for generating sequential insolation data sets. Specifically, the chapter describes the various models and algorithms used to generate the insolation data sets identified in chapter 2. It also addresses the very important aspect of accuracy in the data sets. A portion of the chapter addresses the methods used to generate the well-known (see chapter 2) SOLMET (*SOL*ar *MET*eorological) and ERSATZ data bases. These data bases are currently in common use by the solar energy community. It is important that the users of these data bases understand how they were generated and their accuracy. The chapter also covers algorithms that relate global-horizontal insolation to its components of direct and diffuse insolation and algorithms that convert diffuse-horizontal insolation to insolation on tilted surfaces. As with any technical field, improvements are constantly being sought and produced. Hence, by the time this book is published, improved insolation models and algorithms will most likely be available. The reader should examine the current literature to supplement this information.

Chapter 4 presents a historical overview of mainly the NWS national insolation monitoring network(s). This chapter is unique in that it provides the reader with valuable insights regarding the national monitoring network and the data used to generate the SOLMET and ERSATZ data bases described in chapter 2. Chapter 4 also includes descriptions of other major monitoring networks in the United States. Monitoring networks tend to be dynamic; therefore, the reader should supplement this chapter by investigating current monitoring networks operating across the United States. Sources of information include state climatologists, certain universities, utilities, NWS, and SERI.

Chapter 5 covers the instrumentation used to measure total solar irradiance and spectral solar irradiance. The chapter by itself could serve as a textbook on this subject. It gives an extensive overview of the many different types and manufacturers of solar irradiance instrumentation and calibration procedures and methods and discusses the accuracy of the resultant data. It also covers the very important aspect of national and international standardization. The solar and meteorological communities need to pay strict attention to these subjects to obtain high-quality insolation data.

Chapter 6 presents an overview of the nature of spectral solar irradiance at the earth's surface, some recent models for such, and some examples of measured data. It does not treat these in great detail because the spectral distribution of the insolation resource to solar thermal technologies is not considered as important as the other subjects in this book. However, the chapter does contain a fairly extensive reference list for those wanting more complete information.

Chapter 7 describes the NWS Operational Solar Insolation Forecast System. This system, developed through a joint DOE-NOAA project started in 1979, predicts the daily total global-horizontal insolation for 2 days. This chapter contains contour maps (one for each day) of the daily total insolation, with individual values given for many cities across the United States. As solar energy systems become more common, forecasts of daily insolation will become more widely used to help operate and manage such energy systems.

Chapter 8 is an overview of daylighting models and resources. It provides a summary of DOE-funded efforts and is intended to be independent of the other chapters; however, it describes methods for converting insolation data to illuminance data.

The authors of the individual chapters used terminology, units, and symbols as they deemed appropriate and have defined such within their chapters.

Those interested in more detailed technical information or updates on the current federal efforts in insolation resource assessment can contact the Solar Energy Research Institute/Resource Assessment and Instrumentation Branch/1617 Cole Boulevard/Golden, CO 80401.

Those interested in currently available insolation or meteorological data bases can contact the National Climatic Data Center (NCDC)/Federal Building/Asheville, NC 28801.

2 Insolation Data Bases and Resources in the United States

Raymond J. Bahm

2.1 Introduction

This chapter describes solar radiation data and how such data are archived and presented. It also offers some uses for the data. The chapter contains a listing of computerized data sets, printed data sets, and maps, along with where to obtain the data that are still available. This listing also includes information about the data that a serious user would want to know. Finally, the chapter gives recommendations for future solar radiation data collection programs.

This chapter does not contain foreign data or references to work done outside the United States, and, thus, a great deal of excellent work is not cited here. It is a review of data bases or large collections of data and does not include small data sets.

2.2 Insolation Measurement, Data Processing, Archiving, and Data Presentation

2.2.1 A Definition of Solar Radiation

Solar radiation is energy originating at the sun but measured at the earth's surface. Before solar radiation reaches the earth's surface, it must pass through the atmosphere, where it is absorbed, reflected, refracted, and otherwise changed. Because the intensity of solar radiation outside the earth's atmosphere is essentially constant, the measures of solar radiation at the earth's surface are measures of how the atmosphere affects it.

The sensing surfaces that measure this radiation are less than an inch in diameter, so although the measures are normally used to represent the energy incident over a large area (many square miles), they are actually measured at one point. Since some of the energy measured has been reflected from the ground and then the atmosphere, ground cover in the immediate areas around the sensor can affect the values measured somewhat.

Solar radiation is normally described as having two components: the beam (also called direct and sometimes direct beam) radiation and the diffuse radiation. The beam is that which comes directly from the apparent solar disk without undergoing any reflection or refraction. The diffuse is all other (e.g., from the sky) solar radiation falling on the surface. Together these components are called the total (or global) radiation.

Two instruments are commonly used to measure solar radiation: the pyrheliometer and the pyranometer. The pyrheliometer measures the beam radiation; the pyranometer measures the total radiation on a plane surface—usually horizontal. Historically, a variety of other names have been applied to these instruments, but the type of measure has been typically one of these two. Recent improvements, such as temperature compensation, have made these instruments more accurate and easier to use. For more information on instruments, see chapter 5.

This chapter refers to measurement of electromagnetic radiation with a wavelength from 0.3 to 3.0 μm (micrometers); thus, it excludes far infrared, far ultraviolet, x rays, gamma rays, and particles from the sun.

2.2.2 A Definition of Insolation

The term insolation is defined as incident solar radiation, the solar radiation energy incident on a surface. Thus, when a measure of solar radiation is given, a surface (plane) orientation must also be specified. Two orientations are commonly used—that of a horizontal surface and that of a plane normal to the radiation.

Henceforth, this chapter will use the term insolation when referring to solar radiation incident on a surface.

2.2.3 Data Recording Mechanisms

The earliest measurements were simply instantaneous values recorded by hand. By the 1950s, strip chart recorders were in common use, and essentially all the solar data from 1952 through 1976 measured by the National Weather Service (NWS) were recorded on these recorders and integrated manually to obtain hourly or daily values.

In 1977, the strip chart recorders in the NWS Solar Radiation Monitoring Network (hereafter referred to as NWS Network—see chapter 4) were replaced by electronic instruments that were sometimes unreliable; hence, the data recorded between 1977 and now can be patchy, and may not be as useful as earlier data.

2.2.4 Data Archiving Methods

The earliest data are archived in handwritten or printed form. Data from the NWS Network (see chapter 4) between 1952 and 1976 were archived on 80-column IBM cards—hence, the names Card Deck 280 and Card Deck 480. These data were later placed on 9-track computer magnetic tape in "card-image" format (80 characters per record). These card decks or tapes contain only the insolation data and no other measures.

The data from the NWS Network from 1977 through 1980 are archived on 9-track magnetic tape in the SOLMET format. Since the data are recorded in 1-minute intervals and processed into hourly values, the option of archiving the 1-minute data was considered.

As of October 1985, the 1-minute recording devices at most of the stations were not functioning well enough for that data to be recorded. Instead, the hourly totals printed by the recording devices were used as manual input to an hourly data base. In addition, maintenance of these systems had been so poor that less than 40% of the data that could have been collected in 1982 were actually collected. Chapter 4 presents the details of such data collection.

2.2.5 Data Presentation Methods and Their Use

The data presentation media tend to be determined by the data volume and their ultimate use. The early data were just single measures published in printed form. The printed format is one that is easy to understand. Later, when the volume of data became large, statistics of the data, usually long-term averages, were published in a printed form. Later, as data users became more sophisticated, the data were made available on punched cards and computer magnetic tape.

Some specialized forms of presentation have been used. The most common of these are maps with isolines of insolation availability on a horizontal surface or on surfaces of different tilts.

2.2.6 Data Presentation Statistics

Historically, insolation data were summarized and published in the form of daily or monthly averages. The most common form was the long-term averages for each month at a given location.

The increased use of electronic digital computers has led to the development of detailed simulation models, increasing the demand for solar radiation data on an hourly basis. Many researchers have demanded actual measures of hourly data to investigate the dynamic performance of systems in a realistic environment.

In some cases, the measures were converted to other measures for presentation. For example, designers of solar water heating systems are most interested in data that represent the performance of collectors on a tilt. In other cases, the insolation data were combined with other data, most notably temperature data, for presentation. The joint distribution of solar radiation and temperature illustrates the combinations of times when the energy is available as it is needed (without storing it).

2.3 Uses of Insolation Data

Today most people are aware that solar energy is potentially a major energy source for the world. Already there are significant applications where solar energy is replacing other energy sources (Bahm, 1982). The most well known of these are domestic water heating, space heating of buildings, and electric power for remote installations and spacecraft.

Some not-so-well-known uses of insolation data are economic planning, building energy performance modeling, atmospheric weather and climate models, weathering of materials, crop growth models, and irrigation models.

Awareness of the importance of measuring the spectral distribution of solar radiation has increased. The need to estimate the solar radiation resource availability for photovoltaic devices has caused many people to look for data that show how the spectral characteristics of atmospheric transmission affect the solar radiation resource availability for photovoltaic devices. Many factors must be considered, including.

1. Most measures of solar radiation availability are made with instruments having equal sensitivity to all parts of the solar spectrum.

2. Photovoltaic devices are sensitive to only a restricted part of the solar spectrum.

3. Different photovoltaic devices have different spectral sensitivities.

4. The spectral transmission of the atmosphere varies with the varying climatic conditions, the major effects caused by clouds, water vapor absorption and absorption, and scattering by atmospheric dust and gaseous molecules.

5. Some of the energy scattered from the solar beam reappears as diffuse solar radiation. This scattering depends strongly on the wavelength and has its most rapid change in the region of the spectrum where the photovoltaic sensitivity is the greatest.

The resulting problems are that

1. The accuracy of the resource estimates for flat-plate photovoltaics using the traditional measures will depend upon the atmospheric conditions at the time of use.

2. The accuracy of the resource estimates for concentrating photovoltaic systems will be even poorer than those for flat-plate systems, and additional measures of the atmospheric characteristics are needed to improve the accuracy of the resource knowledge for these concentrating systems.

2.4 Available Insolation Data Bases for the United States

2.4.1 Insolation Data Base Characteristics

Archiving, disseminating, and using the data are different functions and, thus, often require the data to be in different forms. Archives are designed to hold large volumes of detailed data inexpensively but permanently. Extracting the data from the archives should be fast, easy, and inexpensive. Disseminating data requires a form that is easily transportable and has a format and media compatible with the data users' requirements. The format for using data depends on how the data are to be used and the devices used to manipulate the data. The SOLMET data format (see section 2.4.2.1) attempts to satisfy all of these requirements.

Today, insolation data are archived along with coincidently measured meteorological data on computer-readable media, usually magnetic tape, or in a printed form, such as microfiche.

The data are usually disseminated in their archival form or summarized, usually in printed form. These summary data are either tables or maps. Sometimes the data are presented as graphs.

The following sections list useful solar radiation data sets according to their format. Within each section they are listed with the most commonly used or most useful first.

The source of the data is given where it is known. The user should contact the source indicated before attempting to order the data to ascertain the availability, format, media, and price. A telephone call is recommended.

2.4.2 Computer-Readable Data Sets

The following data bases are available on computer-readable media.

2.4.2.1 SOLMET—TD9724 Measured Data from 27 Stations (NCDC, 1978) This is an hourly data set that covers 23 years for most of the stations. Table 2.1 lists the stations, and figure 2.1 shows a summary of the data format and elements. The period of record is 1952 through 1975 for most cities.

Table 2.1
Listing of the 27 SOLMET station cities

Albuquerque, NM	Great Falls, MT
Apalachicola, FL	Lake Charles, LA
Bismarck, ND	Madison, WI
Boston, MA	Medford, OR
Brownsville, TX	Miami, FL
Cape Hatteras, NC	Nashville, TN
Caribou, ME	New York, NY
Charleston, SC	North Omaha, NE
Columbia, MO	Phoenix, AZ
Dodge City, KS	Santa Maria, CA
El Paso, TX	Seattle-Tacoma, WA
Ely, NV	Stephenville, TX
Fort Worth, TX	Washington, DC
Fresno, CA	

A map of the SOLMET stations (and ERSATZ stations discussed in section 2.4.2.2) is shown in chapter 4. This and the ERSATZ data following are the primary data sources for most of the insolation data used today. Section 2.4.6 and chapter 3 give some of the limitations and errors to be expected. The serious user should read this section and chapters 3 and 4 before deciding to use this data. Be aware that all of the direct-beam data in this data base are modeled; none represent actual measurements.

Obtain from the National Climatic Data Center (NCDC)/Federal Building/Asheville, NC 28801 [(704) 259-0682]. (This was formerly the National Climatic Center.)

2.4.2.2 ERSATZ SOLMET—TD9724 Regression Modeled Data from 222 Stations (NCDC, 1978 This is an hourly data set that covers 23 years for most of the stations. Table 2.2 lists the stations. A map of the ERSATZ stations is given in chapter 4. This and the preceding SOLMET data are the primary sources for most of the insolation data used today. The absolute accuracy of these data is probably on the order of 5–10%. Section 2.4.6 and chapter 4 give some of the limitions and errors to be expected. The serious user should read this section and chapters 3 and 4 before deciding to use these data. Be aware that all of the data in this data base are modeled; none are from actual measurements.

Despite the fact that these are modeled data, they are given the same tape deck number and released in the same format as the SOLMET data. The inexperienced user can easily miss the difference between the measured and ERSATZ data.

This set does not contain any direct-beam insolation data.

Obtain from the NCDC.

2.4.2.3 TMY—TD9734 Typical Meteorological Year Data Set (NCDC, 1981) This data set is a subset of the hourly SOLMET and ERSATZ data sets. It consists of 1 year of data for each of the stations. The year of data was selected so that it is typical of the entire data set. Each year is made up of selected months from the entire data set that are combined to make 1 complete year. Each month is selected because its weighted cumulative distribution is closest to the long-term distribution for that month. The process for selecting the typical year is described in more detail in Hall et al. (1981).

The list of stations for which TMY data are available is the same as those listed in tables 2.1 and 2.2 except for 14 stations, which are listed in table 2.3. Figure 2.2 shows a summary of the data format and the data elements. This is probably the most widely used data set of hourly insolation data.

Obtain from the NCDC.

2.4.2.4 SOLDAY—TD9739 Daily Insolation and Meteorological Data for 26 Stations (NCDC, 1979) This is a daily data set that covers 26 years for most of the stations. Table 2.4 lists the stations. The solar data in this set were derived from the Card Deck 480 data. One difference between these data and the hourly SOLMET is that the temperature corrections were not applied to the instrument readings. Another is that the insolation measurements for these locations were originally recorded only on a daily basis.

Obtain from the NCDC.

2.4.2.5 TD9736 New Network Data 1977–1980 (NCDC, 1977–1980) In 1977, the instrumentation for the entire NWS Network was replaced. The data processing procedures were revised, and a subset of the previously existing recording stations was selected to be a new network. A map showing the locations of these stations is in chapter 4.

The lengths of record of these data are probably too short to assure a representative data sample for most users; however, the individual mea-

TAPE DECK 9724	SOLMET	PAGE NO. 1

IDENTIFICATION							SOLAR RADIATION OBSERVATION										
TAPE DECK #	WBAN STN #	SOLAR TIME				LST TIME	ETR KJ/m^2	RADIATION VALUES KJ/m^2									SUNSHINE MIN
								DIRECT	DIFFUSE	NET	TILTED	GLOBAL			A	B	
		YR	MO	DY	HRMN							OBS	ENG COR	STD YR COR			
9724	XXXXX	XX	XX	XX	XXXX	XXXX	XXXX	1XXXX	1XXXX	1XXXX	1XXXX	1XXXX	1XXXX	1XXXX	1XXXX	1XXXX	XX
001	002	003				004	101	102	103	104	105	106	107	108	109	110	111

FIELD NUMBER

SURFACE METEOROLOGICAL OBSERVATION																											
OB TIME LST	CEILING dam	SKY COND	VSBY hm	WEATHER	PRESSURE kPa		TEMP °C		WIND		CLOUDS																SNOW COVER
											TOTAL	LOWEST			SECOND				THIRD				FOURTH			OPAQUE	
					SEA LEVEL	STATION	DRY BULB	DEW-PT.	DIR deg	SPD m/s		AMOUNT	TYPE	HEIGHT dam	AMOUNT	TYPE	HEIGHT dam	SUM AMT	AMOUNT	TYPE	HEIGHT dam	SUM AMT	AMOUNT	TYPE	HEIGHT dam		
XX	XXXX	1XXXX	XXXX	XXXXXXXX	XXXXX	XXXXX	XXXX	XXXX	XXX	XXXX	XX	XX	XX	XXXX	XX	XX	XXXX	XX	XX	XX	XXXX	XX	XX	XX	XXXX	XX	X
201	202	203	204	205	206		207		208		209																210

TAPE FIELD NUMBER	TAPE POSITIONS	ELEMENT
001	001 - 004	TAPE DECK NUMBER
002	005 - 009	WBAN STATION NUMBER
003	010 - 019	SOLAR TIME (YR, MO, DAY, HOUR, MINUTE)
004	020 - 023	LOCAL STANDARD TIME (HR AND MINUTE)
101	024 - 027	EXTRATERRESTRIAL RADIATION
102	028 - 032	DIRECT RADIATION
103	033 - 037	DIFFUSE RADIATION
104	038 - 042	NET RADIATION
105	043 - 047	GLOBAL RADIATION ON A TILTED SURFACE
106	048 - 052	GLOBAL RADIATION ON A HORIZONTAL SURFACE - OBSERVED DATA
107	053 - 057	GLOBAL RADIATION ON A HORIZONTAL SURFACE - ENGINEERING CORRECTED DATA
108	058 - 062	GLOBAL RADIATION ON A HORIZONTAL SURFACE - STANDARD YEAR CORRECTED DATA
109, 110	063 - 072	ADDITIONAL RADIATION MEASUREMENTS
111	073 - 074	MINUTES OF SUNSHINE
201	075 - 076	TIME OF COLLATERAL SURFACE OBSERVATION (LST)
202	077 - 080	CEILING HEIGHT (DEKAMETERS)
203	081 - 085	SKY CONDITION
204	086 - 089	VISIBILITY (HECTOMETERS)
205	090 - 097	WEATHER
206	098 - 107	PRESSURE (KILOPASCALS)
207	108 - 115	TEMPERATURE (DEGREES CELSIUS TO TENTHS)
208	116 - 122	WIND (SPEED IN METERS PER SECOND)
209	123 - 162	CLOUDS
210	163	SNOW COVER INDICATOR

7/77

Figure 2.1
SOLMET tape deck.

Table 2.2
Listing of the 222 ERSATZ station cities (alphabetically by state)

Birmingham, AL	San Diego, CA
Mobile, AL	San Francisco, CA
Montgomery, AL	Sunnyvale, CA
Adak, AK	Colorado Springs, CO
Annette, AK	Denver, CO
Barrow, AK	Eagle, CO
Bethel, AK	Grand Junction, CO
Bettles, AK	Pueblo, CO
Big Delta, AK	Hartford, CT
Fairbanks, AK	Wilmington, DE
Gulkana, AK	Washington, DC
Homer, AK	Daytona Beach, FL
Juneau, AK	Jacksonville, FL
King Salmon, AK	Orlando, FL
Kodiak, AK	Tallahassee, FL
Kotzebue, AK	Tampa, FL
McGrath, AK	West Palm Beach, FL
Nome, AK	Atlanta, GA
Summit, AK	Augusta, GA
Yakutat, AK	Macon, CA
Prescott, AZ	Savannah, GA
Tucson, AZ	Barbers Point, HI
Winslow, AZ	Hilo, HI
Yuma, AZ	Honolulu, HI
Fort Smith, AR	Lihue, HI
Little Rock, AR	Boise, ID
Arcata, CA	Lewiston, ID
Bakersfield, CA	Pocatello, ID
China Lake, CA	Chicago, IL
Daggett, CA	Moline, IL
El Toro, CA	Springfield, IL
Long Beach, CA	Evansville, IN
Los Angeles, CA	Fort Wayne, IN
Mount Shasta, CA	Indianapolis, IN
Needles, CA	South Bend, IN
Oakland, CA	Burlington, IA
Point Mugu, CA	Des Moines, IA
Red Bluff, CA	Mason City, IA
Sacramento, CA	Sioux City, IA

Table 2.2 (continued)

Goodland, KS	Elko, NV
Topeka, KS	Las Vegas, NV
Wichita, KS	Lovelock, NV
Covington, KY	Reno, NV
Lexington, KY	Tonopah, NV
Louisville, KY	Winnemucca, NV
Baton Rouge, LA	Yucca Flats, NV
New Orleans, LA	Concord, NH
Shreveport, LA	Lakehurst, NJ
Bangor, ME	Newark, NJ
Portland, ME	Clayton, NM
Baltimore, MD	Farmington, NM
Patuxent River, MD	Roswell, NM
Alpena, MI	Truth or Consequences, NM
Detroit, MI	Tucumcari, NM
Flint, MI	Zuni, NM
Grand Rapids, MI	Albany, NY
Houghton, MI	Binghamton, NY
Sault Ste. Marie, MI	Buffalo, NY
Traverse City, MI	Massena, NY
Duluth, MN	New York City (La Guardia), NY
International Falls, MN	Rochester, NY
Minneapolis-St. Paul, MN	Syracuse, NY
Rochester, MN	Asheville, NC
Jackson, MS	Charlotte, NC
Meridian, MS	Cherry Point, NC
Kansas City, MO	Greensboro, NC
Springfield, MO	Raleigh, NC
St. Louis, MO	Fargo, ND
Billings, MT	Minot, ND
Cut Bank, MT	Akron, OH
Dillon, MT	Cincinnati, OH
Glasgow MT	Cleveland, OH
Helena, MT	Columbus, OH
Lewistown, MT	Dayton, OH
Miles City, MT	Toledo, OH
Missoula, MT	Youngstown, OH
Grand Island, NE	Oklahoma City, OK
North Platte NE	Tulsa, OK
Scottsbluff, NE	Astoria, OR

Table 2.2 (continued)

Burns, OR	Kingsville, TX
North Bend, OR	Laredo, TX
Pendleton, OR	Lubbock, TX
Portland, OR	Lufkin, TX
Redmond, OR	Midland-Odessa, TX
Salem, OR	Port Author, TX
Koror Island, PN	San Angelo, TX
Kwajalein Island, PN	San Antonio, TX
Wake Island, PN	Sherman, TX
Allentown, PA	Waco, TX
Avoca, PA	Wichita Falls, TX
Erie, PA	Bryce Canyon, UT
Harrisburg, PA	Cedar City, UT
Philadelphia, PA	Salt Lake City, UT
Pittsburgh, PA	Burlington, VT
San Juan, PR	Norfolk, VA
Providence, RI	Richmond, VA
Columbia, SC	Roanoke, VA
Greenville, SC	Olympia, WA
Greer, SC	Spokane, WA
Huron, SD	Whidbey Island, WA
Pierre, SD	Yakima, WA
Rapid City, SD	Charleston, WV
Sioux Falls, SD	Huntington, WV
Chattanooga, TN	Eau Claire, WI
Knoxville, TN	Green Bay, WI
Memphis, TN	La Crosse, WI
Abilene, TX	Milwaukee, WI
Amarillo, TX	Casper, WY
Austin, TX	Cheyenne, WY
Corpus Christi, TX	Rock Springs, WY
Dallas, TX	Sheridan, WY
Del Rio, TX	
Houston, TX	Other station: Guantanamo Bay, Cuba

Table 2.3
Listing of the 14 station cities from SOLMET or ERSATZ data bases for which TMY data sets do not exist

Barrow, AK	Huntington, WV
Bettles, AK	Kansas City, MO
Burns, OR	Kotzebue, AK
Cleveland, OH	Needles, CA
Dallas (Love Field), TX	Pendleton, OR
Farmington, NM	Wichita, KS
Houghton, MI	Zuni, NM

surements are probably more accurate than those of the earlier 1952–1976 data.

Obtain from the NCDC.

2.4.2.6 New Network Data 1981 to Present Funding for the NWS Network was dropped in 1980. Although some of the stations have still collected data, the data are of uncertain quality, and the data set contains large gaps. No meteorological data are included with the solar data.

The data can sometimes be obtained from the NCDC; they are only recommended for research purposes. See also chapter 4 of this volume.

2.4.2.7 WYEC—Weather Year for Energy Calculations (Crow, 1981; 1984) This set is a subset of the hourly SOLMET and ERSATZ data set and contains one year of data for each of the stations included. It is prepared so that the mean values of the temperature and insolation parameters for each month of data are the same as the long-term means for those parameters. This data set is intended primarily for evaluating the energy consumption performance of large buildings. Table 2.5 lists the stations in this data set.

This data set is new (1984) and will most likely be widely used. It is available in the format as the Test Reference Year (TRY) data set and is intended to supersede TRY. It includes snow cover to provide modelers with the capability of enhancing ground reflectance during periods of snow cover.

The most significant differences between the WYEC data set an TMY are that

1. WYEC focuses on having the data set match the long-term mean monthly values for temperatures and insolation, whereas the TMY is weighted heavily toward typical insolation values.

TAPE DECK	TYPICAL METEOROLOGICAL YEAR	PAGE NO.
9734		6

IDENTIFICATION						SOLAR RADIATION OBSERVATION										
WBAN STN #	SOLAR TIME				LST TIME	ETR KJ/m^2	RADIATION VALUES KJ/m^2									SUNSHINE MIN
	YR	MO	DY	HRMN			DIRECT	DIFFUSE	NET	TILTED	GLOBAL OBS	GLOBAL ENG COR	GLOBAL STD YR COR	A	B	
XXXXX	XX	XX	XX	XXXX	XXXX	XXXX	iXXXX	iXXXX	iXXXX	iXXXX	iXXXX	iXXXX	iXXXX	iXXXX	iXXXX	XX
002	003				004	101	102	103	104	105	106	107	108	109	110	111

FIELD NUMBER

SURFACE METEOROLOGICAL OBSERVATION														
OB TIME LST	CEILING dam	SKY COND	VSBY hm	WEATHER	PRESSURE kPa SEA LEVEL	PRESSURE kPa STATION	TEMP °C DRY BULB	TEMP °C DEW-PT.	WIND DIR deg	WIND SPD m/s	CLOUDS TOTAL	CLOUDS OPAQUE	SNOW COVER	
XX	XXXX	iXXXX	XXXX	XXXXXXXX	XXXXX	XXXXX	XXXX	XXXX	XXX	XXXX	XX	XX	X	XXXXXXXXX
201	202	203	204	205	206		207		208		209		210	211

TAPE FIELD NUMBER	TAPE POSITIONS	ELEMENT
002	001 - 005	WBAN STATION NUMBER
003	006 - 015	SOLAR TIME (YR, MO, DAY, HOUR, MINUTE)
004	016 - 019	LOCAL STANDARD TIME (HOUR AND MINUTE)
101	020 - 023	EXTRATERRESTRIAL RADIATION
102	024 - 028	DIRECT RADIATION
103	029 - 033	DIFFUSE RADIATION
104	034 - 038	NET RADIATION
105	039 - 043	GLOBAL RADIATION ON A TILTED SURFACE
106	044 - 048	GLOBAL RADIATION ON A HORIZONTAL SURFACE —OBSERVED DATA
107	049 - 053	GLOBAL RADIATION ON A HORIZONTAL SURFACE —ENGINEERING CORRECTED DATA
108	054 - 058	GLOBAL RADIATION ON A HORIZONTAL SURFACE —STANDARD YEAR CORRECTED DATA
109,110	059 - 068	ADDITIONAL RADIATION MEASUREMENTS
111	069 - 070	MINUTES OF SUNSHINE
201	071 - 072	TIME OF COLLATERAL SURFACE OBSERVATION (LST)
202	073 - 076	CEILING HEIGHT (DEKAMETERS)
203	077 - 081	SKY CONDITION
204	082 - 085	VISIBILITY (HECTOMETERS)
205	086 - 093	WEATHER
206	094 - 103	PRESSURE (KILOPASCALS)
207	104 - 111	TEMPERATURE (DEGREES CELSIUS TO TENTHS)
208	112 - 118	WIND (SPEED IN METERS PER SECOND)
209	119 - 122	CLOUDS
210	123	SNOW COVER INDICATOR
211	124 - 132	BLANK (UNUSED)

Figure 2.2
Typical meteorological year tape deck.

Table 2.4
Listing of the 26 SOLDAY station cities

Astoria, OR	Los Angeles, CA
Atlanta, GA	Midland, TX
Blue Hill, MA	North Little Rock, AR
Boise, ID	Oklahoma City, OK
Burlington, VT	Portland, ME
Cleveland, OH	Rapid City, SD
Glasgow, MT	Salt Lake City, UT
Grand Junction, CO	San Antonio, TX
Greensboro, NC	Sault Ste. Marie, MI
Indianapolis, IN	Spokane, WA
Lakeland, FL	St. Cloud, MN
Lander, WY	Tallahassee, FL
Las Vegas, NV	Tampa, FL

2. TMY focuses on a definition of the typical that is based on the distributions of the hourly values of the various parameters, whereas the WYEC uses the monthly mean values of the data.

Obtain from the American Society of Heating, Refrigerating, and Air-Conditioning Engineers/Publication Sales Department/1791 Tullie Circle NE/Atlanta, GA 30329 [(404) 636-8400].

2.4.2.8 TD9706—Test Reference Year (NCDC, 1976) This is an hourly data set and has been widely used for evaluating the energy performance of large buildings. It was originally developed for comparing the performance of different air-conditioning systems in the same city, not for comparing different cities. It is available for 60 cities throughout the United States (see table 2.6). The parameters in this data set are dry-bulb, wet-bulb, and dew-point temperatures; wind direction and speed; station pressure; weather (type); total sky cover; and global horizontal insolation (where available).

Obtain from the NCDC.

2.4.2.9 TD1440—Airways Surface Observation (TDF-14) (NCDC, 1983) This is an hourly data set of meteorological observations for about 300 locations and does not contain insolation data. It is the source of most of the meteorological data for the SOLMET data set.

Obtain from the NCDC.

Table 2.5
Listing of the 51 cities included in the WYEC data base

First tape (22 cities)	U.S. stations
Albuquerque, NM	Amarillo, TX
Bismarck, ND	Atlanta, GA
Boise, ID	Birmingham, AL
Brownsville, TX	Boston, MA
Charleston, SC	Cheyenne, WY
Chicago, IL[a]	Denver, CO
Cleveland, OH	Des Moines, IA
Dayton, OH	Detroit, MI
Dallas-Fort Worth, TX	Dodge City, KS
El Paso, TX	Great Falls, MT
Lake Charles, LA	Indianapolis, IN
Las Vegas, NV	Kansas City, MO
Los Angeles, CA	Little Rock, AR
Madison, WI	Minneapolis, MN
Medford, OR	Oklahoma City, OK
Miami, FL	Phoenix, AZ
Nashville, TN	Pittsburgh, PA
New York, NY	Portland, ME
Omaha, NE	Portland, OR
Seattle, WA	Raleigh, NC
Tallahassee, FL	Salt Lake City, UT
Washington, DC	St. Louis, MO
Second tape (29 cities)	San Antonio, TX
Canadian Stations	Tampa, FL
Edmonton, AL	
Montreal, QB	
Toronto, ON	
Vancouver, BC	
Winnipeg, MB	

a. Chicago is on the tape even though it is missing from the documentation.

Table 2.6
Test reference years (TRY) for 60 cities

Tape 1		Tape 2		Tape 3	
Station	TRY	Station	TRY	Station	TRY
Birmingham, AL	1965	Chicago, IL	1974	Phoenix, AZ	1951
Washington, DC	1957	Dodge City, KS	1971	Fresno, CA	1951
Jacksonville, FL	1965	Boston, MA	1969	Los Angeles, CA	1973
Miami, FL	1964	Portland, ME	1965	Sacramento, CA	1962
Tampa, FL	1953	Minneapolis, MN	1970	San Diego, CA	1974
Atlanta, GA	1975	Columbia, MO	1968	San Francisco, CA	1974
Lake Charles, LA	1966	Kansas City, MO	1968	Boise, ID	1966
New Orleans, LA	1958	St. Louis, MO	1972	Indianapolis, IN	1972
Jackson, MS	1964	Omaha, NE	1966	Louisville, KY	1972
Raleigh, NC	1965	Albany, NY	1969	Detroit, MI	1968
Philadelphia, PA	1969	Buffalo, NY	1974	Great Falls, MT	1956
Charleston, SC	1955	New York, NY	1951	Bismarck, ND	1970
Memphis, TN	1964	Cleveland, OH	1969	Albuquerque, NM	1959
Nashville, TN	1972	Oklahoma City, OK	1951	Cincinnati, OH	1957
Brownsville, TX	1953	Tulsa, OK	1973	Medford, OR	1966
Fort Worth, TX	1975	Amarillo, TX	1968	Portland, OR	1960
Houston, TX	1966	El Paso, TX	1967	Pittsburgh, PA	1957
San Antonio, TX	1960	Lubbock, TX	1955	Salt Lake City, UT	1948
Norfolk, VA	1951	Burlington, VT	1966	Seattle-Tacoma, WA	1960
Richmond, VA	1969	Madison, WI	1974	Cheyenne, WY	1974

2.4.2.10 TD9788—Historical Sunshine Data (NCDC, 1983) This is a data base of monthly and annual percentages of possible sunshine for 240 U.S. stations. The period of record is from 1891 to 1980 or shorter.

Obtain from the NCDC.

2.4.2.11 TD9744—Input Data for Solar Systems (Cinquemani, Owenby, and Baldwin, 1978) This data set contains monthly mean values of global insolation and mean daily temperature data for 248 stations. It is also published in printed form (see section 2.4.3.3).

Obtain from the NCDC.

2.4.2.12 Data from the West Associates Solar Monitoring Network (Patapoff and Yinger, 1976–1980) This data base contains 15-minute insolation and temperature data for about 50 locations in the western United

States. This network was established in 1976, and data from 13 stations were first published for 1976. The number of stations increased each year until 1980, when data collection was discontinued. This data base contains 15-minute measured beam solar radiation data as well as global data. Not all parameters were measured at each site. Table 2.7 lists the stations.

Obtain from the NCDC.

2.4.2.13 Card Deck 280—TD9725 Hourly Solar Data (NCDC, n.d.1) This is the original insolation data set for the U.S. Solar Network 1952–1976. The majority of these data are included on the SOLMET data set for the 27 stations with measured data. Those stations not included were those with little data or data judged to be of poor quality.

This data set is available on magnetic tape as 80-character card images and is primarily of historical interest.

Obtain from the NCDC.

2.4.2.14 Card Deck 480—TD9726 Daily Solar Data (NCDC, n.d.2) This data set contains daily total global insolation data for over 100 stations but does not include any temperature or other meteorological data. These data have varying periods of record and many have calibration drift errors and other errors that have not been corrected. Some Canadian stations are included in the data set.

This set is included here because it contains data from a number of locations not included in the other computer data bases. However, the data in this data base have not been subjected to the review and rehabilitation processes that the foregoing data sets have. Thus, quality of these data is less certain. Because of this, the Card Deck 480 data are primarily of historical or research interest.

This data set is available on magnetic tape as 80-character card images.

Obtain from the NCDC.

2.4.2.15 Aerospace Insolation Data Base (Randall, 1976) This data base is a compendium of insolation data from the Card Deck 480 and meteorological data sets that have not been corrected. This data set is primarily of historical or research interest and is no longer available.

2.4.2.16 Other Solar Radiation Measurements Eight university solar energy meteorological research and training stations collected various insolation and related data beginning in 1977. The data collected included global horizontal, beam, and diffuse horizontal, and data on various tilted

Table 2.7
WEST Associates insolation monitoring stations

- Arizona Public Service
  - Gila Bend
  - Parker
  - Phoenix
  - Saguaro
- Imperial Irrigation District
  - Imperial
- Los Angeles Department of Water and Power
  - Inglewood
  - Los Angeles
  - San Pedro
  - Sun Valley
  - West Los Angeles
  - West Valley
- Nevada Power Company
  - Boulder City (DRI)
  - Las Vegas
- Public Service Co. of Colorado
  - Alamosa
  - Cheyenne
  - Denver 1 (Southeast)
  - Denver 2 (Holly)
  - Pueblo
- Public Service Co. of New Mexico
  - Albuquerque
- Salt River Project
  - Page
  - St. Johns
- San Diego Gas and Electric
  - Alpine
  - Carlsbad
  - Chula Vista
  - El Cajon
  - Escondido
  - Ramona
  - San Diego
  - Spring Valley
- Sierra Pacific Power
  - Reno
- Southern California Edison
  - Alhambra
  - Arrowhead
  - Barstow
  - Blythe
  - Eldorado
  - El Segundo
  - El Toro
  - Huntington Beach
  - Laguna Bell
  - Lancaster
  - Mandalay
  - Moorpark
  - Palm Springs
  - Pardee
  - Rialto
  - Ridgecrest
  - Victorville
  - Villa Park
  - Visalia
  - Walnut
  - Yucca Valley
- Tucson Electric Power Company
  - Tucson

Table 2.8
Circumsolar telescope data collection sites and periods of record

Site	Motivation	Dates
Albuquerque, NM	CRTF tests	11/79[b] ...
Albuquerque, NM	CRTF,[a] high-desert climate	5/76–10/79
Argonne, IL	Cool cloudy climate	8/77–10/78
Atlanta, GA	ACTF,[c] humid climate	6/77[b] ...
Barstow, CA	Mojave, 10-MW[e] pilot plant	5/77–10/79
Berkeley, CA	Rehabilitate and upgrade	11/78–10/79
Berkeley, CA	Storage	11/79[b] ...
Boardman, OR	Boeing heliostat test	2/77–5/77
China Lake, CA	Mojave desert climate	7/76–5/77
Colstrip, MT	NOAA Atmospheric tests	5/77
Edwards, CA	JPL PDTF,[d] Mojave climate	11/79[b] ...
Ft. Hood, TX	TES,[e] warm cloudy climate	7/76–8/77

a. Central Receiver Test Facility at Sandia Laboratories.
b. Ongoing; contact LBL for more information on the period-of-record.
c. Advanced Component Test Facility at Georgia Tech.
d. Jet Propulsion Laboratory Parabolic Dish Test Facility.
e. Total Energy System (proposed).

surfaces. One feature is that some of the data is on a 1-minute time scale (SERI, 1978). These data are primarily of research interest.

Obtain from the Solar Energy Research Institute (SERI)/1617 Cole Boulevard/Golden, CO 80401 [(303) 231-1815].

A special instrument was designed by Lawrence Berkeley Laboratory to measure the circumsolar radiation. The data collected allow a detailed analysis of the circumsolar radiation and its variation with some other atmospheric measurements. Measurements were taken for a few months at several different locations in different climates. Table 2.8 lists these data sites and the period of record (Evans et al., 1980; Grether et al., 1979, 1980). These data are primarily of research interest.

Obtain from the Lawrence Berkeley Laboratory/One Cyclotron Road/University of California/Berkeley, CA 94720.

2.4.3 Tabular Printer Data Sets

2.4.3.1 Monthly Summaries of Solar Radiation Data (NCDC, 1977–1980)

The hourly measurements from the new U.S. Solar Monitoring Network, 1977–1980, were tabulated and published monthly. One booklet covers the

Table 2.9
Listing of cities that have stations in the NWS Solar Radiation Monitoring Network (1976–1985)

Albuquerque, NM	Lake Charles, LA
Bismarck, ND	Lander, WY
Blue Hill, MA	Las Vegas, NV
Boise, ID	Los Angeles, CA
Boulder, CO	Madison, WI
Brownsville, TX	Medford, OR
Burlington, VT	Miami, FL
Caribou, ME	Midland, TX
Columbia, MO	Montgomery, AL
Desert Rock, NV	Nashville, TN
Dodge City, KS	Omaha, NE
El Paso, TX	Phoenix, AZ
Ely, NV	Pittsburgh, PA
Fairbanks, AK	Raleigh, NC
Fresno, CA	Salt Lake City, UT
Grand Junction, CO	San Juan, PR
Great Falls, MT	Seattle-Tacoma, WA
Guam, PN	Tallahassee, FL
Honolulu, HI	Washington, DC (Sterling, VA)

entire network for each month. Four volumes, one for each year, exist (48 monthly booklets in all). See table 2.9 and chapter 4.

Obtain from the NCDC.

2.4.3.2 Microfiche Presentations from the Solar Radiation Energy Resource Atlas of the United States (SERI, 1981) This publication identifies a variety of statistical summary presentations developed from the SOLMET and ERSATZ data bases. These microfiche are ordered separately from the *Atlas* itself. They are ordered by station and are available for all but a few of the SOLMET stations. Table 2.10 gives a brief list of these presentations.

Obtain from the NCDC.

2.4.3.3 Insolation Data Manual (Knapp, Stoffel, and Whitaker, 1980) This publication contains long-term monthly means of insolation, temperature, degree-days, and global insolation fractional transmittance (K_t) for 248 NWS stations in the United States. This information was drawn primarily from the SOLMET and ERSATZ data bases and was previously

Table 2.10
Statistical microfiche presentations described in the *Solar Radiation Energy Resource Atlas of the United States*

Type I presentation tables for SOLMET stations and Type II for ERSATZ stations

- Joint distribution of mean daily dry-bulb and solar radiation on a horizontal surface by year
- Joint distribution of mean daily dry-bulb and mean daily wet-bulb by year
- Mean daily global horizontal solar radiation by month and year
- Mean daily direct above selected threshold values by month and year
- Mean daily diffuse horizontal solar radiation by month and year
- Mean daily global solar radiation on south-facing surfaces at selected tilts and above selected thresholds by month and year
- Mean daily direct solar radiation on south-facing surfaces tilted at latitude and vertical by month and year
- Mean daily global solar radiation on a north-facing surface tilted at latitude by month and year
- Mean daily global solar radiation on east-, southeast-, southwest-, and west-facing surfaces tilted at latitude and vertical by month and year
- Mean daily opaque sky cover by month and year
- Normal daily mean temperature by month and year
- Mean daily wind speed by month and year
- Mean daily wet-bulb temperature by month and year
- Mean daily dry-bulb temperature by month and year
- Mean nighttime temperature by month and year

Type I presentations—distributions—SOLMET stations only

- Cumulative frequency distributions of global solar radiation on surfaces tilted at horizontal, vertical, and at latitude by month
- Cumulative frequency distribution of direct solar radiation on a surface tilted at latitude by month
- Cumulative frequency distribution of direct normal solar radiation by month

Type I presentations—run length graphs—SOLMET stations only

- Global and direct normal radiation days 1–100 and 100–365 by year

Type III presentations—cumulative frequency distributions—ERSATZ stations only

- Daily total solar radiation by month
- Wind velocity mean and maximum by month
- Wet-bulb mean, minimum, and maximum by month
- Dry-bulb mean, minimum, and maximum by month

Type IV presentations—summary statistics used in the TMY selection process—ERSATZ stations only

---

published under the title *Input Data for Solar Systems* (Cinquemani, Owenby, and Baldwin, 1978). These data are also available on computer magnetic tape (see section 2.4.2.11).

This publication also has maps of K_t.

Obtain from the Superintendent of Documents/U.S. Government Printing Office/Washington, DC20402, or obtain in microfiche from the National Technical Information Center (NTIS)/U.S. Department of Commerce/5285 Port Royal Road/Springfield, VA 22161.

2.4.3.4 Direct Normal Solar Radiation Data Manual (Knapp and Stoffel, 1982) This is an addendum to the *Insolation Data Manual* and contains long-term, monthly mean, daily totals for 235 NWS stations. These data are based on the modeled direct-normal data from the SOLMET and ERSATZ data bases, not on measured direct-normal data. This publication also has maps of the beam radiation.

Obtain from the NTIS.

2.4.3.5 Estimates of Available Solar Radiation and Photovoltaic Energy Production for Various Tilted and Tracking Surfaces throughout the United States, based on PVFORM, a Computerized Performance Model (Menicucci and Fernandez, 1986) This is a tabular listing of the solar radiation available to collectors in a variety of tilted and tracking schemes. Data are presented for 38 stations on a quarterly basis. This project used the improved Perez model (see chapter 3). These data probably represent the best available on a quarterly basis.

Obtain from the NTIS.

2.4.3.6 Monthly Average Solar Radiation on Inclined Surfaces for 261 North American Cities (Klein, Beckman, Duffie, 1977) This publication presents temperature, heating degree-days, K_t, and insolation data.

Obtain from the University of Wisconsin/Solar Energy Laboratory/Madison, WI 53706.

2.4.3.7 Other Listings of Insolation Data The following citations are generally superseded by later publications listed in this chapter: Randall and Whitson (1977), Boes et al. (1976, 1977), and Watt (1978).

Obtain from the NTIS.

2.4.3.8 National Weather Service Publications A variety of NWS publications include solar radiation data. These are

- *Climatological Summaries Monthly and Annual* (NCDC, 1983),
- *Climatography of the United States* (NCDC, 1983),
- *Climatic Atlas of The United States (1968)* (NCDC, 1968), and
- *Monthly Weather Review.*

The *Monthly Weather Review* was initially published by NWS, and their first insolation measurements were published here. More recently, this has become a publication of the American Meteorological Society.

Obtain from the NCDC.

2.4.3.9 Publications by State and Regional Groups A number of publications of insolation data measurements by different groups exist. The most current and useful of these are

West Associates Solar Resource Evaluation Project—Solar Energy Measurements at Selected Sites Throughout the Southwest (Patapoff and Yinger, 1976–1980). There are five reports, one for each of the years 1976–1980. Obtain from the NCDC.

The Northeast Utilities Solar Resource Evaluation Project—Solar Energy Measurement During 1974–1983 (Goodrich, Buisick, and Janowiec, 1982). Obtain from the Northeast Utilities/P.O. Box 270/Hartford, CT 06101.

The California Solar Data Manual (Berdahl et al., 1978). Obtain from the NTIS.

Montana Solar Data Manual (Fowlkes, 1982). All of the data in this publication was measured on a tilted surface with photovoltaic silicon sensors. Obtain from the Montana Department of Natural Resources and Conservation/32 S. Ewing/Helena, MT 59715.

Pacific Northwest Solar Radiation Data (McDaniels et al., 1983). Obtain from the Center for Environmental Research/School of Architecture and Allied Arts/University of Oregon/Eugene, OR 97403.

2.4.3.10 Printed Documents Listing Other Insolation Data Bases Many government agencies and other groups have collected insolation measurements where, sometimes, the instruments have not been calibrated and maintained properly. Without proper installation and maintenance, it is common to see biased errors of 20% or greater. When using data from any source, be sure first to determine that the data are satisfactory for the purpose.

The *Solar Radiation Data Directory* (SERI, 1983) lists data from a variety of sources.

Obtain from the SERI.

2.4.4 Maps of Insolation Data

2.4.4.1 Solar Radiation Energy Resource Atlas of the United States (SERI, 1981) This publication is a major compendium of insolation data in a reduced form and is based on data contained in the SOLMET and ERSATZ data bases. This *Atlas* presents the distribution of mean daily insolation for each month geographically. It also presents, in a similar way, cloud cover, temperature, and humidity data and contains some graphs of run length, beam intensity, and cumulative frequency distribution of solar radiation availability.

This document is folio size 22″ × 16″, which makes it difficult to use. All the information in this could be easily printed in an 8″ × 10″ format which would be much easier to use and to store.

Table 2.11 lists the contents. Figures 2.3 and 2.4 are examples of maps of the direct normal and global insolation on a south-facing surface.

Obtain from the Superintendent of Documents/U.S. Government Printing Office/Washington, DC 20402.

2.4.4.2 Quarterly Maps of Insolation Quarterly average values of insolation were prepared for a number of different surfaces, both fixed and tracking by Boes et al. (1977). Contour maps were prepared from these data for total horizontal and beam radiation, for surfaces titled at 45° and 90° from the horizontal towards the south, and for diffuse radiation on a horizontal surface.

Obtain from the NTIS.

Table 2.11
Contents of the *Solar Radiation Energy Resource Atlas of the United States*

Text
Preface
Acknowledgments
Solar radiation energy resources
Meteorological data
Contour mapping
Relationship between ERSATZ and SOLMET stations
Solar radiation units
References

Table 2.11 (continued)

Global solar radiation contour maps (annual and monthly)
- Average daily solar radiation on a horizontal surface
- Average daily solar radiation on a south-facing surface: tilt = latitude
- Average daily solar radiation on a south-facing surface: tilt = latitude; above threshold of 0.180 MJ/m^2
- Average daily solar radiation on a south-facing surface: tilt = latitude; above threshold of 0.360 MJ/m^2
- Average daily solar radiation on a south-facing surface: tilt = latitude; above threshold of 0.720 MJ/m^2
- Average daily solar radiation on a south-facing surface: tilt = latitude; above threshold of 1.440 MJ/m^2
- Average daily solar radiation on a south-facing surface: tilt = latitude + 15 degrees
- Average daily solar radiation on a south-facing surface: tilt = latitude − 15 degrees
- Average daily solar radiation on a south-facing surface: tilt = vertical

Direct normal solar radiation contour maps (annual and monthly)
- Average daily direct normal solar radiation
- Average daily direct normal solar radiation above threshold of 0.180 MJ/m^2
- Average daily direct normal solar radiation above threshold of 0.360 MJ/m^2
- Average daily direct normal solar radiation above threshold of 0.720 MJ/m^2
- Average daily direct normal solar radiation above threshold of 1.440 MJ/m^2
- Average daily direct normal solar radiation on a south-facing surface: tilt = latitude
- Average daily direct normal solar radiation on a south-facing surface: tilt = vertical

Diffuse solar radiation contour maps (annual and monthly)
- Average daily diffuse solar radiation on a horizontal surface

Weather contour maps (annual and monthly)
- Average daily opaque sky cover
- Normal daily average temperature
- Normal daily maximum temperature
- Normal daily minimum temperature
- Normal total heating degree-days based on 18.3°C
- Normal total cooling degree-days based on 18.3°C
- Average daytime dry-bulb temperature
- Average nighttime dry-bulb temperature
- Average daily wet-bulb temperature

Solar radiation graphical plots for 26 SOLMET stations
- Run length plots (1 year)
- Bar graphs (monthly)
- Cumulative frequency distribution plots (monthly)

Appendix A: microfiche presentations (see table 2.8)

Appendix B: methodology used to generate atlas data

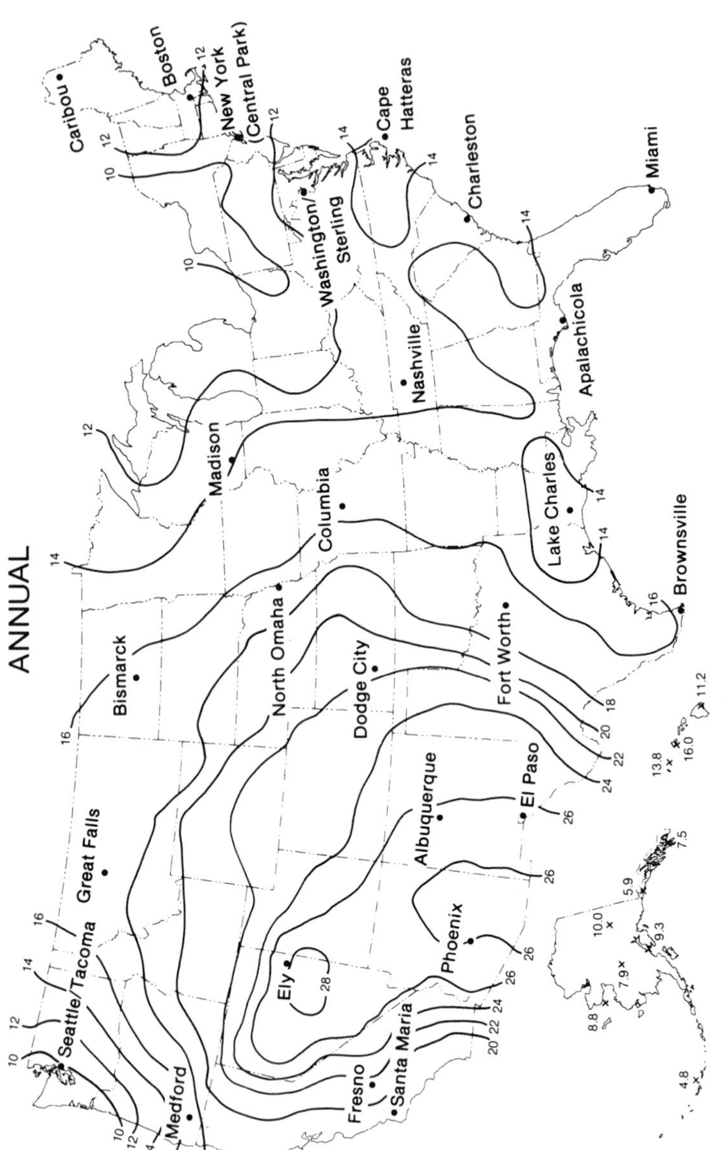

Figure 2.3
Average daily direct normal solar radiation for the United States (annual and 12 monthly figures), in MJ/m^2 (Mega-Joules per square meter).

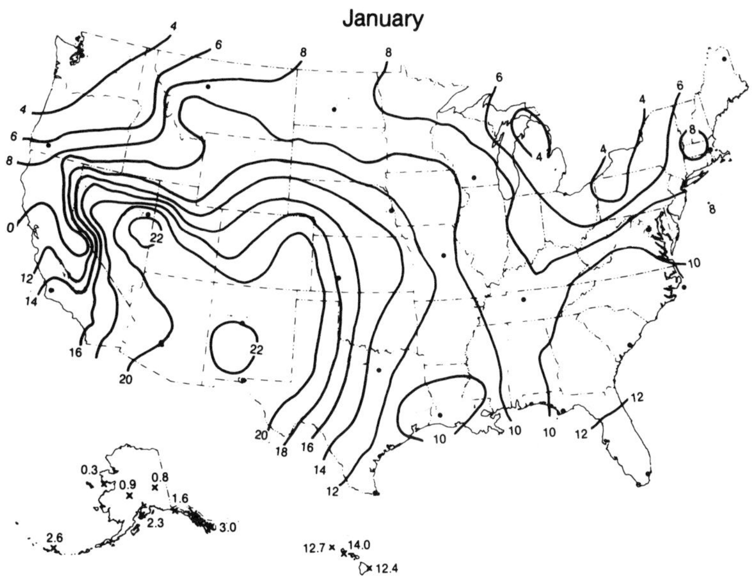

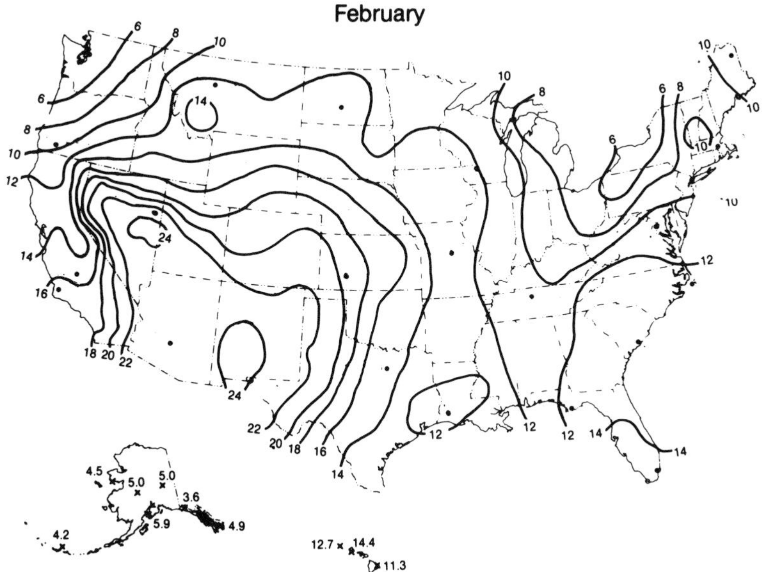

Figure 2.3 (continued)

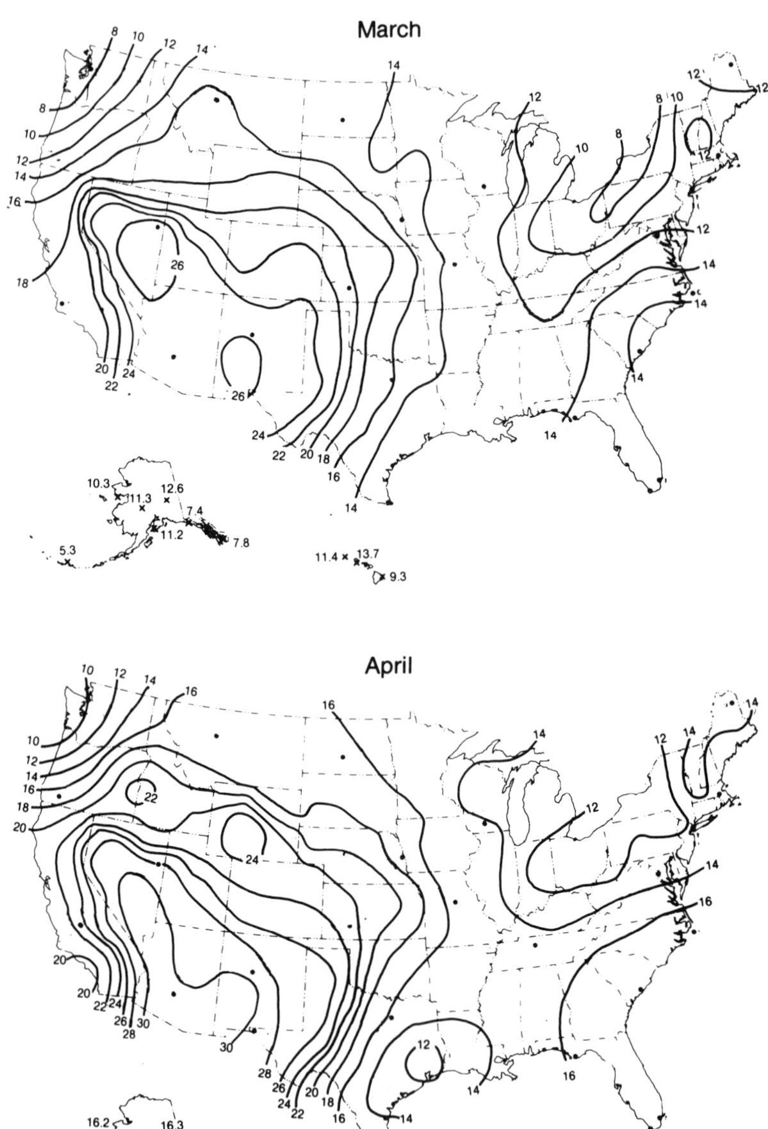

Figure 2.3 (continued)

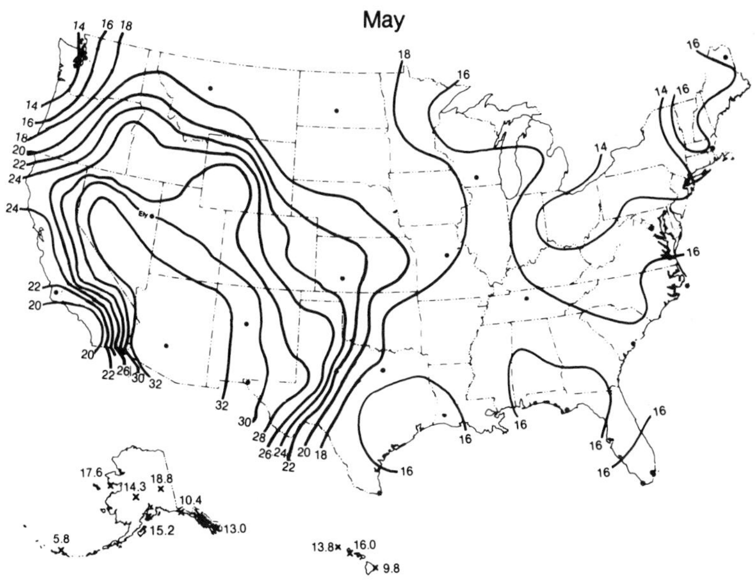

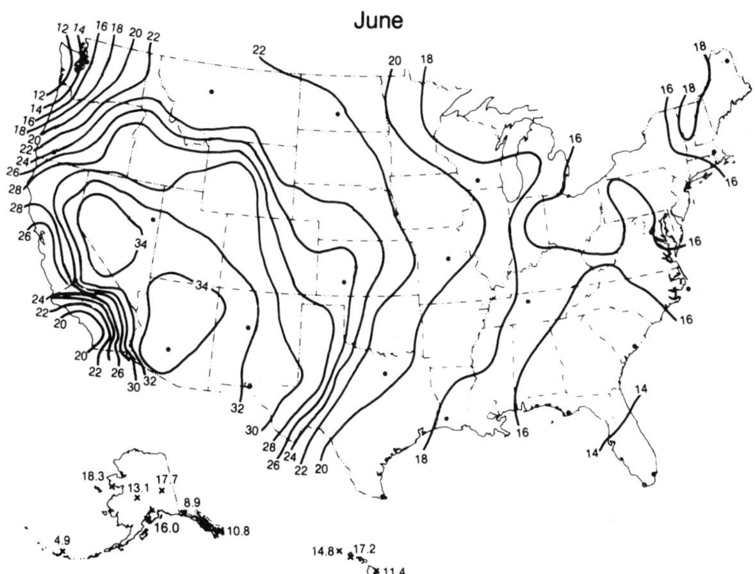

Figure 2.3 (continued)

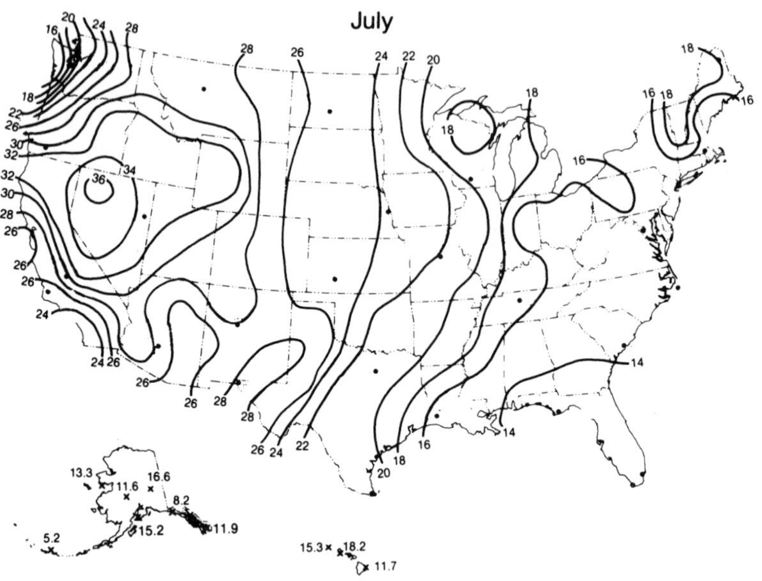

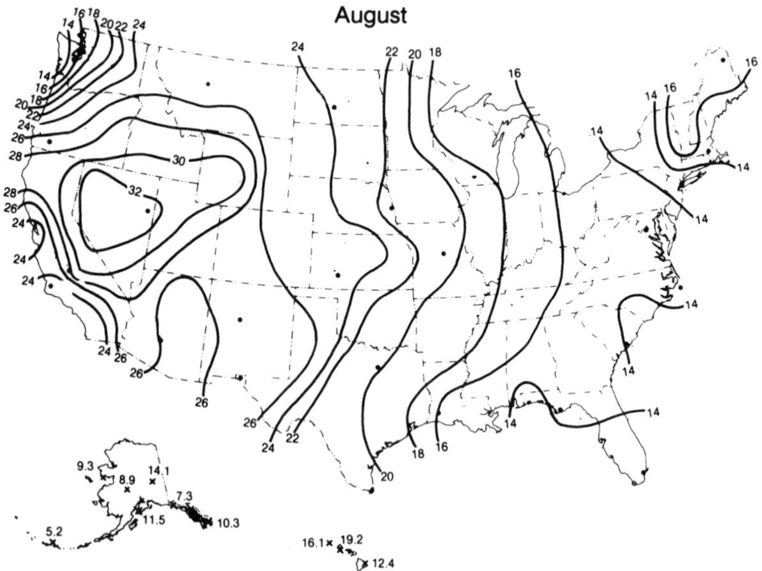

Figure 2.3 (continued)

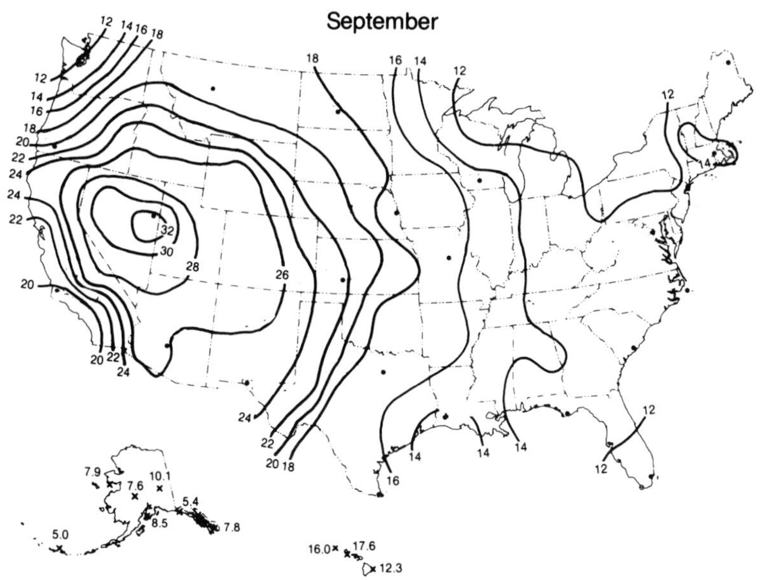

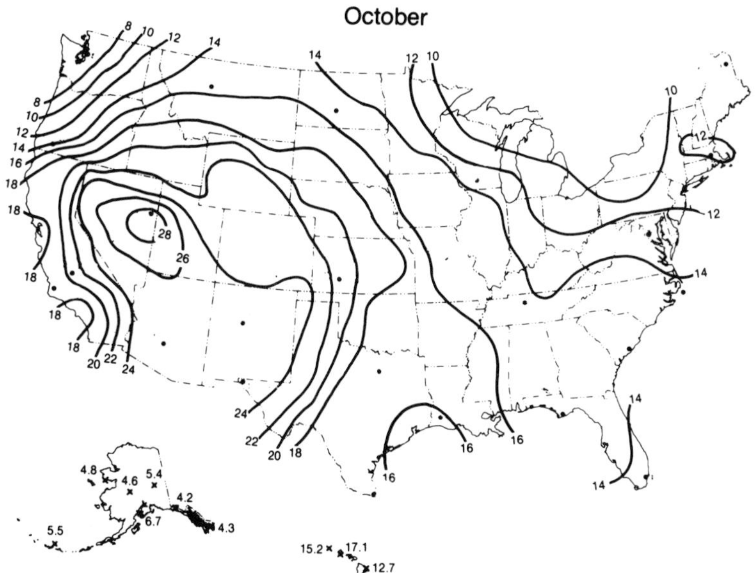

Figure 2.3 (continued)

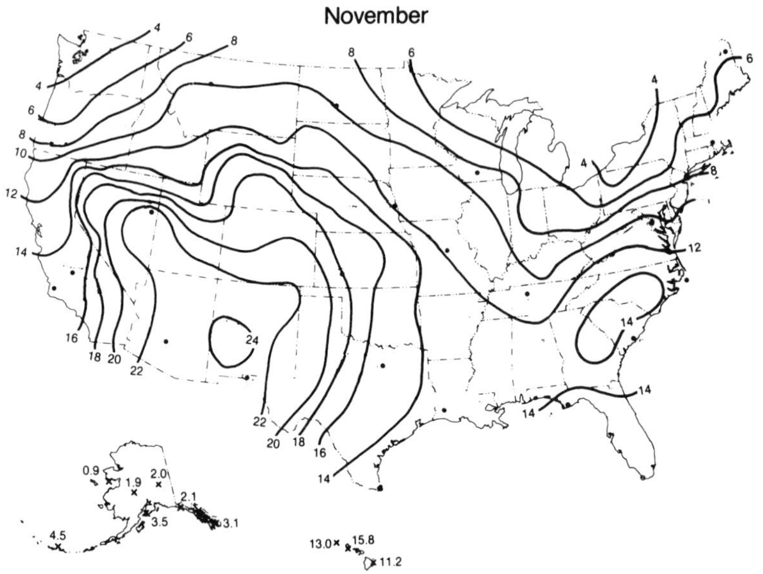

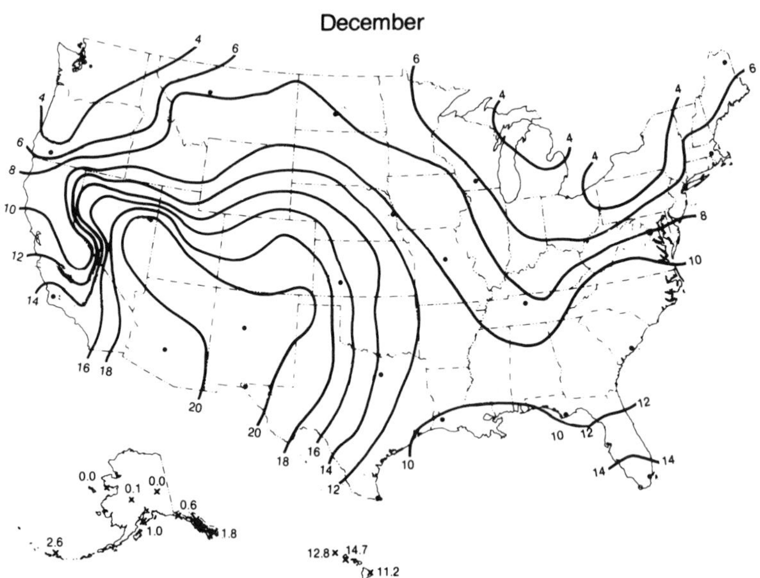

Figure 2.3 (continued)

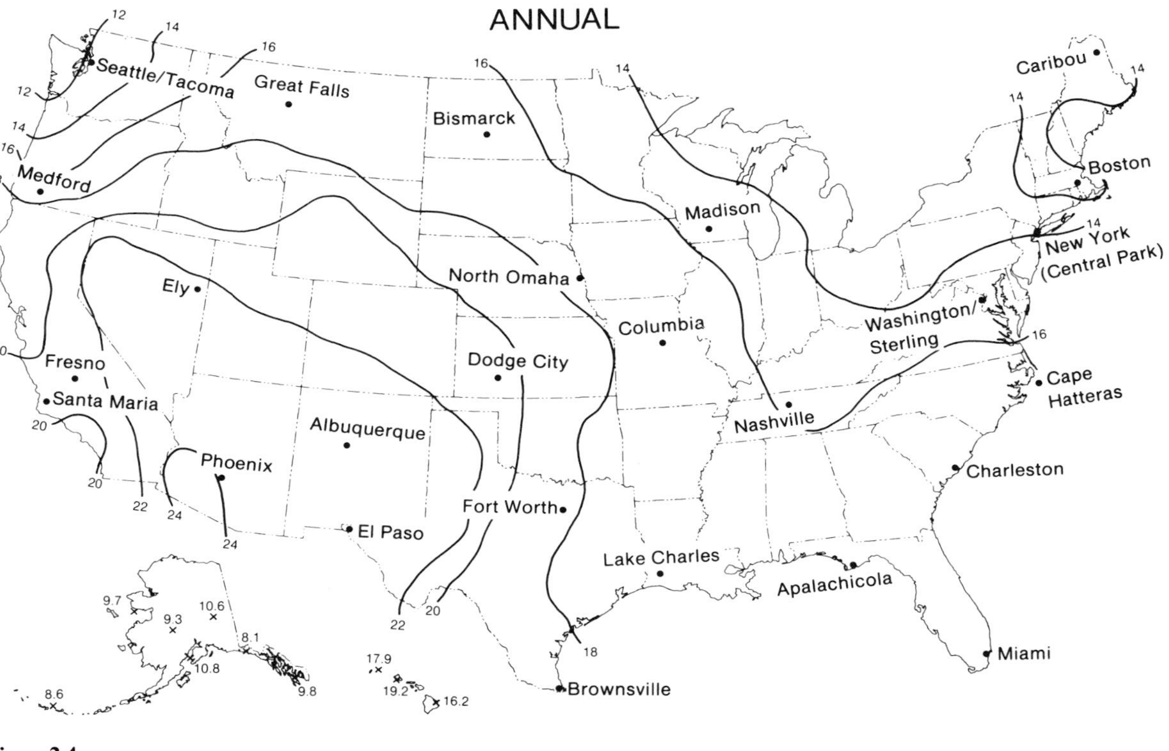

Figure 2.4
Average daily global solar radiation on a south-facing surface, tilt = latitude (annual and 12 monthly figures), in MJ/m^2.

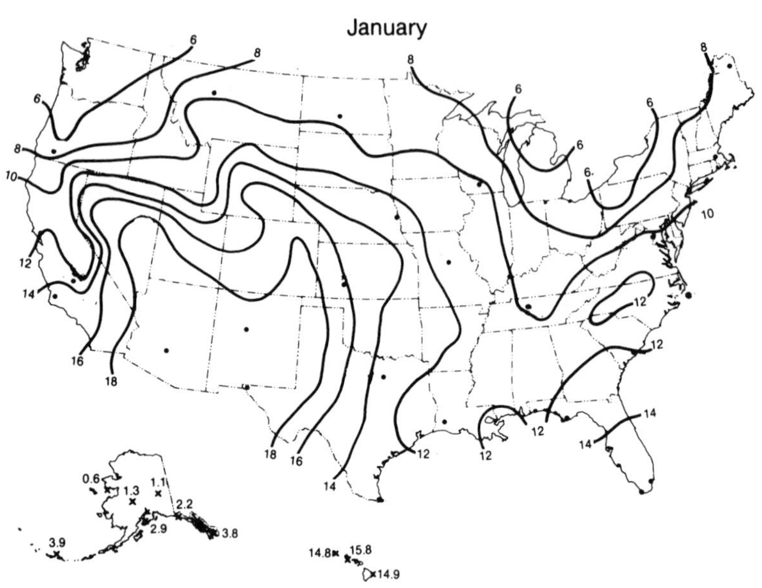

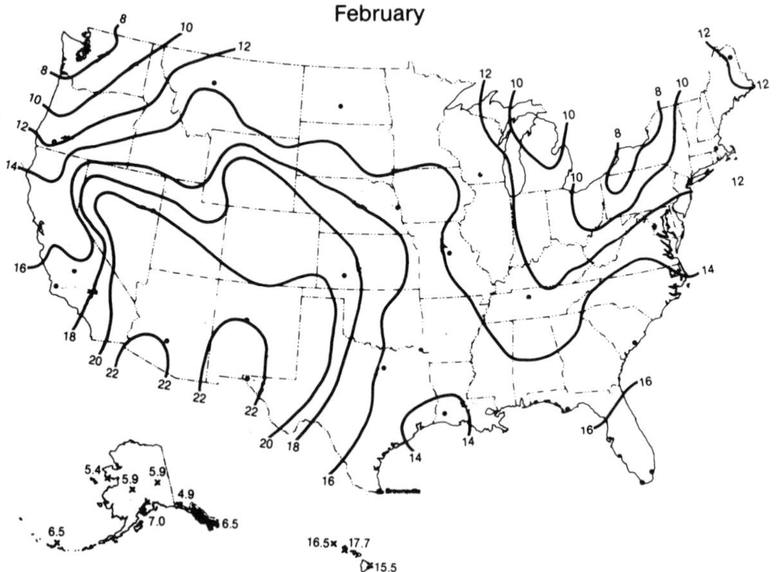

Figure 2.4 (continued)

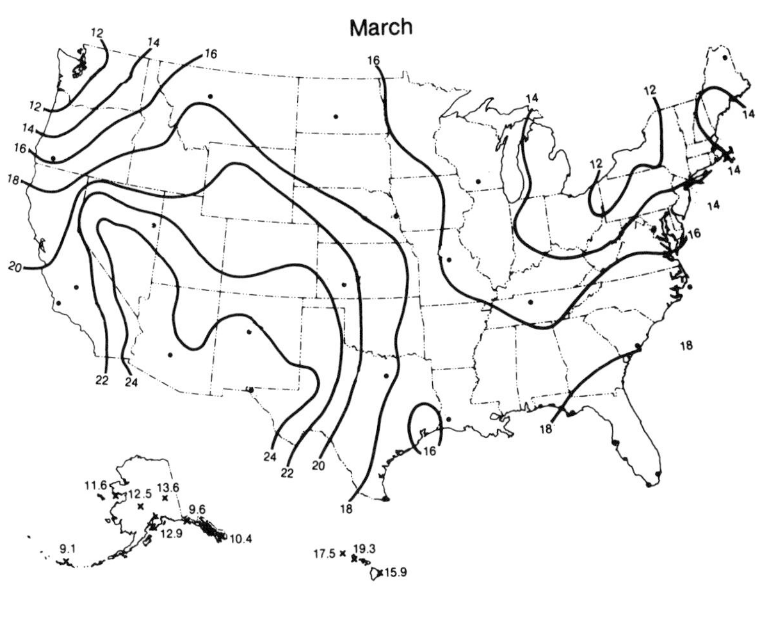

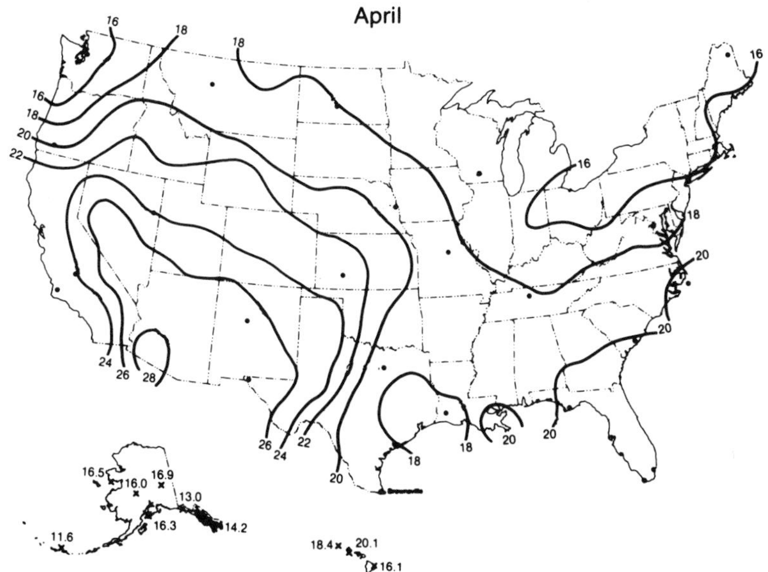

Figure 2.4 (continued)

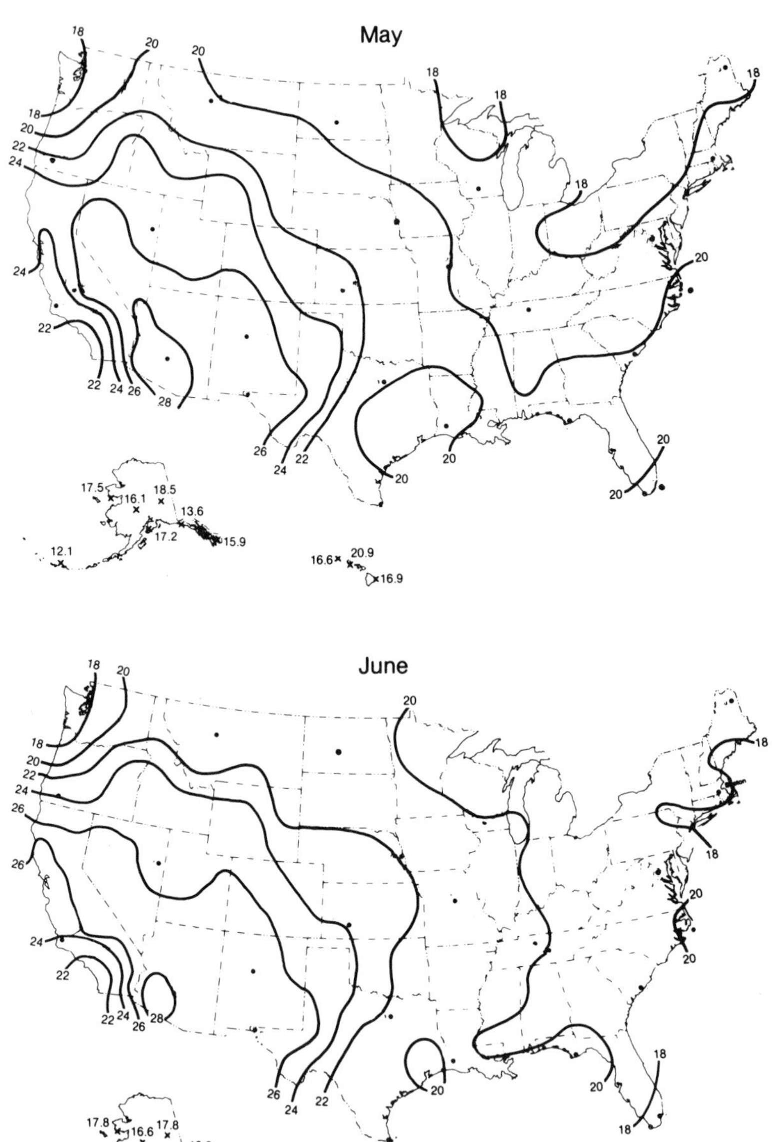

Figure 2.4 (continued)

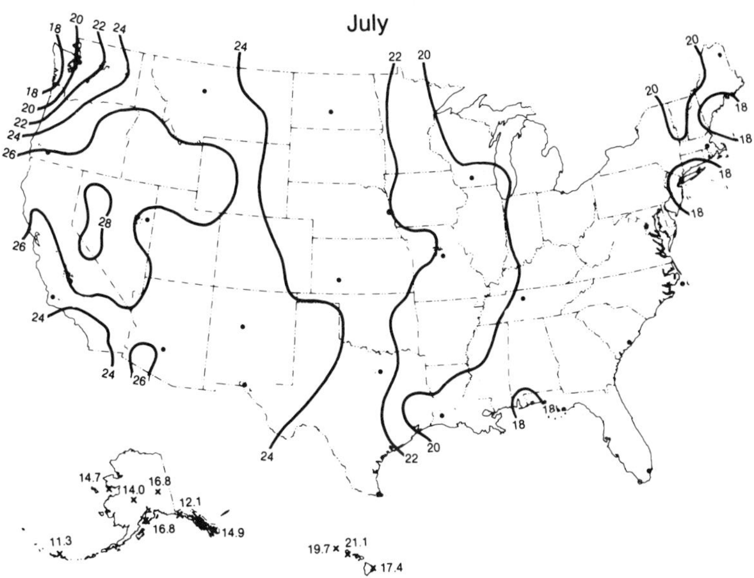

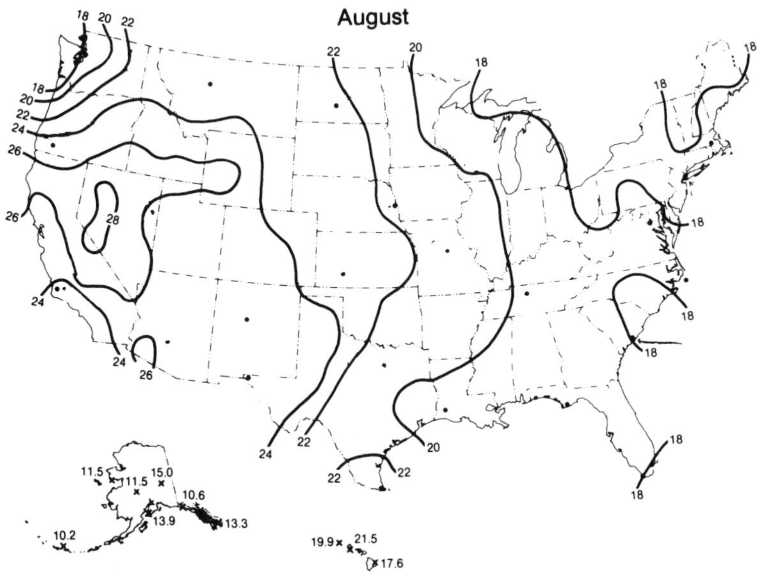

Figure 2.4 (continued)

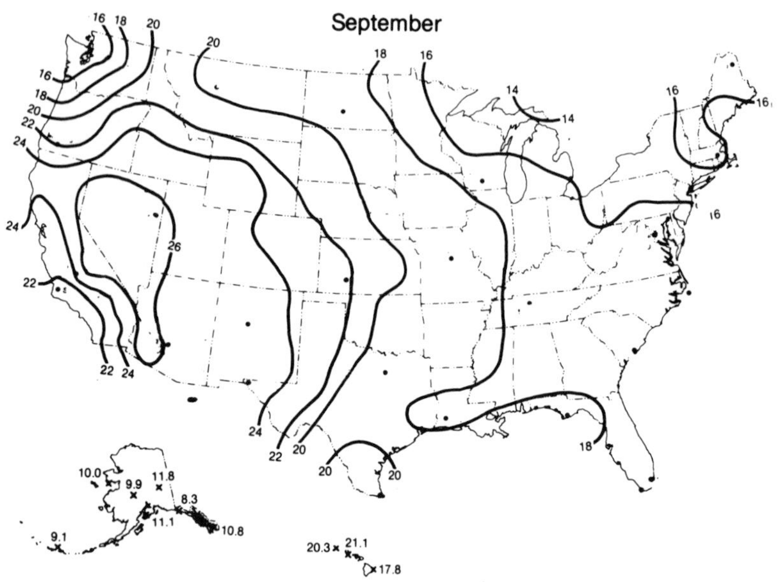

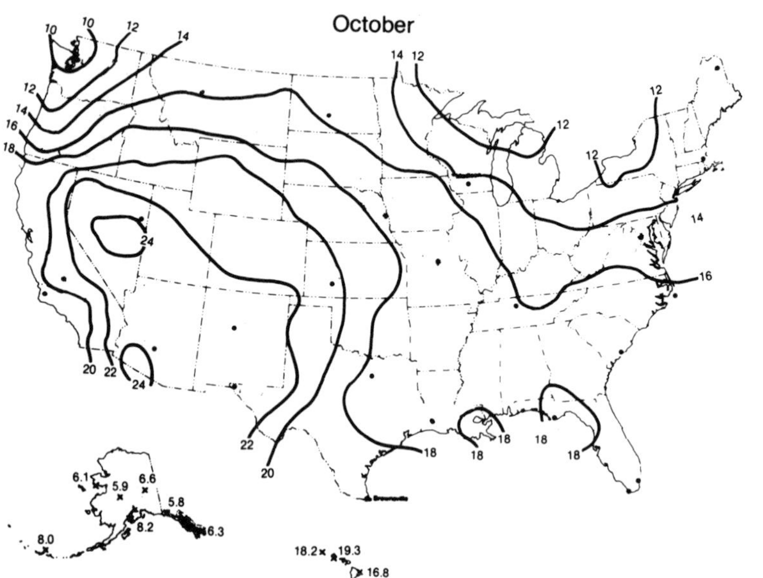

Figure 2.4 (continued)

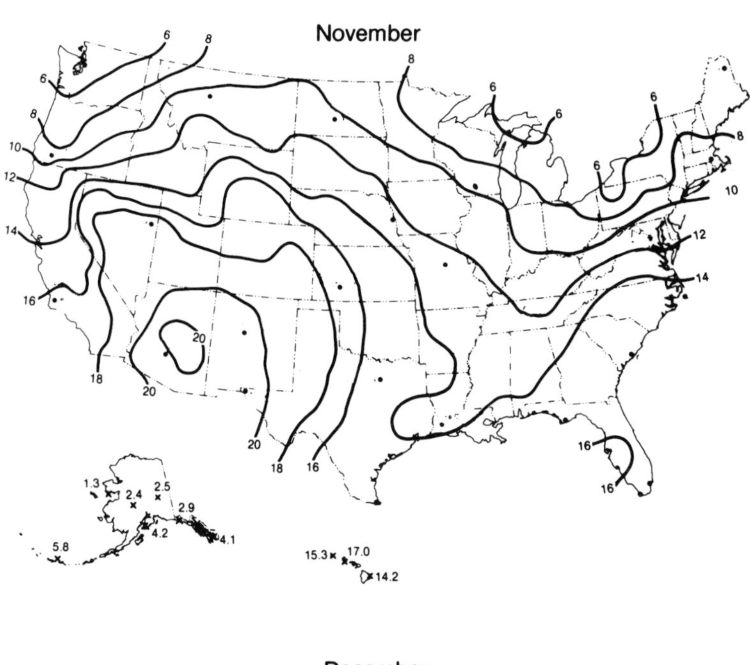

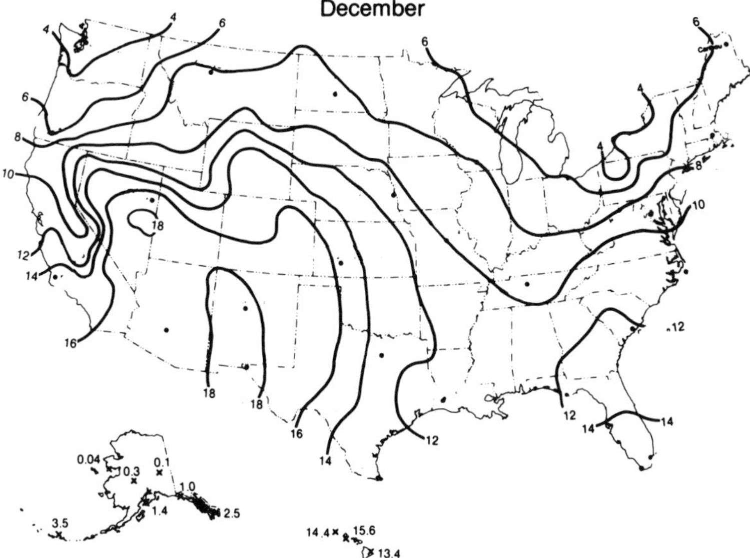

Figure 2.4 (continued)

2.4.4.3 "Monthly Maps of Mean Daily Insolation for the United States" (Bennett 1965) These are probably the first maps showing the variation of the solar resource over the United States.

2.4.5 Printed Documentation of Computer Data Bases

This section briefly discusses the printed documents describing the computerized data bases and identifies some of the problems that make these data sets difficult to work with.

2.4.5.1 SOLMET Volume 1—User's Manual (TD9724) Hourly Solar Radiation—Surface Meteorological Observation (NCDC, 1978) This document describes the SOLMET and the ERSATZ data base format and lists the SOLMET and ERSATZ stations and their locations.

The principal difference between the data formats of the SOLMET and ERSATZ data sets is the tape blocking factor. The SOLMET tapes are blocked 3,912 characters (24 hourly records) per physical block. The ERSATZ tapes are blocked 7,824 characters (48 hourly records) per physical block. Each tape is a serial file of records. All of the records have the same format.

Although I have found no case where the documentation is actually wrong, there are several areas in which the user can easily run into difficulty. The following are problem areas:

1. Since the blocking factors are different, two different programs must be prepared for processing the SOLMET and ERSATZ data sets.
2. In the data for the nighttime hours, the solar radiation fields are sometimes filled with nines (meaning missing data) instead of zeros (meaning no solar radiation).
3. After 1964, some of the stations only have observations every 3 hours for the meteorological parameters, but the solar radiation data are presented for all hours. Since the insolation data for the ERSATZ stations are based on models that use these meteorological values as input, the other 2 hours are not based on observations but instead are based on a linear interpolation.
4. Even though the format indicates beam data are present, they are missing from the ERSATZ stations.
5. No indication of the data accuracy is provided.

2.4.5.2 SOLMET Volume 2—Final Report (Quinlan, 1979) This document describes the process used to rehabilitate the SOLMET data and to generate the ERSATZ data. Chapter 4 presents an overview of this process.

2.4.5.3 SOLDAY User's Manual (TD9739) Daily Solar Radiation—Surface Meteorological Data (NCDC, 1979) This document describes the SOLDAY data base format and lists the SOLDAY stations and their locations. It also describes the rehabilitation process used. Figure 2.5 shows the layout of the data fields.

2.4.5.4 TMY User's Manual (Hall et al. 1981) This document describes the TMY data base format and the selection process. Figure 2.2 shows the layout of the data fields.

The following are problem areas:

1. In the data for the nighttime hours, the solar radiation fields are sometimes filled with nines (meaning missing data) instead of zeros (meaning no solar radiation).
2. Even though the format indicates beam data are present, they are missing from the ERSATZ stations.
3. No indication of the data accuracy is provided.
4. The documentation does not give the station latitude, longitude, or elevation; nor are these data on the tape. The user must obtain the data elsewhere. These values are almost always needed with the data.

2.4.6 Errors and Limitations of the Data

2.4.6.1 Limiatations as a Result of Measurement Accuracy Most of the insolation data are based on measurements by a pyranometer, which has a wide field of view (measures radiation from an entire hemisphere). Since there is no accepted standard reference instrument, the calibration must be transferred from a narrow field-of-view instrument. The accuracy of this calibration and the varying conditions between calibration and field measurements limit the practical absolute accuracy of the instrument to about 2%. One cannot therefore expect any greater accuracy in the recorded data.

Virtually all of the direct data in the available data bases is not measured but is derived from a model. The model introduces additional error, and, thus, the direct solar radiation data are likely to be less accurate. Further, the diffuse is usually computed by subtracting the beam from the global, allowing for potentially larger errors in the diffuse data or data from radiation on a tilt. This topic is covered in greater detail in chapter 5.

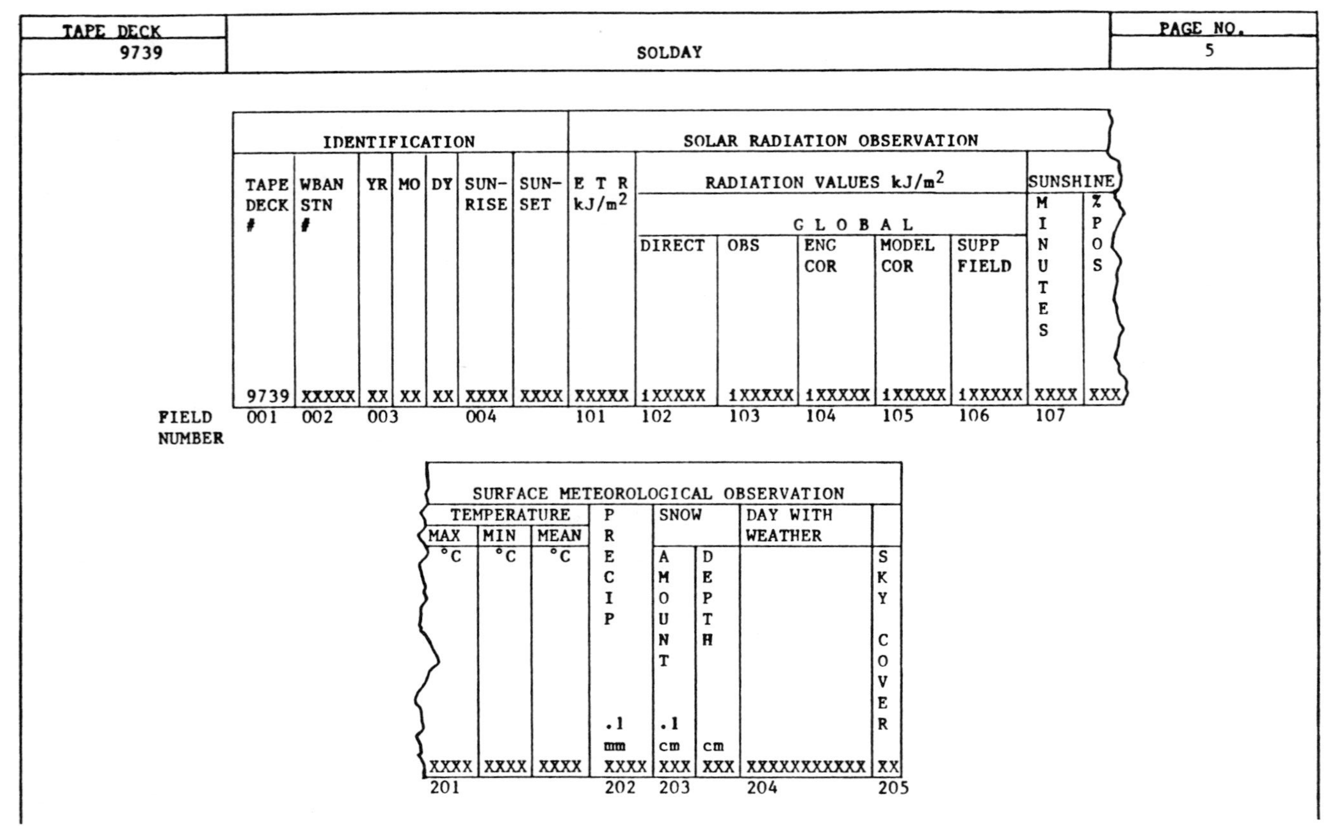

TAPE DECK		PAGE NO.
9739	SOLDAY	5

	IDENTIFICATION							SOLAR RADIATION OBSERVATION							
									RADIATION VALUES kJ/m²					SUNSHINE	
										GLOBAL					
	TAPE DECK #	WBAN STN #	YR	MO	DY	SUN-RISE	SUN-SET	E T R kJ/m²	DIRECT	OBS	ENC COR	MODEL COR	SUPP FIELD	MINUTES	% POS
	9739	XXXXX	XX	XX	XX	XXXX	XXXX	XXXXX	1XXXXX	1XXXXX	1XXXXX	1XXXXX	1XXXXX	XXXX	XXX
FIELD NUMBER	001	002	003			004		101	102	103	104	105	106	107	

SURFACE METEOROLOGICAL OBSERVATION							
TEMPERATURE			PRECIP	SNOW		DAY WITH WEATHER	SKY COVER
MAX	MIN	MEAN		AMOUNT	DEPTH		
°C	°C	°C	.1 mm	.1 cm	cm		
XXXX	XXXX	XXXX	XXXX	XXX	XXX	XXXXXXXXXXX	XX
201			202	203		204	205

TAPE FIELD NUMBER	TAPE POSITIONS	ELEMENT
001	001-004	TAPE DECK NUMBER
002	005-009	WBAN STATION NUMBER
003	010-015	DATE (YEAR, MONTH, DAY)
004	016-023	SUNRISE/SUNSET (LOCAL STANDARD TIME-HOUR AND MINUTE)
101	024-028	EXTRATERRESTRIAL RADIATION
102	029-034	DIRECT RADIATION
103	035-040	GLOBAL RADIATION ON A HORIZONTAL SURFACE-OBSERVED DATA
104	041-046	GLOBAL RADIATION ON A HORIZONTAL SURFACE-ENGINEERING CORRECTED DATA
105	047-052	GLOBAL RADIATION ON A HORIZONTAL SURFACE-MODEL CORRECTED DATA
106	053-058	ADDITIONAL RADIATION MEASUREMENT
107	059-065	SUNSHINE (MINUTES AND PERCENT OF POSSIBLE)
201	066-077	TEMPERATURE (TENTHS OF DEGREES C)
202	078-081	PRECIPITATION (TENTHS OF MILLIMETERS)
203	082-087	SNOW (SNOWFALL AMOUNT IN TENTHS OF CENTIMETERS; SNOW DEPTH IN WHOLE CENTIMETERS)
204	088-098	DAY WITH WEATHER
205	099-100	SKY COVER (TENTHS)

Figure 2.5
SOLDAY tape deck.

2.5 Needs for the Future

The more we learn about solar radiation, the more we become aware of the need for better knowledge and the gaps in the present knowledge. This section presents some areas where more work is needed.

A priority is to reestablish a high-quality solar radiation monitoring network. This network should collect continuous data for a minimum of 10 years, although longer would be better. This network should include instrumentation to establish the spectral content of the terrestrial solar radiation.

A comprehensive reference or text is needed on the subject of solar radiation measurement and the use of those data. Although some good references exist, much has been learned since they were published, and that information is scattered. This new text should include the following:

1. information on how to use the data,
2. a clear and detailed definition of the measures,
3. a complete history of instrument types, errors, corrections, etc.,
4. practical accuracies to be expected from the data,
5. current knowledge of various models used with the data:

- global horizontal to beam,
- global and beam to diffuse, etc., and

6. identification of known data discrepancies.

An excellent text (Iqbal, 1983) is available on the theory of solar radiation, but one is still needed on the measurement process.

Better calibration methods are needed. The current capabilities for calibration accuracy are roughly

1. Pyrheliometers—absolute instruments—0.5% under typical circumstances, 0.2% in very favorable situations.

2. Pyrheliometers—field instruments—1–2%; however, the real limitation here is the instruments' performance in the field and the relation of the measurement conditions to the conditions at the time of calibration. Another limitation is the training of the staff to take care of the instrument. Lack of care accelerates the problems.

3. Pyranometers—transfer of calibration from a pyrheliometer to a pyranometer is about 2%. However, the real problem here is the variability of sky conditions, the distribution of the diffuse brightness over the sky dome at the time of calibration and the time of measurement, and the relative intensities of the beam and the diffuse energy. Since these parameters are quite variable, depending upon sky conditions, the measure itself—total global horizontal radiation—cannot be precisely quantified. Hence, there must remain a considerable amount of statistical variability in the measure just because of its imprecise definition with respect to other standards. To obtain more accurate measurements it will be necessary to define new measures, such as a combination of direct beam and an array of values of the diffuse as a function of position on the sky dome.

Better measurements of the spectral content of solar radiation are needed. Some of the use of these data, if they were available, would be to

1. evaluate the resource availability for flat-plate and concentrating photovoltaics as a function of the spectrally selective scattering of the beam caused by atmospheric dust and other particulates, and
2. monitor the air pollution at various levels of the atmosphere by spectral measures of the solar radiation.

Statistical methods are needed that allow data users to understand data quality in terms of the performance of the solar energy systems that they are designing. Although these methods may exist, they are not commonly expressed in terms data users understand.

Finer geographic resolution of measured data is needed. Most system designers today must interpolate between points of measurement.

Most users of insolation data require associated weather data, especially temperature. Data bases need to be generated that include temperature measurements and other meteorological parameters with the solar radiation data.

Better methods are needed for describing the nonuniformity of the brightness of the sky dome to allow better specification of the energy available on surfaces at a variety of orientations.

More comprehensive specifications of instruments for measuring solar radiation are needed. This includes the specification of idealized detectors for each type of measure currently made and the deviation of each instrument from this ideal.

New instruments for measures, such as the nonuniformity of the sky dome, are needed.

A comprehensive philosophy, structure, and organization of the measurement process is needed. Historically, insolation measurement has been related to the individual experimenters and their respective interests. Rather than a single goal, there has been a variety of projects. The currently available data are still largely measurement oriented rather than user oriented—a focus that is needed. The rest of this section contains suggestions for such a program philosophy.

A primary goal of a resource assessment program should be to provide a standardized method for predicting the availability of the solar resource at any given place over any given time period and to improve the accuracy of that prediction. Such a procedure would necessarily include forecasting, measurement, interpolation, modeling, and a knowledge of errors and uncertainty and their effect on the estimate. Such a procedure should

1. give values or sets of values where combined occurrence is important,
2. give an error band for those values (and an uncertainty),
3. give the expected variation or distribution about those values,
4. give the same result no matter who uses the procedure,
5. give better values with better or more data,
6. be adopted by, authorized by, or otherwise approved by appropriate standards committees or organizations,
7. be acceptable in a court of law,
8. be traceable to accepted standard measures,
9. be testable and provable,
10. include long-term climate stability factors (e.g., pollution increases),
11. work for broad spectrum data (0.2–4.0 μm), for narrow spectrum (photovoltaic or biological) measures, or be confined to the visible part of the spectrum,
12. work for the variety of measures required: total global horizontal radiation, direct beam radiation, diffuse radiation only, energy through a window, etc.,
13. work for new types of system (e.g., multilayer photovoltaics), and
14. incorporate the results of both measurements and modeling.

The existence of such a procedure will

1. provide a precise definition of data quality,
2. provide a precise framework for specifying a data user's needs,
3. provide a framework for estimating the cost of attaining a specific data quality level,
4. provide a mechanism for evaluating solar system performance at any given location, and
5. provide a clear mechanism for defining intermediate objectives in a resource assessment program.

This procedure does not have to be simple or easy to understand by everyone. It may be desirable to have a single agency or group of agencies provide a common and current data base for all users.

Most users of insolation data do not or cannot foresee a future need for those data far enough in advance of their need to make their own measurements. Only after they recognize the need is a request for the data generated. The planning for and the measurement of solar radiation must be based on extrapolated demands as well as foresight and scientific inquiry, since many years of measurements are normally required to satisfy a need.

Acknowledgments

I wish to thank the editor, Roland Hulstrom, of SERI and the reviewers Valentine S. Szwarc, Loren Crow, and Eugene Clark for their careful review and helpful suggestions. I also wish to thank my wife, Linda, for reviewing the manuscript.

I also wish to acknowledge the assistance and information provided by Loren Crow, a consulting meteorologist, and Frank Quinlan and Richard Cram of the NCDC for their current information on the various data bases.

References

Bahm, R. J. 1982. "Radiometry—the data." *Advances In Solar Energy* 1(1):1–18.

Bennett, I. 1965. "Monthly maps of mean daily insolation for the United States." *Solar Energy* 9(3):145.

Berdahl, P., D. Grether, M. Martin, and M. Wahlig. 1978. *California Solar Data Manual.* LBL-5971. Berkeley, CA: Lawrence Berkeley Laboratory.

Boes, E. C., H. E. Anderson, I, J. Hall, R. R. Prairie, and R. T. Stromberg. 1977. *Availability of Direct, Total, and Diffuse Solar Radiation to Fixed and Tracking Collectors in the USA.* SAND77-0885. Albuquerque, NM: Sandia National Laboratories.

Boes, E. C., I. J. Hall, R. R. Prairie, R. T. Stromberg, and H. E. Anderson. 1976. *Distribution of Direct and Total Solar Radiation Availabilities for the USA*. SAND76-0411. Albuquerque, NM: Sandia National Laboratories.

Cinquemani, V., J. R. Owenby, and R. G. Baldwin. 1978. *Input Data for Solar Systems.* Asheville, NC: National Climatic Data Center.

Crow, L. 1981. "Development of hourly data for weather year for energy calculations (WYEC)." *ASHRAE Journal* 23(10).

Crow, L. 1984. "Weather year for energy calculation (WYEC) data tapes—now complete." *ASHRAE Journal* 26(6).

Evans, D. B., D. F. Grether, A. Hunt, and M. Wahlig. 1980. *The Spectral Character of Circumsolar Radiation.* LBL-10802. Berkeley, CA: Lawrence Berkeley Laboratory.

Fowlkes, C. W. 1982. *Montana Solar Data Manual.* Helena, MT: Montana Department of Natural Resources and Conservation.

Goodrich, R. W., K. C. Buisick, and C. A. Janowiec. 1982. *The Northeast Utilities Solar Resource Evaluation Project—Solar Energy Measurement During 1974–1981.* Hartford CT: Northeast Utilities.

Grether, D. F., D. Evans, A. Hunt, and M. Wahlig. 1979. *Measurement and Analysis of Circumsolar Radiation.* LBL-10243. Berkeley, CA: Lawrence Berkeley Laboratory.

Grether, D. F., D. Evans, A. Hunt, and M. Wahlig. 1980. *Measurement and Analysis of Circumsolar Radiation.* LBL-11645, Berkeley, CA: Lawrence Berkeley Laboratory.

Hall, I. J., R. R. Prairie, H. E. Anderson, and E. C. Boes. 1981. "Generation of typical meteorological years for 26 SOLMET stations." Appendix to *TMY User's Manual.* Asheville, NC: National Climatic Data Center.

Iqbal, M. 1983. *An Introduction to Solar Radiation.* New York: Academic Press.

Klein, S. A., W. A. Beckman, and J. A. Duffie. 1977 (rev. Aug. 1978). *Monthly Average Solar Radiation on Inclined Surfaces for 261 North American Cities.* EES 44-2. Madison, WI: University of Wisconsin, Solar Energy Laboratory, Engineering Experiment Station.

Knapp, C. L., T. L. Stoffel, and S. D. Whitaker. 1980. *Insolation Data Manual.* SERI/SP-755-789. Golden, CO: Solar Energy Research Institute.

Knapp, C. L., and T. L. Stoffel, 1982. *Direct Normal Solar Radiation Data Manual.* SERI/SP-281-1658. Golden, CO: Solar Energy Research Institute.

McDaniels, D. K., F. Vignola, J. Hull, P. Erickson, P. Ryan, and D. Fong. 1983. *Pacific Northwest Solar Radiation Data.* Eugene, OR: Solar Energy Center, University of Oregon.

Menicucci, D. F., and J. P. Fernandez. 1986. *Estimates of Available Solar Radiation and Photovoltaic Energy Production for Various Tilted and Tracking Surfaces throughout the United States Based on PVFORM, a Computerized Performance Model.* SAND85-2775. Albuquerque, NM: Sandia National Laboratories.

National Climatic Data Center (NCDC). n.d.1. *Card Deck 280 Solar Radiation Hourly Record.* Asheville, NC: Data Processing Division, NCDC.

National Climatic Data Center (NCDC). n.d.2. *Card Deck 480 Solar Radiation—Summary of Day.* Asheville, NC: Data Processing Division, NCDC.

National Climatic Data Center (NCDC). 1968. *Climatic Atlas of the United States.* Asheville, NC: NCDC.

National Climatic Data Center (NCDC). 1976. *ASHRAE Test Reference Year Reference Manual (TD9706)*. Asheville, NC: NCDC.

National Climatic Data Center (NCDC). 1977–1980. *Monthly Summaries of Solar Radiation Data*. Volumes 1–4. Nos. 1–12. Asheville, NC: NCDC.

National Climatic Data Center (NCDC). 1978. *SOLMET Volume 1 User's Manual (TD9724) Hourly Solar Radiation—Surface Meteorological Observations*. Asheville, NC: Environmental Data and Information Service, NCDC.

National Climatic Data Center (NCDC). 1979. *SOLDAY User's Manual (TD9739) Daily Solar Radiation—Surface Meteorological Data*. Asheville, NC: Environmental Data and Information Service, NCDC.

National Climatic Data Center (NCDC). 1981. *Typical Meteorological Year User's Manual (TD9734) Hourly Solar Radiation—Surface Meteorological Observations*. Asheville, NC: Environmental Data and Information Service, NCDC.

National Climatic Data Center (NCDC). 1983. *Selective Guide to Climatic Data Sources, Key to Meteorological Records Documentation*. Asheville, NC: NCDC.

Patapoff, N. W., and R. W. Yinger. 1976–1980. *The WEST Associates Solar Resource Evaluation Project—Solar Energy Measurements at Selected Sites throughout the Southwest*. Rosemead, CA: Southern California Edison.

Quinlan, F. T., ed. 1979. *SOLMET Volume 2—Final Report*. Asheville, NC: Environmental Data and Information Service, NCDC.

Randall, C. M. 1976. *Insolation Data Base*. ATR-76(7523-11)-9. El Segundo, CA: Aerospace Corporation.

Randall, C. M., and M. E. Whitson, Jr. 1977. *Hourly Insolation and Meteorological Data Bases Including Improved Direct Insolation Estimates*. ATR-78(7592)-1. El Segundo, CA: The Aerospace Corporation.

Solar Energy Research Institute (SERI). 1978. *Solar Energy Meteorological Research and Training Site Program Second Annual Report*. SERI/SP-290-1478. Golden, CO: SERI.

Solar Energy Research Institute (SERI). 1981. *Solar Radiation Energy Resource Atlas of the United States*. SERI/SP-642-1037. Golden, CO: SERI.

Solar Energy Research Institute (SERI). 1983. *Solar Radiation Data Directory*. SERI/SP-281-1973. Golden, CO: SERI.

Watt, Engineering, Ltd. 1978. *On the Nature and Distribution of Solar Radiation*. HCP/T2552-01. Washington, DC: U.S. Department of Energy.

3 Insolation Models and Algorithms

Charles M. Randall and Richard Bird

3.1 Introduction

Insolation data are used within the solar energy community for designing and sizing solar energy systems. The required insolation values frequently are not available directly from measurements at the site of interest. These values must then be inferred from auxiliary data using an estimation model (algorithm). In this chapter we review the various types of insolation estimation models, with particular emphasis on those models based on the widely used SOLMET/ERSATZ data bases (Quinlan, 1977, 1979) and tables (SERI, 1981; Knapp and Stoffel, 1982).

Insolation modeling is the process by which required insolation values are obtained from some other observable quantities. Most practical solar energy system design applications require at least some insolation modeling. The reasons are numerous: Insolation values are often required for a surface with an orientation different from the surface used for reference in the available data; insolation observations for the system location are either nonexistent or inadequate; available insolation observations for a site do not include some insolation component of interest. In this chapter we concentrate on describing models for formulating sequential insolation data sets such as the SOLMET/ERSATZ data set. The terms insolation and solar radiation are used interchangeably.

Our purpose is to review and summarize the various models and algorithms developed to relate the various components of insolation with each other and to estimate unavailable data. These topics have been discussed in many books and a variety of journals. Rather than try to review this extensive literature comprehensively, we shall concentrate on those models used to formulate the widely used U.S. insolation data bases (chapter 2) and to evaluate the accuracy of these methods and models.

Nomenclature

AZ	azimuth of the sun
CCF	simplified ASHRAE cloud cover factor
CLD	cloud factor
D	direct (direct normal) insolation
Dec	declination of the sun

D_o	solar constant
D_x	extraterrestrial direct solar radiation
dy	day of the year
EL	elevation of sun above horizon
ET	equation of time: (apparent solar time) − (mean solar time)
GR_t	ground-reflected insolation
HR	hour angle of sun
K_d	fractional transmittance of direct insolation
K_t	fractional transmittance of the global horizontal insolation
k	conversion from degrees to radians = 0.01754329°/rad
Lng	longitude of site
Lnt	longitude of standard time meridian
LAT	latitude of site
LST	local standard time
OPQ	opaque cloud cover
P	coefficient
Q	global (global-hemispheric) insolation (global-horizontal)
Q_B	global-horizontal insolation
Q_x	extraterrestrial global radiation
R	distance from the earth to the sun
$\bar{R}$	ratio of daily average insolation on titled surface to that on horizontal surface
R_b	ratio of average direct normal insolation on tilted to horizontal for each month
R_o	mean distance from south to sun, 1.496×10^8 km
S	diffuse radiation from sky from SERI (BIRD) model
$\bar{S}$	monthly average daily diffuse insolation
SRCS	clear sky irradiance
TST	true solar time
z	solar zenith angle
$\langle \rangle$	average value of the arguments over the time interval of interest

β	tilt of surface from horizontal
ρ	ground reflectance (albedo)
θ	incidence angle on a tilted surface
ϕ	azimuth angle of the sun

3.2 Insolation Modeling Approaches

3.2.1 Insolation Fundamentals

A common basis of understanding is required to discuss the various insolation models. We first present an overview of the relationships relevant to the later discussion of specific models. Similar information is in the introductory material in many atmospheric science and solar energy engineering textbooks (Boes, 1981; Sayigh, 1977; Stine and Harrigan, 1985; Paltridge and Platt, 1976; Iqbal, 1983).

3.2.1.1 Insolation Measurements Solar radiation reaches the surface of the earth directly from the sun and indirectly after being scattered or reflected one or more times, as shown in figure 3.1. Iqbal (1983) summarizes the various processes that attenuate and scatter solar radiation as it passes through the atmosphere.

Since some solar energy applications use only the radiation coming directly from the sun and others can use radiation from the entire sky, insolation is usually measured in two complementary ways. These two parameters could, in principle, be any two of the following: direct, global, and diffuse insolation.

Direct or direct normal insolation D is the solar radiation (W/m^2) incident on a surface perpendicular to the sun's rays. The field of view of the measuring instrument, a pyrheliometer, is limited, so it only measures radiation coming directly from the sun. Ideally, the field of view would be limited to the apparent angular diameter of the sun (about 0.5°); but this imposes an accuracy requirement on the solar tracker, needed to keep the pyrheliometer pointed at the sun, that is too stringent for practical solar trackers. Consequently, a somewhat larger field of view of about 5.5° is commonly used.

Global or global-hemispheric insolation Q is the total (i.e., direct and diffuse sky) solar radiation (W/m^2) on a horizontal surface. The field of view of the measuring instrument, a pyranometer, is 2π sr, which includes

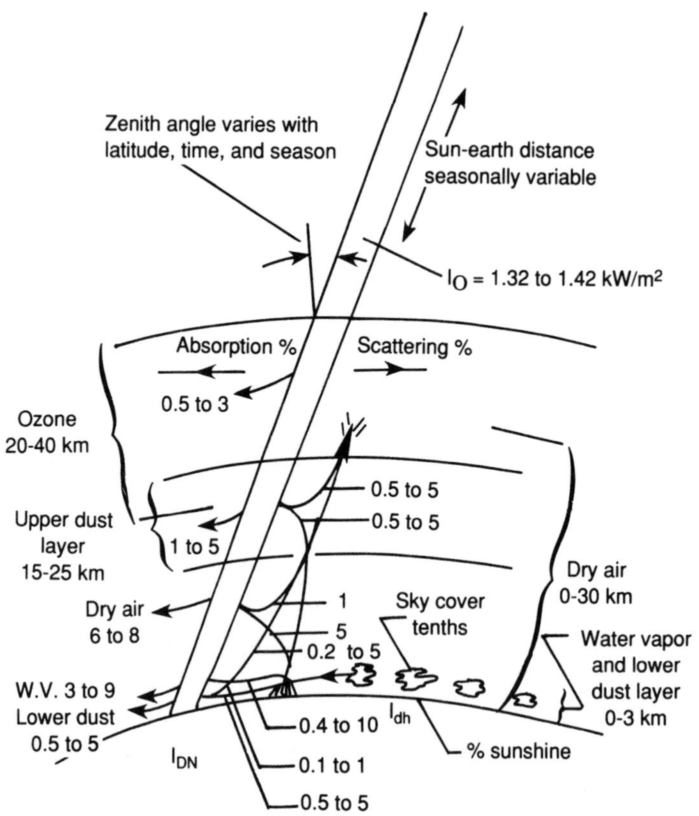

Figure 3.1
Nominal range of clear sky absorption and scattering of incident solar energy. Values are typically for one air mass. Sources: Watt (1978) and Stine and Harrigan (1985).

the entire upward hemisphere. The sensor responds to radiation coming directly from the sun as well as the diffuse radiation coming from the rest of the sky S, where

$$Q = D\cos(z) + S \tag{3.1}$$

and z is the solar zenith angle (i.e., the angle between the solar disk and the zenith).

The diffuse solar radiation coming from the sky is sometimes measured using a pyranometer with an occultation device to prevent the direct solar radiation from reaching the sensor. Using the occultation disk or shadow band requires that the observation be corrected to compensate for the part of the sky also obscured. This causes such techniques to be less accurate than direct or global measurement (chapter 5).

Additional specialized solar radiation measurements are sometimes required to support particular projects. The actual solar radiation distribution within the pyrheliometer's 5.5° field of view is a function of the details of the forward scattering characteristics of the atmospheric aerosols and may be critically important in highly concentrating solar energy systems. The Lawrence Berkeley Laboratory designed and deployed four instruments to make routine measurements of this circumsolar radiation (Grether et al., 1981; Watt, 1980). These measurements are the basis for empirical analytical models of the sun's angular brightness distribution (Vittitoe and Biggs, 1981).

Studies of the transient response at the Solar One power tower required coincident measurements at high time resolution over the area the heliostat field would occupy (Randall, Whitson, and Johnson, 1978; Randall, Johnson, and Whitson, 1980). Solar energy systems operating with flat-plate collectors tilted at an angle will often have a pyranometer mounted at the same tilt.

3.2.1.2 Geometric Effects In the absense of clouds, the sun moving across the sky during the day varies the insolation significantly by changing the direct beam's path length through the atmosphere and the projection angle of the sun on the horizontal reference surface. If the latitude, longitude, and local standard time (including the date) of the location are known, we can compute the position (azimuth and elevation) of the sun. First we need to determine for the day in question the sun's distance from the earth and its position in the spherical celestial coordinate system specified by declination (analogous to terrestrial latitude) and the equation of time. The equation of time is the angular distance between the true position of the sun and the position of the mean sun. These quantities for any given day change slightly from year to year and can be obtained exactly from a published ephemeris, such as *The Astronomical Almanac* or *Astronomical Phenomena*, for any given year. Either of these volumes is available from the U.S. Government Printing Office. Approximate formulas, which ignore the variations from year to year, are often suitable for solar energy system calculations. Various formulas have been presented in the literature. The following are provided by Spencer (1971) and are expressed in terms of an angle:

$$\mathrm{TH} = 2\pi \mathrm{dy}/365 \text{ (rad)}, \tag{3.2}$$

$$(R/R_o)^2 = 1.000110 + 0.034221\cos(\mathrm{TH}) + 0.001280\sin(\mathrm{TH}) + 0.000719\cos(2\,\mathrm{TH}) + 0.000077\sin(2\,\mathrm{TH}), \tag{3.3}$$

$$\mathrm{ET} = 0.0172 + 0.4281\cos(\mathrm{TH}) - 7.3515\sin(\mathrm{TH}) - 3.3495\cos(2\,\mathrm{TH}) - 9.3619\cos(2\,\mathrm{TH}), \tag{3.4}$$

$$\mathrm{Dec} = 0.39637 - 22.91326\cos(\mathrm{TH}) + 4.02543\sin(\mathrm{TH}) - 0.38720\cos(2\,\mathrm{TH}) + 0.05197\sin(2\,\mathrm{TH}) - 0.15453\cos(3\,\mathrm{TH}) + 0.08479\sin(3\,\mathrm{TH}), \tag{3.5}$$

where

R = the distance from the earth to the sun (km),

R_o = the mean distance from the earth to the sun, 1.496×10^8 km (Allen, 1973),

ET = the equation of time, the difference between the hour angle of the actual solar position and the mean solar position (min) = (apparent solar time) – (mean solar time),

dy = the day of the year, counting from January 1 as day 0 to December 31 as day 364, and

Dec = the declination of the sun (deg).

The maximum error in the declination is $<0.03°$ if all terms are kept. If the last two terms are omitted, the maximum error is still only 0.2°. The maximum error in ET is about 35s. The maximum error in $(R/R_0)^2$ is less than 0.0001. Walraven (1978) provides a FORTRAN program that contains more accurate relationships that account for the variations from year to year and provide results to within 0.01°.

Once the position of the sun is known, we can compute the azimuth and elevation of the sun:

$$\mathrm{TST} = \mathrm{LST} + (\mathrm{Lnt} - \mathrm{Lng})/15 + \mathrm{ET}/60, \tag{3.6}$$

$$\mathrm{HR} = (\mathrm{TST} - 12)15, \tag{3.7}$$

$$\mathrm{EL} = \arcsin[\sin(k\,\mathrm{Lat})\sin(k\,\mathrm{Dec}) + \cos(k\,\mathrm{Lat})\cos(k\,\mathrm{Dec})\cos(k\,\mathrm{HR})], \tag{3.8}$$

$$z = 90 - \mathrm{EL}, \tag{3.9}$$

$$\mathrm{AZ} = \arcsin[\cos(\mathrm{Dec})\sin(\mathrm{HR})/\cos(\mathrm{EL})], \tag{3.10}$$

where

TST = true solar time (h),

LST = local standard time (h),

Lnt = longitude of the standard time meridian (deg),

Lng = longitude of the site (deg),

HR = hour angle of the sun (deg),

EL = elevation of the sun above the horizon (deg),

k = conversion from degrees to radians = 0.01745329 rad/deg,

AZ = solar azimuth measured from south (deg), and

Lat = latitude of the site (deg).

The solar azimuth can exceed 90°, so the algorithm for determining AZ must include provisions to recognize this. This computer subroutine of Walraven (1978) provides an example of such a procedure.

We can now compute the extraterrestrial direct solar radiation D_x (i.e., radiation at the top of the earth's atmosphere):

$$D_x = D_o(R_o/R)^2. \tag{3.11}$$

The parameter D_o is the solar constant, which various investigators have measured. Thekaekara and Drummond (1971) recommend a value of 1.353 kW/m^2 for D_o. This value was corrected upward by 1.8% based on comparisons by Fröhlich et al. (1973), resulting in a solar constant of 1.377 kW/m^2 on an absolute scale. The SOLMET/ ERSATZ data base (Quinlan, 1977) was prepared using the solar constant of 1.377 kW/m^2 (Quinlan, 1979). Measurements from space indicate that the solar output does vary by a few hundredths of a percent over the course of a year (Willson, Hudson, and Woodard, 1984; Willson, 1984). Although such variations are of great interest to solar physicists (White, 1977), these variations are insignificant when compared with how the atmosphere affects the amount of insolation.

We can now derive the extraterrestrial global radiation Q_x:

$$Q_x = D_x \cos(z). \tag{3.12}$$

3.2.2 Approaches to Modeling Atmospheric Effects

3.2.2.1 Physical Effects and Associated Models Weather phenomena associated with the atmosphere add a high degree of complexity to determining insolation values beyond the simple formulas presented in the previous sections. This precludes a complete yet practical computation of the insolation from first principles. Various approaches were developed to deal with this complexity. In this section we describe in general terms the different approaches used. Although there are numerous examples of each (Iqbal, 1983), we are only reviewing a few here. An overview is presented of the specific models used to prepare some of the widely used insolation data bases developed in recent years.

The gases in the clear atmosphere selectively absorb radiation at some wavelengths characteristic of the specific gases (chapter 6). Ozone and oxygen limit the shortwave transparency of the atmosphere. Water vapor, carbon dioxide, and ozone absorb intensely in limited regions in the longer-wave infrared region (beyond about 0.7 μm). In addition, the molecules of the atmosphere scatter radiation. The strength of this scattering depends inversely on the fourth power of the wavelength; therefore, the scattering is much stronger with short wavelengths. This causes the blue appearance of the clear sky. Figure 3.2 illustrates the sun's spectrum through a clear atmosphere, identifying the absorption bands of the atmospheric gases.

Aerosols cause additional changes in the amount of solar radiation reaching the ground by absorbing some radiation but mostly by scattering the radiation. The spectral structure (chapter 6) associated with the aerosol

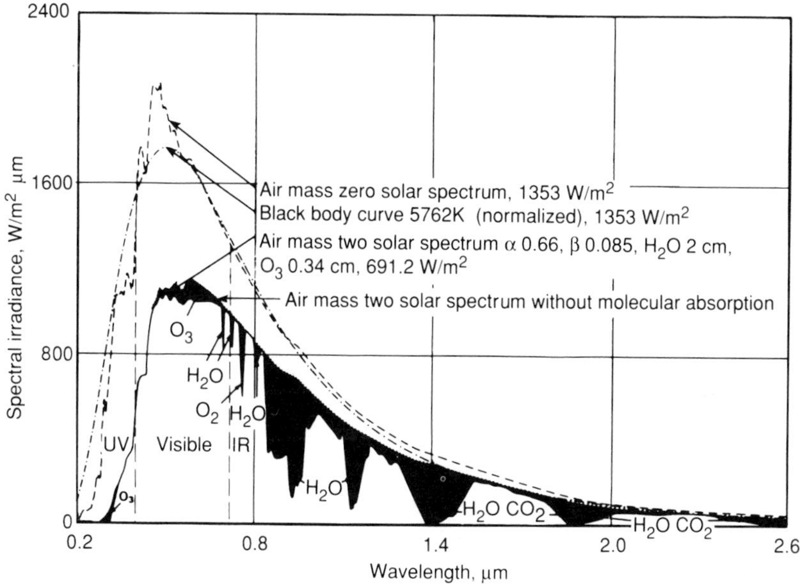

Figure 3.2
Solar spectral irradiance for several different air masses. Source: Sayigh (1977).

effects is much smoother than that for molecular absorption, as indicated by the atmosphere appearing to be white when the aerosol (e.g., dust) content is significant. The solar radiation under nominally clear conditions is then that of the sun modified by a combination of aerosol and molecular effects and the path length through the atmosphere. The relative importance of each is determined by the specific meteorological conditions present at the time (Iqbal, 1983).

Clouds may be thought of as a layer or layers of aerosols sufficiently thick that the radiation reaching the earth's surface is all diffuse (i.e., no direct component). The diffuse radiation may not be uniform over the sky because of variations in cloud thickness or the absence of clouds in some portion of the sky. Modeling the effects of clouds is extremely difficult, particularly in the case of partial cloud cover. There can even be cloud-induced increases in radiation caused by reflections from cloud edges.

The absorption of radiation by the molecules in the earth's atmosphere is not spectrally continuous (chapter 6), but rather occurs at tens of thousands of discrete wavelengths governed by the discrete quantum energy states available to the molecule. At each of these wavelengths, the atmospheric transmittance decreases exponentially with increasing amounts of the gas in the path. Catalogs of the required spectral line parameters are available (Rothman et al., 1982), and the mathematical relationships required to use them are well-known. However, the computational effort required to use all of this detailed information is enormous and not warranted or practical for most solar energy applications. Lower-spectral-resolution models of molecular absorption were formulated originally because there were no computational resources and no detailed spectral line parameters (Elsasser and Culbertson, 1960). A number of approaches were taken in formulating these so-called "band models." Today, one of the most widely used model is the semiempirical model of McClatchey et al. (1972), which forms the basis for the series of LOWTRAN computer codes prepared by the U.S. Air Force Geophysics Laboratory (AFGL). SERI researchers modified the LOWTRAN code available in the late 1970s for solar energy applications by including an extraterrestrial solar spectrum (chapter 6); the resulting code is known as SOLTRAN. More recent versions of the AFGL code, beginning with LOWTRAN6, also contain a solar spectrum option and can compute the direct solar radiation (Kneizys et al., 1983).

Aerosols will in general absorb and scatter radiation. Since a photon can be scattered more than once, a complete computation of the radiation distribution in a scattering atmosphere can be very time consuming, even if the scattering properties of the individual aerosol particles are known. The scattering and absorption properties of the individual aerosol particles can also be modeled based on well-known physical principles expressed in mathematical terms, usually associated with the name Mie (van de Hulst, 1981). Assuming the particles are spherical, these individual particle effects can be computed from a description of the size distribution of the particles and a knowledge of their refractive index. Mie-scattering techniques form the basis for computing the aerosol parameters available in the more recent LOWTRAN codes. Early versions of the LOWTRAN code considered aerosols only as a loss mechanism and did not include radiation scattered into the beam. LOWTRAN6 (Kneizys et al., 1983) was the first version to include single scattering into the beam and would be appropriate for computing diffuse insolation only under very clear conditions. LOWTRAN7 (Kneizys et al., in press) includes a more comprehensive two-stream multiple scattering model. Dave, Halpern, and Braslau (1975) have applied more detailed multiple scattering techniques to some solar energy applications, but the computational effort is usually not justified because of the lack of accurate data (i.e., aerosol characteristics) on which to base the computations for a specific location and time.

In summary, the effect of the clear atmosphere is to modify both the total amount and the spectral distribution of the radiant energy available from the sun. A more detailed discussion of these spectral effects is in chapter 6. Application of these fundamental physical models for solar spectral irradiance to solar energy system simulations is restricted by two problems: (1) The complexity of the complete mathematical description of all the phenomena requires significant computational resources, and (2) the required comprehensive description of the meteorological conditions is not usually available. Fortunately, the weather-induced spectral insolation variation can be ignored for many solar energy applications, so that simpler total solar energy models have great utility. SERI researchers (Bird and Hulstrom, 1981) have derived such a model based on spectrally integrated coefficients from SOLTRAN (Bird and Hulstrom, 1980) and compared it with a number of older, physically based models. However, the user of these simplified models should retain an awareness of their limitations and the situations where the spectral variations may be important, e.g., for photovoltaic systems.

3.2.2.2 Statistical Correlations The straightforward estimation of insolation using a statistical correlation between the desired insolation parameters and one or more commonly measured meteorological parameters is the approach frequently used to estimate insolation. The most commonly used meteorological parameters are minutes of sunshine or some measure of cloud cover. The usual procedure is to derive the correlation coefficients from some limited data base, containing both measured insolation values and values of whatever meteorological parameters are being correlated. The derived correlation coefficients are then used to estimate insolation values from the observed meteorological parameters for some time period, location, or both for which the desired insolation parameters are not available. Since there is nothing inherent in this procedure to assure that the correlations are appropriate to any other location or time, the literature is filled with studies of some insolation parameter or parameters correlated with one or more meteorological parameters. No attempt is made here to review that literature. Several specific examples are cited in describing the formulation of insolation data bases for the United States, which relied heavily on statistical correlation models.

Statistics can be powerful in providing useful insolation values for evaluating solar energy systems. However, understanding the physical principles that give rise to the observed correlations is extremely important in selecting optimal parameters for correlation and in including all of the appropriate meteorological parameters in the correlation. As an example, the reflection of radiation from snow-covered ground to clouds affects the observed value of global insolation under cloudy conditions. Including this physical insight in selecting parameters to correlate with the insolation improves the accuracy of estimation procedures (Hay, 1976). Removing as many deterministic effects as possible is also important. For this reason the global insolation measure usually correlated is the percentage of possible insolation. This is obtained by dividing the observed global insolation by the extraterrestrial global insolation [equation (3.12)] rather than by expressing the global insolation directly in energy units, with include the solar projection angle factor.

3.2.2.3 Other Modeling Approaches When neither insolation values nor meteorological data sufficient for applying some correlation-based estimation algorithm are available, one can usually obtain insolation estimates either by analogy with some other area having a similar climate or by adopting the values for some nearby locations. The adequacy of this approach depends on the spatial variability of insolation.

The availability of meteorological satellites, which produce images of the earth at frequent intervals, can be used to determine insolation at any location within the satellite's field of view. [Bahm (1981) provides an overview of the progress with this approach.] A satellite image can determine the extent of cloud cover at a location, and since cloud cover can partially determine the amount of insolation, one can estimate the insolation. However, the satellite cannot determine the thickness of the cloud layer, and thus the estimate may be inaccurate. Also, it is difficult for satellites to differentiate between clouds and ice or snow. Finally, the amount of data that must be processed can be enormous, so plan carefully when routinely applying these techniques or develop and use simplified techniques.

3.2.2.4 Thermal Sky Radiance Models Thermal radiation emitted to the sky at infrared wavelengths, particularly during a clear night, provides a thermal energy sink that can be used to cool buildings. Although this is not strictly an insolation model, we shall mention some of the work done in this area. The recent LOWTRAN models (Kneizys et al., 1983) can compute spectral sky radiance for any radiation path. The sky spectral radiance is calculated for a representative number of atmospheric paths. These radiances are then summed over the infrared spectral region from about 3 to 25 μm. The radiances for each path, properly weighted by a projection angle factor, appropriate to the orientation of the atmospheric path to the radiating surface are then combined. Although this procedure includes most of the relevant physical relationships for clear conditions, it involves extensive computations based on data for the atmosphere above the surface as well as surface conditions. Because it is complex, this approach is not very useful for parametric studies; however, simplified models are available that are more appropriate to parametric energy studies (Centeno, 1982).

This cooling resource was also measured at selected U.S. locations (Martin and Berdahl, 1984b). Berdahl and Fromberg (1982) and Martin and Berdahl (1984a) used these measurements to prepare and validate models that more immediately meet the needs of the energy system designer.

3.3 Global Insolation Models

When the emphasis on solar energy applications began in the early 1970s, reliable global insolation data bases were lacking for the United States that covered a statistically significant period of time (20 years). The need for such data required using various models to create the required data bases. In this section, we describe the models used to prepare the present global insolation data bases for the United States (chapter 2). Although the literature describes numerous similar models for other countries and specific locations, we shall not discuss them here.

3.3.1 Historical Models

Liu and Jordan (1960) were among the first investigators to develop systematic modeling procedures to infer values of hourly global and direct insolation from the available global-horizontal insolation data. These authors recognized that understanding the performance of solar energy systems depended on knowing the variation of insolation during the day, although most of the climatic data available then provided only mean daily global-horizontal insolation values. Based on detailed hourly insolation values for a few locations, they developed models to infer mean hourly insolation values from the daily values. These hourly values were then used to develop improved system performance measures. These models used as variables the fractional transmittance of the extraterrestrial direct or global insolation:

$$K_d = D/D_x, \tag{3.13}$$

$$\langle K_d \rangle = \langle D \rangle / \langle D_x \rangle, \tag{3.14}$$

$$K_t = Q/Q_x, \tag{3.15}$$

$$\langle K_t \rangle = \langle Q \rangle / \langle Q_x \rangle, \tag{3.16}$$

where

K_d = fractional transmittance of the direct insolation,

K_t = fractional transmittance of the global-horizontal insolation, and

$\langle\ \rangle$ = average value of the arguments over the time interval of interest (e.g., hours or days).

Data for applying these techniques were provided for 80 representative locations throughout North America (Liu and Jordan, 1963). These data were subsequently provided in the American Society of Heating, Refrigerating and Air-Conditioning Engineers (ASHRAE) handbooks (Jordan and Liu, 1977) and continue to be widely used. The insolation values included in those summaries are based largely on the measured values available at the time, some of which were subsequently shown to have significant errors.

Bennett (1965) prepared maps of mean monthly values of daily global insolation for the contiguous 48 states using the same basic insolation data. To provide more spatial detail, he obtained additional insolation values based on multiple linear regression from observed sunshine and site elevation data. Löf (Löf, Duffie, and Smith, 1966) used similar techniques to provide worldwide maps of insolation. Löf used cloud cover data in addition to sunshine and measured insolation values. Budyko (1963) derived worldwide insolation maps using similar techniques.

The Aerospace Corporation in 1973 needed serially complete hourly global insolation data for the southern California area. This information was used to provide a standard data base to compare various approaches for converting solar thermal energy. Leonard and Randall (1974) reviewed the available hourly data and, in the case of the Inyokern data, corrected for instrument drift and an error in converting local time to true solar time. The data were insufficient to reflect the dramatic climatic differences present in this limited geographic region, so they investigated correlations of insolation with cloud cover observations. Total and opaque sky cover are measures of cloudiness and are routinely observed hourly at most meteorological stations. Southern California insolation values correlated highly with these measures of cloud cover, as Bennett (1969) had demonstrated for other areas of the country. The fractional transmittance K_t had the highest correlation with opaque sky cover. Leonard and Randall (1974) then developed regression relationships between cloud cover and insolation from existing southern California insolation data. These relationships were then used to generate entire hourly data bases for some locations and to estimate missing data values for the rest. These data bases covered 1962 and 1963 and provided serially complete hourly surface meteorological values and insolation values for the entire 2 years. The techniques originally developed as a part of this Aerospace Corporation study of Southern California were eventually used to prepare 34 data sets as the studies were extended to include first the southwest (Hall et al., 1974) and eventually the entire contiguous 48 states (Randall, 1976).

Similar requirements for a standard insolation data set to compare solar photovoltaic conversion systems led Boes and his colleagues at Sandia National Laboratories to prepare similar data bases (Boes, 1975; Boes et al., 1976, 1977). They independently developed correction procedures for suspected inaccuracies in the raw data and developed direct insolation estimation procedures to be discussed later.

3.3.2 SOLMET

3.3.2.1 Description The requirement for a reliable standard insolation data base and the documented problems with the old data led to the development of several new data bases. The U.S. government initiated two parallel activities under the sponsorship of the Energy Research and Development Agency (ERDA), which later became the U.S. Department of Energy (DOE). First, a new network of improved insolation measuring and recording instruments was deployed (see chapter 4). Second, the existing hourly data were rehabilitated. The rehabilitation process, in fact, required a significant amount of modeling, which is summarized here. Quinlan (1979) describes these procedures in detail.

The original expectations of rehabilitating the hourly insolation data, based solely on applying carefully reviewed instrument calibrations, proved to be impossible. Some critical instrument calibrations were never carried out, and calibration procedures had sometimes used artificial light in an integrating sphere and at other times natural light, leading to an indeterminate uncertainty in the relative calibration of various sensors. It was decided that one could more accurately *model* the transmittance of the clear atmosphere than determine these calibration uncertainties. Therefore, the data were normalized to modeled global-horizontal insolation values for a specific location under clear sky conditions at local solar noon (Quinlan, 1979).

Using this procedure, researchers eventually corrected the data covering more than 20 years from 26 locations in the contiguous 48 states. The locations, known as SOLMET stations, are indicated by asterisks in figure 3.3. The first step in treating the data was to establish all known calibration factors for the various sensors used. The global insolation value for noon on each clear day was then obtained from the original strip-chart recordings using these calibration factors. The extraterrestrial radiation for the date and time was then used to normalize these values. Figure 3.4 shows 15-day averages of these mean clear noon K_t values for Bismarck, ND.

The basic clear-sky transmittance model used was that of Hoyt (1978, 1979). He used climatic average input data for each site to compute the clear solar noon transmittance for global radiation for each day of the year. When Hanson (1979) compared the clear-sky results from this model with a limited amount of insolation data from Bismarck, ND, Albuquerque, NM, and Raleigh, NC, he felt some slight water vapor and air mass dependent adjustments in the model were justified. He proceeded to develop empirical correction terms to Hoyt's model based on the limited data from these three locations. The result of these procedures was a table of clear solar noon values of K_t for each day of the year for each of the 26 locations considered. Figure 3.5 shows and example for Bismarck, ND.

The ratio of the modeled clear solar noon value to the observed value was then used to correct all observed values from that station for that day. For days without a clear solar noon, the correction factor was determined by interpolation from the closest adjacent days with clear noons. This procedure could only be used over a limited time span of 60 days or less. If clear noons were not available for more than 60 days, the observations for which correction coefficients were unavailable were treated as missing data. For a few locations—for example, Miami, FL, where prevailing weather pat-

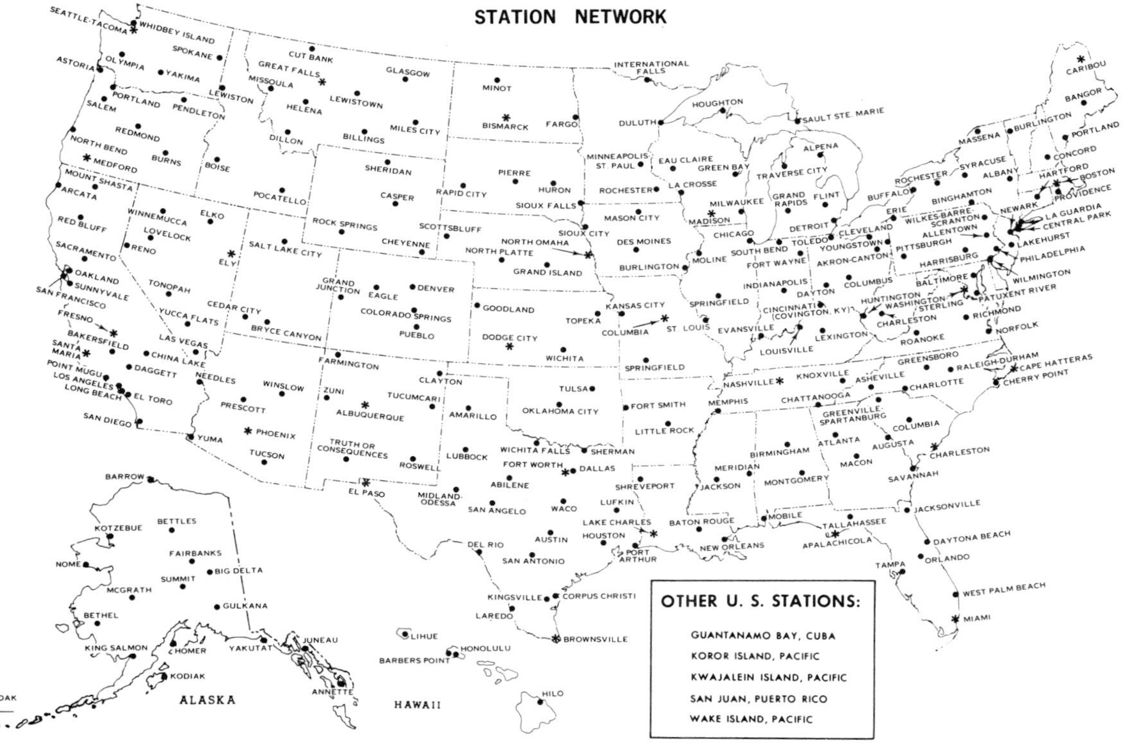

Figure 3.3
Locations for which SOLMET/ERSATZ data tapes are available from the National Climatic Center.

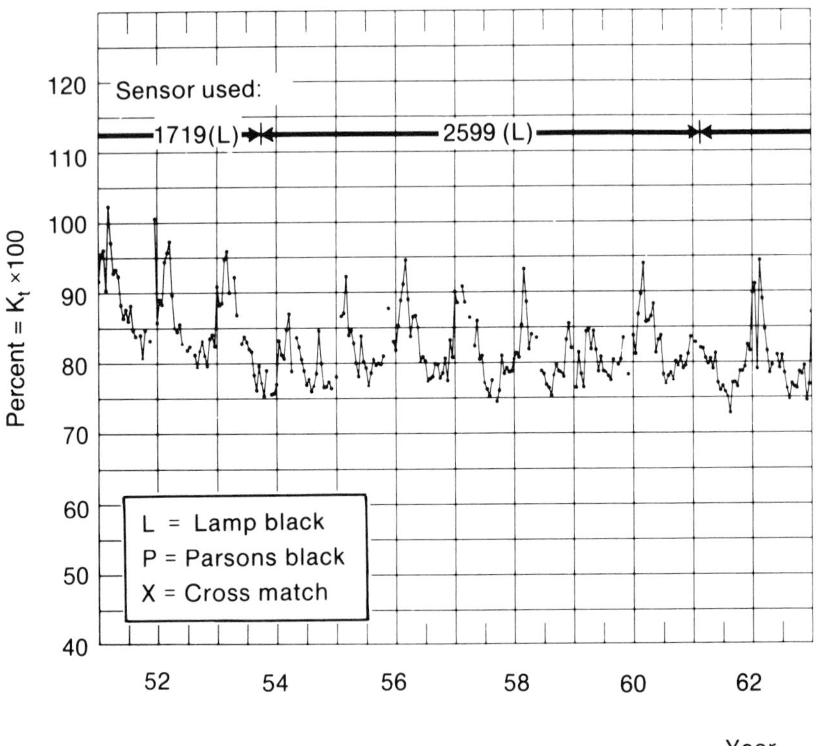

Figure 3.4
An example of observed clear solar noon values (Bismarck, N). Source: Quinlan (1979).

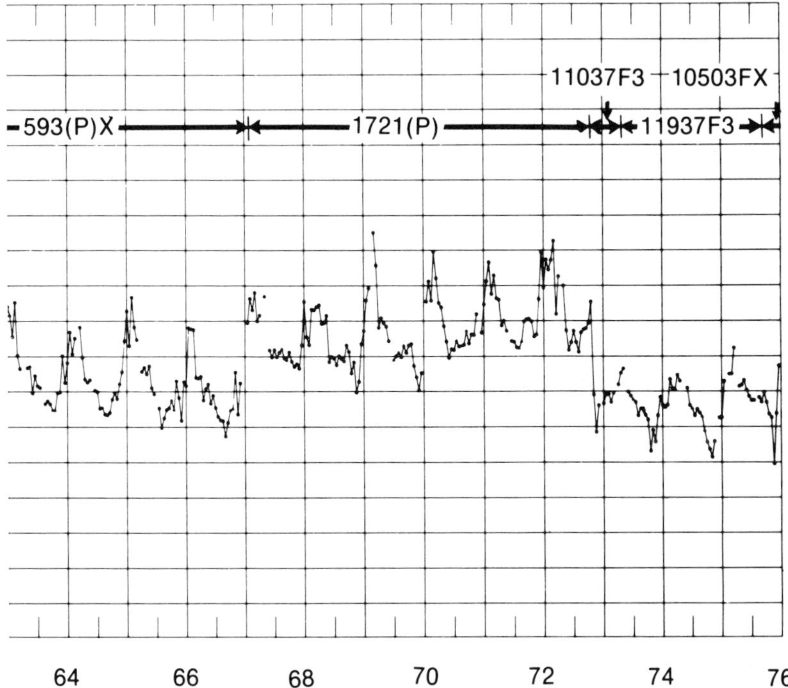
11037F3
10503FX
593(P)X
1721(P)
11937F3
64
66
68
70
72
74
76

1. .772	56. .818	111. .822	166. .798	221. .787	276. .792	331. .778
2. .773	57. .819	112. .822	167. .798	222. .787	277. .792	332. .777
3. .774	58. .820	113. .822	168. .798	223. .787	278. .792	333. .777
4. .774	59. .820	114. .821	169. .797	224. .787	279. .792	334. .776
5. .775	60. .821	115. .821	170. .797	225. .787	280. .792	335. .775
6. .776	61. .822	116. .820	171. .797	226. .786	281. .792	336. .775
7. .776	62. .822	117. .820	172. .797	227. .786	282. .792	337. .774
8. .777	63. .823	118. .819	173. .796	228. .786	283. .792	338. .774
9. .778	64. .824	119. .819	174. .796	229. .787	284. .792	339. .773
10. .779	65. .824	120. .818	175. .796	230. .787	285. .791	340. .773
11. .779	66. .825	121. .818	176. .796	231. .787	286. .791	341. .773
12. .780	67. .826	122. .818	177. .795	232. .787	287. .791	342. .772
13. .781	68. .826	123. .817	178. .795	233. .788	288. .791	343. .772
14. .782	69. .827	124. .817	179. .795	234. .788	289. .791	344. .772
15. .783	70. .828	125. .816	180. .795	235. .788	290. .791	345. .771
16. .784	71. .828	126. .816	181. .794	236. .788	291. .791	346. .771
17. .784	72. .829	127. .815	182. .794	237. .789	292. .791	347. .771
18. .785	73. .829	128. .815	183. .794	238. .789	293. .791	348. .771
19. .786	74. .830	129. .814	184. .793	239. .789	294. .791	349. .771
20. .787	75. .830	130. .814	185. .793	240. .789	295. .791	350. .771
21. .788	76. .830	131. .813	186. .793	241. .789	296. .790	351. .770
22. .788	77. .830	132. .813	187. .792	242. .790	297. .790	352. .770
23. .789	78. .830	133. .812	188. .792	243. .790	298. .790	353. .770
24. .790	79. .830	134. .812	189. .792	244. .790	299. .790	354. .770
25. .791	80. .830	135. .811	190. .791	245. .790	300. .790	355. .770
26. .792	81. .830	136. .811	191. .791	246. .790	301. .790	356. .770
27. .793	82. .829	137. .810	192. .791	247. .790	302. .790	357. .770
28. .794	83. .829	138. .810	193. .790	248. .790	303. .789	358. .770
29. .795	84. .829	139. .810	194. .790	249. .791	304. .789	359. .770
30. .796	85. .829	140. .809	195. .790	250. .791	305. .789	360. .770
31. .796	86. .829	141. .809	196. .789	251. .791	306. .789	361. .771
32. .797	87. .829	142. .808	197. .789	252. .791	307. .789	362. .771
33. .798	88. .829	143. .808	198. .789	253. .791	308. .789	363. .771
34. .799	89. .829	144. .807	199. .789	254. .791	309. .789	364. .772
35. .800	90. .828	145. .807	200. .789	255. .791	310. .788	365. .772
36. .801	91. .828	146. .807	201. .789	256. .791	311. .788	
37. .802	92. .828	147. .806	202. .789	257. .791	312. .788	
38. .803	93. .828	148. .806	203. .789	258. .792	313. .788	
39. .804	94. .828	149. .805	204. .789	259. .792	314. .788	
40. .805	95. .827	150. .805	205. .789	260. .792	315. .788	
41. .806	96. .827	151. .805	206. .789	261. .792	316. .788	
42. .807	97. .827	152. .804	207. .789	262. .792	317. .788	
43. .808	98. .827	153. .804	208. .789	263. .792	318. .788	
44. .809	99. .827	154. .803	209. .789	264. .792	319. .788	
45. .810	100. .826	155. .803	210. .789	265. .792	320. .787	
46. .811	101. .826	156. .802	211. .789	266. .792	321. .786	
47. .811	102. .826	157. .802	212. .789	267. .792	322. .785	
48. .812	103. .825	158. .802	213. .788	268. .792	323. .784	
49. .813	104. .825	159. .801	214. .788	269. .792	324. .783	
50. .814	105. .825	160. .801	215. .788	270. .792	325. .782	
51. .814	106. .824	161. .800	216. .788	271. .792	326. .782	
52. .815	107. .824	162. .800	217. .788	272. .792	327. .781	
53. .816	108. .824	163. .799	218. .788	273. .792	328. .780	
54. .817	109. .823	164. .799	219. .788	274. .792	329. .779	
55. .817	110. .823	165. .799	220. .788	275. .792	330. .779	

Figure 3.5
True solar noon transmission values for each day of the year for Bismarck, N.D. Source: Quinlan (1979).

terns often cause cloudiness near noon—researchers had to discard significant amounts of data.

The resulting series of corrected hourly global-horizontal insolation values often had significant gaps because of either equipment malfunctions or the inability to define a clear solar noon correction factor. These data gaps were unacceptable for driving system simulations, so Cotton (1979) developed an additional model to estimate global horizontal insolation from surface meteorological observations. This model is basically an empirical regression model with parameters derived from the serially incomplete global insolation data sets resulting from the rehabilitation steps described in the previous three paragraphs. This model estimates the global insolation as the product of clear sky and cloud factors:

$$Q = \text{SRCS} \times \text{CLD}, \tag{3.17}$$

where

SRCS = clear sky irradiance and

CLD = cloud factor.

The clear-sky insolation model is a regression of the clear-sky values from the corrected observation set on powers of the cosine of the zenith angle (Cotton, 1979). The cloud factor is a multiple regression of the observed cloudiness on minutes of sunshine, opaque sky cover, precipitation, and sky condition. The specific combination of these cloud-indicating variables used at any site depended on the combination of variables available for that site. Cotton (1979) used minutes of sunshine and opaque sky cover if both were available. If not, he used whichever one was available. If neither was available, then he used the sky condition observation.

Although surface meteorological observations were made and manually recorded hourly, to save money only the data for every third hour were transcribed to the computer-compatible tapes, which are the source of the meteorological part of the SOLMET/ERSATZ data base. Meteorological values used with the Cotton model were obtained by interpolation for those observation hours not transcribed.

A serially complete set of global insolation values was provided by using the Cotton model to fill in the values missing from the normalization of insolation observations by modeled clear solar noon insolation. This serially complete set of insolation values was then combined with hourly surface meteorological measurements to provide a complete data set known as the solar radiation surface meteorological observations (SOLMET), TD-9724 (Quinlan, 1977). SOLMET data sets were prepared for 26 locations. For most of these locations, the data sets cover a period of 20 or more years. These data are available from the National Climatic Data Center in Asheville, NC (chapter 2). This chapter refers to these data as the rehabilitated, or SOLMET, data set, although even this data set includes a significant amount of modeling.

The surface meteorological observations, from which the Cotton model estimates insolation, were available for 222 weather stations in the United States in addition to those 26 having insolation observations. To indicate the insolation resource at additional locations, the Cotton model was used to generate an ERSATZ data base (chapter 2) for these additional locations, which are shown with dots in figure 3.3.

The procedures used to generate the SOLMET/ERSATZ data base were described to educate the user of these data. These data make up the best insolation data base available in the United States covering 10 years or more. Even for the 26 rehabilitated locations, however, the insolation values are not directly measured but rather the result of significant manipulation. This manipulation almost certainly affects the statistical distributions of the data and may also affect the mean insolation values.

3.3.2.2 Evaluations of the SOLMET/ERSATZ Data Base The data manipulations required to form the SOLMET/ERSATZ insolation data base, as outlined in the preceding section, suggest that the data base contents be compared with recent measurements obtained using carefully calibrated instruments. In this section, we discuss several such studies.

Hall et al. (1980) evaluated the SOLMET global horizontal rehabilitation results by comparing the Hanson clear solar noon model with the observed insolation values obtained with NWS's new insolation network (discussed in chapter 4) during 1977. These values came from nine locations: Albuquerque, Bismarck, Brownsville, TX, Dodge City, KS, Madison, WI, Medford, OR, Miami, Nashville, TN, and Seattle, WA. Of the 90 station-months available for analysis, the mean monthly clear solar noon values were within 5% of the observed values for 74 station-months (82% of the station-months studied).

Turner and Mujahid (1984) compared the performance of the Cotton model with a model of their own for estimating two years of data from

Blytheville, AR. They found the average monthly difference between the predicted and the observed to be −3.0%.

Hall et al. (1980) also evaluated the Cotton regression model (for estimating global insolation from cloud cover information) by comparing estimates made with this model with observed values obtained with the new NWS network at the same locations previously cited, except they used Caribou, ME, instead of Madison. Of the 91 station-months available for analysis, the monthly median daily global insolation difference was less than 7.5% for 78 of the station-months. The uncertainty is greater for hourly values, particularly those away from noon. For the noon hours, the median error was less than 7.5% for 77 of the 91 station-months. For hours 1,600 and 1,700, however, only 50 of the 91 station-months had monthly median hourly errors less than 7.5%. Quite large errors can occur. For hours 1,600 and 1,700, the largest 75th percentile monthly error was 40.9% (August, Miami), and the largest magnitude 25th percentile monthly error was −97.5% (December, Caribou). The tables in Hall et al. (1980) contain a more detailed presentation of the results.

Rapp (1979) has compared the SOLMET clear solar noon values (on which the rehabilitation depends) with observations taken at 11 locations between 1972 and 1976 with Eppley PSP pyranometers (chapter 5). He reviewed the observed data and the calibration carefully to assure reliability. The 11 locations were Albuquerque, Bismarck, Caribou, El Paso, TX, Ely, NV, Fresno, CA, Madison, Nashville, Phoenix, AZ, Raleigh, and Seattle. Tables 3.1 and 3.2 (Rapp, 1979) compare the observed clear solar noon transmittance with the values computed from the models of Hoyt and Hanson, respectively, by presenting the transmittance differences (note this is not the percent difference used by Hall). Comparing observations with the Hoyt model indicates that 88 of the months considered agree to within 5% (67% of the months studied). Comparing observations with the Hanson-modified Hoyt model indicates that 105 of the months considered agree to within 5% (79% of the months studied). Note that the improved overall performance of the Hanson model, as indicated by the greater number of months within 5% transmittance difference and the lower overall mean difference, was obtained at the expense of larger errors for some locations and months (January, February, and March at Seattle). Rapp concluded that Hanson's correlation of corrections to Hoyt's model with water vapor and air mass was not justified by the limited data available to Hanson. The improvements resulted simply because the cor-

Table 3.1
Difference between transmission values (% of ETR) at midmonth at solar noon (calculated by Hoyt—measured by Eppley PSP)

	Month												
Location	J	F	M	A	M	J	J	A	S	O	N	D	Year
ABQ	5.6	4.5	4.8	3.9	4.8	4.6	4.5	4.4	5.4	6.4	8.4	7.5	
ELP	1.5	0.9	1.1	−0.5	1.5	1.9	2.2	2.8	2.8	3.8	5.6	2.8	
ELY	3.0	3.0	2.8	2.2	2.0	2.5	2.6	4.1	6.1	6.4	6.1	5.5	
NSH	−2.0	−0.8	−0.4	−3.0	−3.0	−3.0	−2.3	−1.6	−0.5	0.5	1.3	−1.1	
SEA	−6.0	−8.0	−7.8	−7.5	−5.4	−2.2	−0.9	−0.8	−2.1	−3.2	−4.4	−5.6	
CAR	−1.0	0.2	1.5	0.2	0.4	1.0	1.1	1.9	3.0	4.7	5.6	1.5	
BIS	2.5	3.0	4.2	1.6	1.0	3.2	3.6	4.1	7.6	9.6	5.7	3.0	
RAL	3.8	3.5	2.8	2.6	3.0	0.0	2.2	3.0	3.0	4.8	4.9	4.0	
FRE	4.5	3.8	4.0	4.6	3.7	5.2	6.4	6.9	9.5	12.5	10.3	7.3	
MAD	3.2	3.3	2.2	1.0	0.7	1.0	1.4	2.4	3.4	6.0	8.8	7.9	
PHO	4.4	4.2	4.5	3.1	4.3	4.3	4.6	4.8	6.3	7.3	8.8	6.8	
Mean deviation	1.8	1.6	1.8	0.7	1.2	1.7	2.5	2.7	4.1	5.3	5.6	3.6	2.7
Mean absolute deviation	3.5	3.2	3.2	2.7	2.8	2.6	3.0	3.2	4.5	5.9	6.4	4.8	3.8

Source: Rapp (1979).

Table 3.2
Difference between transmission values (% of ETR) at midmonth at solar noon (calculated by Hanson—measured by Eppley PSP)

	Month												
Location	J	F	M	A	M	J	J	A	S	O	N	D	Year
ABQ	1.7	2.4	3.3	2.9	3.7	3.7	2.5	1.8	3.3	4.0	5.3	3.0	
ELP	−2.2	−1.5	−0.8	−1.8	−0.2	0.1	−1.0	−0.3	−0.4	1.2	1.8	−1.8	
ELY	−1.3	0.5	1.3	1.2	1.2	1.5	2.5	2.5	4.4	4.2	2.2	0.5	
NSH	−7.8	−4.4	−3.9	−6.4	−5.9	−6.9	−6.8	−6.1	−5.7	−4.1	−4.3	−7.8	
SEA	−16.9	−14.9	−12.4	−11.2	−8.6	−5.3	−4.2	−5.0	−7.0	−9.6	−13.5	−7.9	
CAR	−10.3	−5.6	−2.4	−2.4	−2.3	−2.2	−2.6	−2.1	−1.4	−1.1	− 2.9	−9.1	
BIS	−6.1	−2.3	0.6	−1.1	−1.4	−0.1	−0.3	−0.1	3.6	4.4	−1.5	7.6	
RAL	−2.1	−0.8	−0.4	−0.4	0.1	−4.2	−2.5	−1.8	−1.4	0.4	−0.5	−2.5	
FRE	−1.8	−0.3	0.6	1.9	1.0	2.8	3.5	4.0	6.0	5.5	4.6	1.0	
MAD	−4.2	−1.4	−1.3	−1.7	−2.5	−2.7	−2.4	−2.3	−1.1	0.4	2.3	−0.4	
PHO	−0.8	0.4	1.9	0.6	2.1	2.0	0.0	0.3	2.6	3.5	4.2	1.0	
Mean deviation	−4.8	−2.5	−1.2	−1.7	−1.2	−1.0	−1.1	−0.8	0.2	0.8	−0.2	−1.6	−1.2
Mean absolute deviation	5.0	3.1	2.7	2.9	2.5	2.8	2.6	2.4	3.4	3.5	3.9	3.9	3.2

Source: Rapp (1979).

rections always reduced the transmittance. The large errors observed by Rapp for Caribou and Seattle in the winter were due to these corrections. By considering all of the observed clear solar noon transmittance values, Rapp suggests one might construct an equally good clear solar noon model by simply taking the mean of the observed transmittance values.

The studies of Rapp and Hall et al. suggest the same qualitative conclusions. The mean annual error introduced by the clear solar noon calibration procedure is only a few percent. For any given location and month, however, it can be as much as 10% or more. Rapp's results suggest a somewhat greater uncertainty in the clear solar noon calibration. The data sets are different, however, so definitive conclusions are not possible. Hall's data were from new instruments that had recently been calibrated; however, he was not able to obtain a uniform sampling of months. Rapp's data uniformly covered the year and had a slightly larger geographic distribution but were based on good instruments from the network in an era when it was not well maintained. To this basic uncertainty, one must add differences in the sensors' responses for hours other than noon and the additional uncertainty introduced by climatic trends in the meteorological parameters on which the Hoyt and Hanson models were based.

The monthly mean daily global insolation values from the SOLMET data set can be compared with observations from the new NOAA network for the 7 years, 1977–1980, for which the data were published (NCC, 1980). Table 3.3 provides the observed monthly mean daily global insolation (kJ/m^2) from the new NOAA betwork. Blank spaces indicate that observations were insufficient to compute a monthly mean value for all the years considered. The values in table 3.3 were obtained by taking the mean of all of the reported monthly means. As a consequence, the values reported in table 3.3 may be the mean of from 1 to 4 monthly values. Table 3.4 compares the global insolation observations for those locations that also have rehabilitated data in the SOLMET data base by providing the percent differences of the mean SOLMET values from the new observed values. Table 3.5 compares the observed global insolation values and values obtained from the Cotton model for those locations having ERSATZ values.

The mean difference between the recent observations and the rehabilitated SOLMET values in table 3.4 is 2.5% (SOLMET greater than observation). The median difference is +3.5%. Of the 219 station-months considered, 68% of the cases have differences from −5.0% to +9.6%. If the differences were normally distributed, then this range of differences

Table 3.3
Mean observed daily global insolation (kJ/m^2), NOAA network 1977–1980

		Month											
Station name	Number	J	F	M	A	M	J	J	A	S	O	N	D
Albuquerque, NM	23050	10031	14421	18540	24633	26086	28326	27273	24759	20521	17327	11708	10371
Bismarck, ND	24011	6400	8826	13887	17067	22054	24264	23572	19383	14878	9524	6017	4788
Blue Hill, MA	14753		11910		17331								4763
Boise, ID	24131	5395	7710	13515	19654	22463	26937	27518	22505	16683	12725	6235	4635
Boulder, CO	94018	8056	11162	15291	19192	20875	24076	22869	19934	18918	13502	8882	7514
Brownsville, TX	12919	8577	11916	14987	18265	20353	23124	24877	21001	18454	15477	11673	8809
Burlington, VT	14742	5153	8665	11638	15523	21066	18776	21322	16071	12884	7060	4860	4024
Caribou, ME	14607	5420	9712	13689	13939	17764	19199	18124	16531	11829	6588	4705	3727
Columbia, MO	3945	7194	10415	13058	15238	20910	24071	23382	19976	17054	12890	7353	6256
Desert Rock, NV	3161				25362	27894	30834				17249		
Dodge City, KS	13985	8813	12691	16341	19523	21193	25768	25780	22374	19392	15693	9274	8152
El Paso, TX	23044	11253	15990	20689	25572	27669	28435	26328	24318	19924	17569	13518	11654
Ely, NV	23154	8576	11957	15785	22129	24220	28674	27418	23175	20400	15072	10131	8175
Fairbanks, AK	26411			7793	14340	19394	19912	20173	13545			449	223
Fresno, CA	93193	6120	11036	15793	21318	26511	28786	27374	25668	19912	14867	10190	6244
Grand Junction, CO	23066	7135	11510	15754	20216	24440	31038	26385	24625	21677	15696	9779	7804
Great Falls, MT	24143	5638	8284	12499	17741	18742	24353	23875	19501	14794	9551	5728	4016
Guam, PI	41415	17372	20195	20145	22439	21851	20030	16786	17438	17168	14308	15935	15982
Honolulu, HI	22521	14330	16731	21189	23127	23292	23438	25107	23969	21074	17985	15944	14603
Indianapolis, IN	93819	5966	9127	10525	14262	21206	22315	21367	16673			5104	4832
Lake Charles, LA	3937	7262	11095	13216	19281	21051	22927	20337	19610	15982	15991	10395	9484
Lander, WY	24021	7758	11700	16768	21728	23035	27275	25392	21735	19301	13722	8382	6425

Table 3.3 (continued)

Station name	Number	Month J	F	M	A	M	J	J	A	S	O	N	D
Las Vegas, NV	23169	8578	13778	17171	24746	26813	30235	28123	25147	21363	16283	11686	9284
Los Angeles, CA	23174	8298	13169	19046	21483	21916		27368	24255	18882	14522	11825	9594
Madison, WI	14837	6122	10575	12300	16184	21836	23121	23797	15891	14339	8853	5562	5270
Medford, OR	24225	4955	7691	13428	17826	22251	26244	27952	23018	17617	11944	5299	3779
Miami, FL	12839	13146	14716	19071	20255	19980	21633	21009	18172	16321	14425	13293	11177
Midland, TX	23023	10546	15294	19203	25833	25282		27310	23542	17431	16609	10528	10454
Montgomery, AL	13895	7874	11966	14103	20551	21857	24106	20357	20619	15707	16210	10059	9499
Nashville, TN	13897	6469	10397	12831	17160	19408	22838	20174	18815	15216	13532	7899	6498
Omaha, NE	94918	7353	10058	13544	16143	20248	23669	23645	18704	17149	11202	7491	6203
Phoenix, AZ	23183	9504	14389	18617	24664	27769	29436	26777	25758	21414	16411	12217	9692
Pittsburgh, PA	94823	5347	11064	11020	13024	16014	20412	18527	14666	14289	9472	5724	4432
Raleigh, NC	13722	7787		15109	18571	19268	19681	21112	19485	14693	15082	7994	6565
Salt Lake City, UT	24127	5959	9799	14322	19712	21174	28164	27009	22336	18559	13031	7424	5671
San Juan, PR	11641		17687	19439	19057	17004	18889	21845	18360	16815		11934	12650
Seattle-Tacoma, WA	24233	3589	4713	10612	15317	20740	23124	21328	16851	11899	8200	3916	2146
Sterling, VA	93734	6049	10639	13901	17221	19303	20489	20334	17705	15233	11176	7580	6438
Tallahassee, FL	93805	9639	12940	15953	20299	21910	22498	20413	18368	12711	15927	11574	9557

Table 3.4
Percent difference of SOLMET global insolation data from observed global insolation, NOAA network 1977–1980

Station name	Number	Month J	F	M	A	M	J	J	A	S	O	N	D
Albuquerque, NM	23050	15.0	5.6	8.2	2.7	10.4	7.3	3.6	5.0	9.0	1.3	9.9	1.5
Bismarck, ND	24011	−17.2	−0.3	−4.5	−3.0	−4.9	−3.7	5.1	9.9	3.3	8.2	−4.3	−11.6
Blue Hill, MA	14753		−32.4		−13.2								−4.0
Brownsville, TX	12919	20.8	8.1	10.4	7.9	7.5	3.8	0.9	9.6	4.2	5.5	2.5	11.1
Caribou, ME	14607	−12.2	−15.4	−6.1	15.1	0.8	3.9	10.4	3.0	5.8	18.6	−11.6	−5.4
Dodge City, KS	13985	6.5	0.3	2.5	9.6	11.9	3.9	1.1	4.3	−1.3	−5.9	9.4	1.9
El Paso, TX	23044	13.5	5.1	4.7	4.9	6.7	7.1	5.6	6.6	13.2	5.9	4.4	0.4
Ely, NV	23154	8.4	8.3	15.5	3.0	8.3	−0.5	1.3	9.2	7.7	6.0	3.8	0.3
Fresno, CA	93193	21.8	4.1	12.5	11.4	6.3	7.7	11.3	7.1	13.1	9.1	−1.0	4.4
Great Falls, MT	24143	−15.4	−1.3	6.3	−4.8	11.9	−2.1	10.7	12.5	5.7	9.9	−1.4	−5.0
Lake Charles, LA	3937	13.8	3.3	12.8	−7.6	−0.3	−2.5	−0.2	−4.1	5.5	−2.0	0.1	−15.6
Madison, WI	14837	−4.5	−13.7	4.8	−1.9	−9.4	−4.4	−7.7	22.0	2.8	16.8	2.9	−16.2
Medford, OR	24225	−6.8	8.8	−4.3	4.3	3.7	−1.5	0.5	4.6	2.4	−6.7	8.0	1.1
Miami, FL	12839	−8.7	1.3	−4.6	4.2	4.7	−10.4	−4.7	1.8	1.3	2.5	−4.5	3.5
Nashville, TN	13897	1.7	−10.1	−0.1	2.1	6.7	−2.5	6.4	4.8	4.3	−6.6	2.2	−9.1
Omaha, NE	94918	−2.1	0.7	2.4	9.6	5.0	1.8	1.1	12.8	−9.1	6.4	−2.4	−6.5
Phoenix, AZ	23183	22.0	8.4	10.6	8.4	9.4	5.6	5.0	1.0	6.8	9.0	6.9	9.1
Seattle-Tacoma, WA	24233	−17.3	19.2	−9.2	−4.2	−6.2	−11.6	19.6	8.9	9.5	−9.2	−2.3	11.7
Sterling, VA	93734	7.3	−13.0	−8.1	−3.9	1.0	5.3	1.4	3.7	−0.2	1.9	−2.5	−15.2

Table 3.5
Percent difference of ersatz insolation data from observed global insolation, NOAA network 1977–1980

Station name	Number	Month J	F	M	A	M	J	J	A	S	O	N	D
Boise, ID	24131	2.1	23.6	9.5	5.5	15.0	3.8	7.8	10.8	18.2	1.5	14.4	7.0
Burlington, VT	14742	−15.1	−20.5	−8.3	−5.2	−15.2	4.5	−8.4	4.2	−1.2	19.0	−12.5	−20.1
Columbia, MO	3945	−3.5	−4.7	2.5	13.6	2.0	−1.5	2.7	6.7	−3.5	−3.1	8.5	−5.2
Fairbanks, AK	26411			−1.8	−5.5	−6.2	−0.2	−13.2	−6.3			87.3	−87.4
Grand Junction, CO	23066	25.9	10.3	11.9	11.5	10.5	−5.0	6.0	0.6	−4.0	−2.8	6.5	6.4
Honolulu, HI	22521	−6.6	−5.3	−13.1	−11.9	−5.0	−2.0	−9.5	−6.9	−2.5	−2.8	−9.9	−12.0
Indianapolis, IN	93819	−5.7	−7.1	11.9	11.3	−9.7	−5.0	−4.1	11.9			28.8	−2.2
Las Vegas, NV	23169	29.4	10.3	20.5	6.4	12.0	4.3	4.5	6.3	8.2	7.3	5.4	7.6
Los Angeles, CA	23174	26.7	4.6	−3.5	3.1	6.7		−4.3	−2.7	1.1	2.9	−3.7	0.4
Midland, TX	23023	16.4	2.6	8.7	−3.7	9.1		−0.7	6.5	20.1	4.0	26.8	8.5
Montgomery, AL	13895	8.3	−3.9	7.9	−4.5	−1.5	−7.1	2.6	−3.9	6.0	−11.7	3.3	−14.0
Pittsburgh, PA	94823	−9.9	−35.9	−2.9	14.7	13.5	−2.1	3.5	16.9	−4.0	7.2	0.1	−11.2
Raleigh, NC	13722	1.1		−4.2	0.5	6.5	7.5	−4.6	−6.1	6.4	−16.8	15.3	9.9
Salt Lake City, UT	24127	21.7	14.5	15.2	9.1	26.6	3.2	8.8	14.5	12.7	12.6	20.4	14.0
San Juan, PR	11641		−1.4	4.4	12.6	21.0	9.2	−2.7	13.6	13.0		30.1	10.9
Tallahassee, FL	93805	3.2	−0.2	5.3	1.9	0.3	−5.0	−2.8	3.5	33.3	−6.1	−1.1	−3.5

would correspond to ± 1 standard deviation. Differences in table 3.4 arise from errors in rehabilitating the SOLMET data and any real climatic changes that may have taken place. Note that these differences are representative of the uncertainty that using the SOLMET data might introduce into a real solar energy design program.

The mean difference between the recent observations and the ERSATZ values in table 3.5 is 3.3% (ERSATZ greater than observation). The median difference is +3.1%. Of the 181 station-months considered, 68% of the cases differ from −6.1% to +13.6%. If the difference were normally distributed, then this range of differences would correspond to ± 1 standard deviation. Differences in table 3.5 arise from errors in the rehabilitation of the SOLMET data, uncertainties introduced by the ERSATZ model, and real climatic changes that may have taken place. As pointed out before, these differences are representative of the uncertainty that using the monthly mean ERSATZ data might introduce into a real solar energy design.

More recently, SERI (1984) compared the hourly observations for 4 years (1977–1980) from 31 stations in the new NWS network with two models. One model was the original Cotton clear-sky model modified by the Cotton cloudiness model described previously. For brevity we are calling this the SOLMET/ERSATZ model. The other model was the SERI (Bird and Hulstrom, 1981) clear-sky model used with the same Cotton cloudiness model. For brevity we are calling this the SERI (Bird) model-I. The comparisons were separated into two groups: 19 of the original SOLMET stations and 12 ERSATZ stations. The SOLMET stations permit us to evaluate the Cotton model for the locations for which it was derived but for a different, more recent time period.

Figures 3.6–3.9 show a comparison of the data from the 19 SOLMET stations and these two models. Figure 3.6 shows the percent mean bias error (%MBE) as a function of the month of the year. Throughout most of the year, the SOLMET/ERSATZ model usually overestimates the insolation by a few percentage points. The SERI (Bird) model-I does better during the year except during the fall months. The percent root-mean-square error (%RMSE) indicates the width of the distribution of errors around the mean and depends on the length of time over which the insolation is integrated. The %RMSE comparison of hourly data from the 19 SOLMET stations and the two models is shown as a function of the month in figure 3.7. Note that these errors are from 20% to more than 30%,

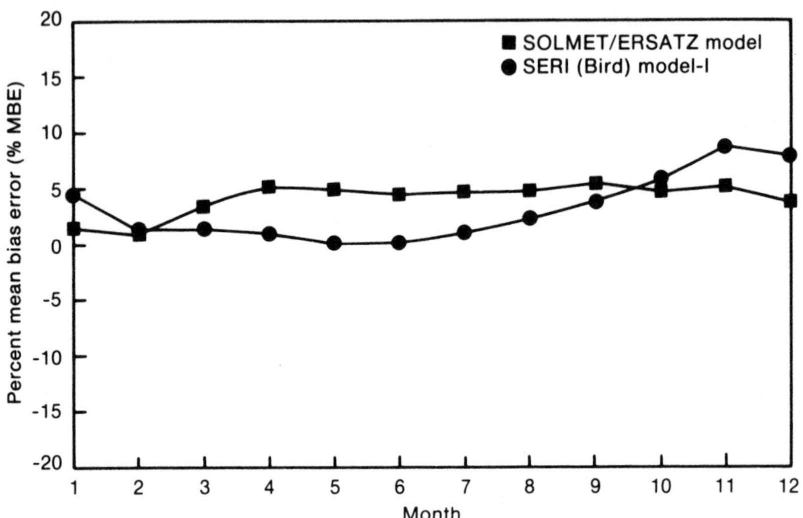

Figure 3.6
Percent mean bias error (%MBE) versus month for the SOLMET/ERSATZ model and the SERI (Bird) model-I for selected (19) SOLMET sites.

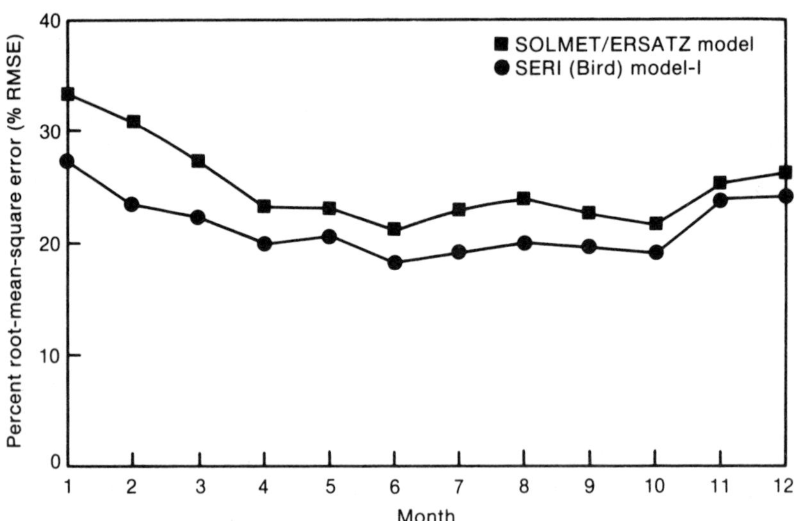

Figure 3.7
Percent hourly root-mean-square error (%RMSE) versus month for the SOLMET/ERSATZ model and the SERI (Bird) model-I for selected (19) SOLMET sites.

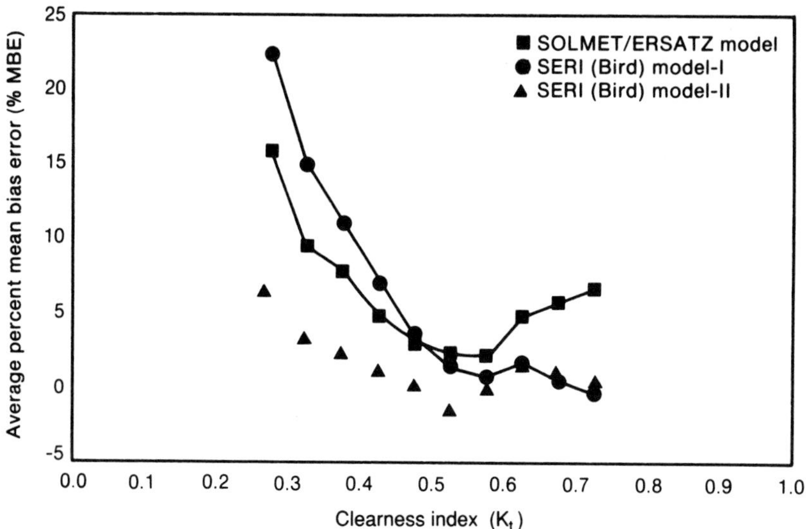

Figure 3.8
Percent mean bias error versus clearness index for the SOLMET/ERSATZ model and the SERI (Bird) model-I and SERI (Bird) model-II for selected (19) SOLMET sites.

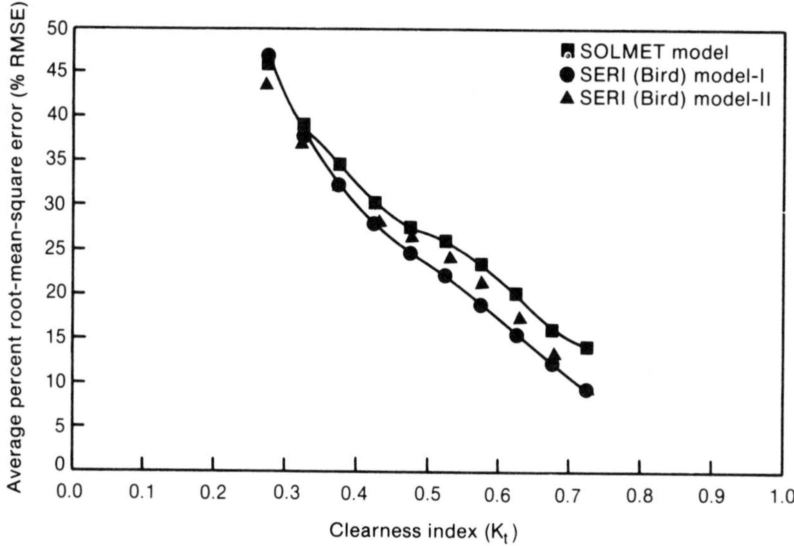

Figure 3.9
Percent hourly root-mean-square error versus clearness index for the SOLMET/ERSATZ model and the SERI (Bird) model-I and SERI (Bird) model-II for selected (19) SOLMET sites.

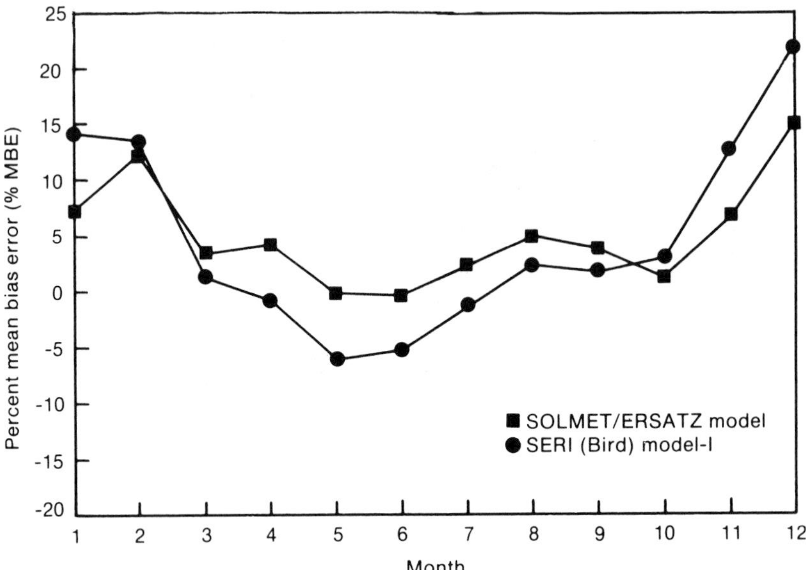

Figure 3.10
Percent mean bias error versus month for the SOLMET/ERSATZ model and the SERI (Bird) model-I for Seattle-Tacoma, WA.

indicating that an insolation estimate for any individual hour is quite likely to have a large error associated with it. Figures 3.8 and 3.9 plot the same %MBE and %RMSE comparisons as a function of the clearness index K_t rather than of the month of the year. The percent errors are significantly greater both in the mean and variation for cloudier conditions, indicating that it is more difficult for insolation models to work well for cloudy conditions. The insolation values, however, are also lower under cloudy conditions, so the absolute energy error does not vary as much with cloudiness.

SERI (1984) researchers have also compared individual station data from which they infer the source of the differences between their own model and the SOLMET/ERSATZ model. Based on these observations, they concluded that the uncertainty of the estimates increases with air mass (i.e., winter northern latitudes), cloudiness, and air pollution. Seattle is an example of a cloudy northern latitude site. Figure 3.10 shows that the %MBE for this site can be as large as 15%.

Figure 3.11–3.14 show a comparison of the data for the 12 ERSATZ stations and the SOLMET/ERSATZ model. The %MBE shown in figure

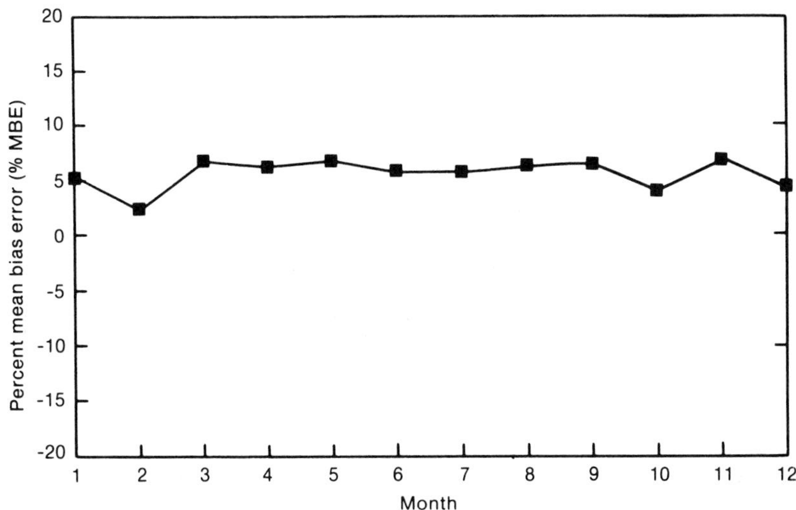

Figure 3.11
Percent mean bias error versus month for selected (12) ERSATZ sites for the SOLMET/ERSATZ model.

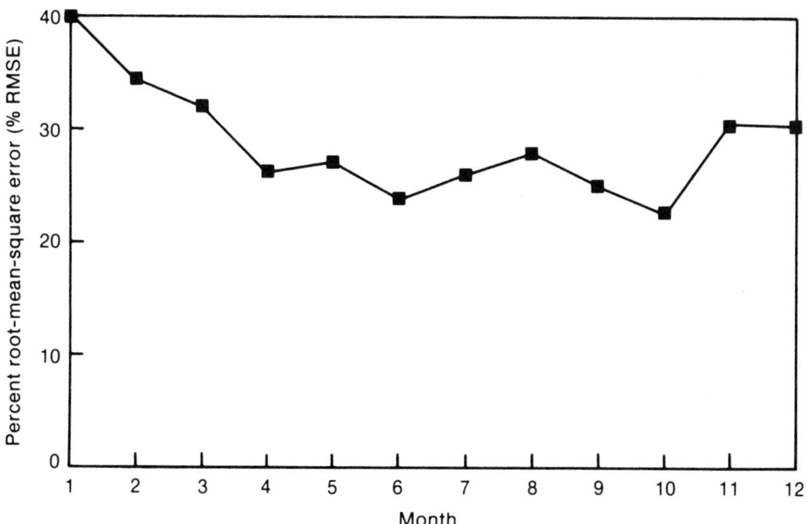

Figure 3.12
Percent hourly root-mean-square error versus month for selected (12) ERSATZ sites for the SOLMET/ERSATZ model.

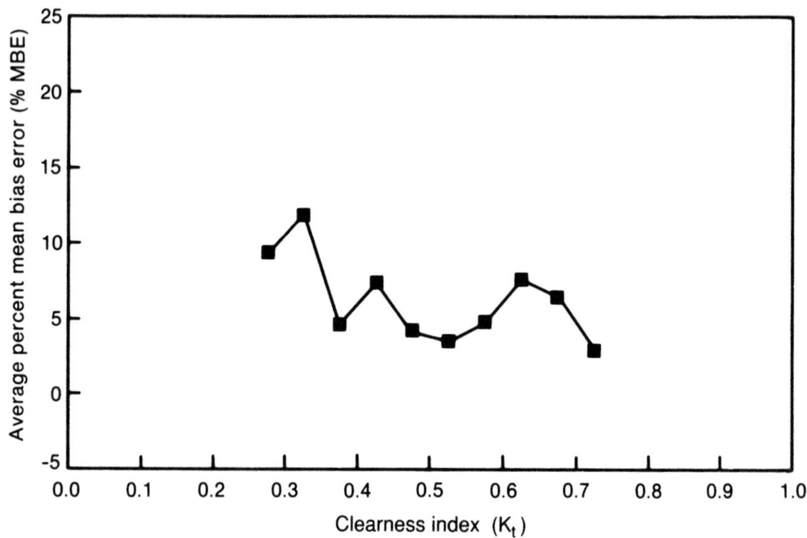

Figure 3.13
Percent mean bias error versus clearness index for selected (12) ERSATZ sites for the SOLMET/ERSATZ model.

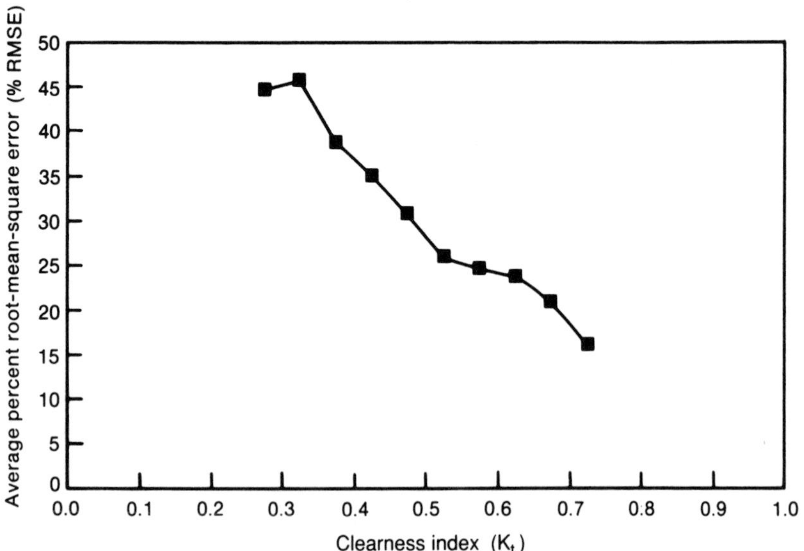

Figure 3.14
Percent hourly root-mean-square error versus clearness index for selected (12) ERSATZ sites for the SOLMET/ERSATZ model.

3.11 is a function of the month of the year. On the average, the SOLMET/ERSATZ model overestimates the insolation by a few percentage points throughout most of the year, just as at the SOLMET locations. The %RMSE indicates the width of the error distribution around the mean and depends on the length of time over which the insolation is integrated. The %RMSE comparison of hourly data from the 12 ERSATZ stations and the SOLMET/ERSATZ model is shown as a function of the month in figure 3.12. Note that these %RMSE are from 20% to 40%, indicating that an insolation estimate for any individual hour is quite likely to have a large error associated with it. Figures 3.13 and 3.14 plot the same %MBE and %RMSE comparisons, respectively, as functions of the clearness index K_t rather than of the month of the year. These results for the ERSATZ stations are similar to those of the SOLMET station. The main differences are that the effect of the air mass correction in the SOLMET clear sky model is less evident and that the error appears to fluctuate more. These differences could be partially caused by the location and number of stations and because these stations had more missing hours of data.

With the data for individual ERSATZ stations, we can examine the extrapolation of the SOLMET/ERSATZ regression models to surrounding regions. Figures 3.15–3.17 present comparisions from three stations to illustrate the varied results obtained. The selected stations are Montgomery, AL, Burlington, VT, and Salt Lake City, UT.

Figure 3.15 shows the %MBE for Montgomery, an example of very good extrapolation results. Nashville was the control station. Figure 3.16 presents the %MBE for Burlington. The magnitude of the error is significantly greater, especially for fall. Caribou was the control station for Burlington. Figure 3.17 gives the results for Salt Lake City from the ERSATZ process; its control station was Ely, NV. The extrapolation from Ely to Salt Lake City is poor (i.e., the %MBEs are greater than 20%).

SERI (1984) researchers also used the 4-year data set from the new NWS network's 18 stations to evaluate four widely used cloud models, including the ASHRAE model of Kimura and Stephenson (1969), the MAC model of Davies and Hay (1978), the Center for Environment and Man (CEM) model of Atwater and Ball (1981), and the Cotton (1979) opaque cloud-cover regression model used in the previous section. This evaluation used the Bird clear-sky model with each of these cloudiness models (Bird and Hulstrom, 1981). The turbidity, precipitable water, and ground albedo values of Hoyt (1979) were used in the clear-sky model.

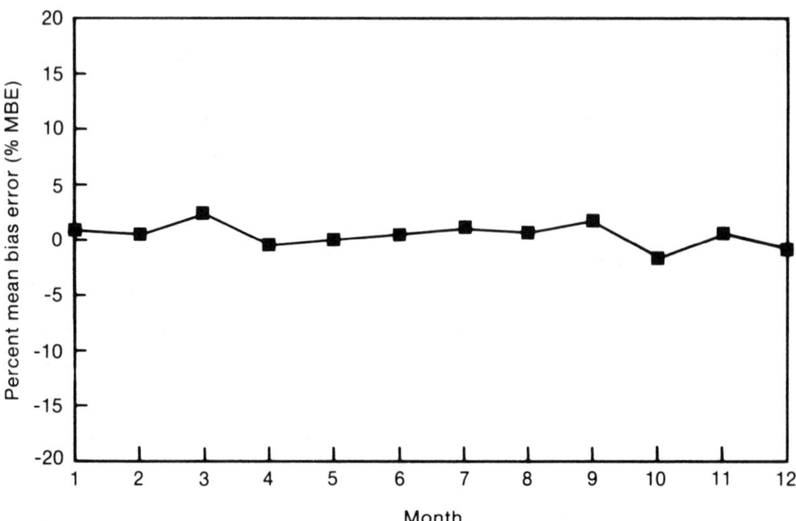

Figure 3.15
Percent mean bias error versus month for Montgomery, AL, for the SOLMET/ERSATZ model.

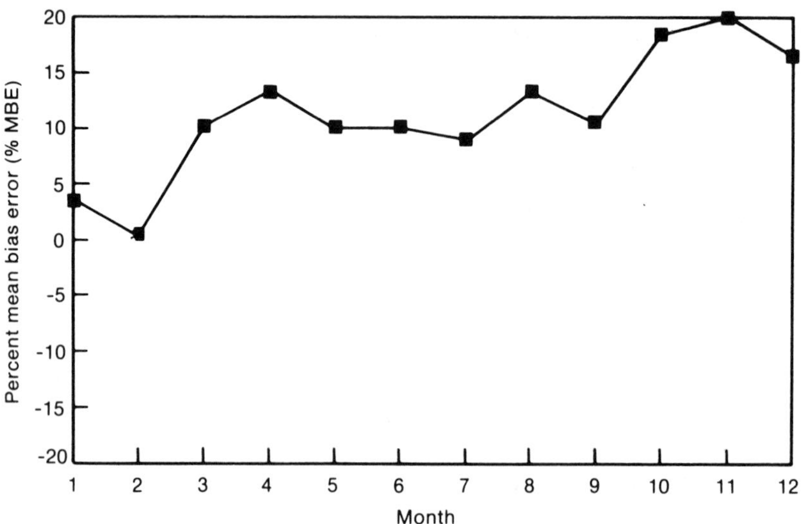

Figure 3.16
Percent mean bias error versus month for Burlington, VT, for the SOLMET/ERSATZ model.

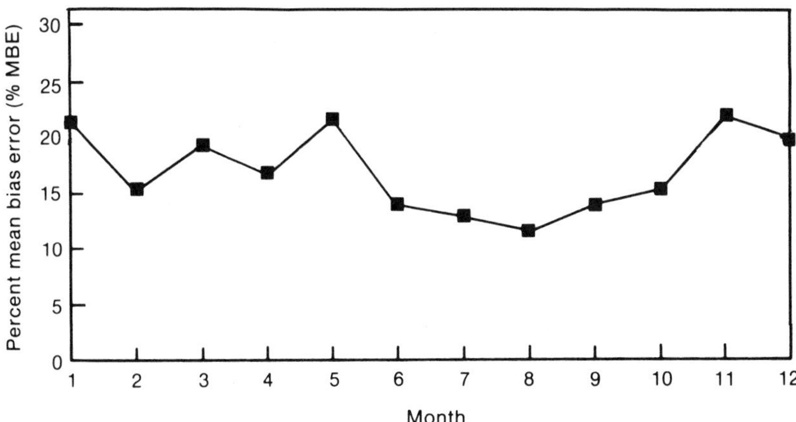

Figure 3.17
Percent mean bias error versus month for Salt Lake City, UT, for the SOLMET/ERSATZ model.

The following general results were obtained for each model for the 18-station group as a whole. The ASHRAE cloud model overestimates the irradiance by ~14% for winter and underestimates by ~4% for summer. The MAC cloud model was very accurate for winter and tended to underestimate the irradiance by ~8% for summer. The CEM cloud model overpredicts 5–20% for all months of the year, winter the greatest error occurring in the winter. The Cotton regression cloud model overpredicts ~9% for winter but is very accurate for the summer.

SERI researchers investigated several modified forms of these cloud models, and they found that a simplified form of the ASHRAE model provided the most accurate results. This model, the SERI (Bird) model-II, was the simplest model to implement. The global insolation on a horizontal surface for this new model is given by

$$Q_{BC} = Q_B(\mathrm{CCF}), \tag{3.18}$$

where Q_B is the global horizontal insolation from the SERI (Bird) clear-sky model, and CCF is the ASHRAE cloud cover factor, which takes the form

$$\mathrm{CCF} = P + S(\mathrm{OPQ}) - R(\mathrm{OPQ})^2, \tag{3.19}$$

where OPQ is the opaque cloud cover in tenths (0.0–1.0). Kimura and Stephenson (1969) give values of P, S, and R for each month; the values are provided in table 3.6.

Table 3.6
Coefficients for the SERI (Bird) model-II cloud model for each month

	J	F	M	A	M	J	J	A	S	O	N	D
P	1.05	1.05	1.02	1.0	1.0	1.0	1.0	1.0	1.0	1.02	1.04	1.04
S	0.06	0.09	0.12	0.19	0.26	0.33	0.32	0.31	0.30	0.21	0.12	0.03
R	0.82	0.83	0.84	0.91	0.99	1.06	1.07	1.07	1.08	0.99	0.91	0.82

Table 3.7
Error statistics for each month for the Bird clear-sky model combined with the opaque-cloud ASHRAE model results obtained from comparing measured data at 18 sites for 4 years

Month	%MBE	Hourly %RMSE	Daily %RMSE
1	1.15	29.73	18.16
2	0.95	25.77	14.65
3	−1.34	24.07	13.14
4	−2.38	21.72	11.02
5	−2.43	22.73	11.21
6	0.19	19.80	9.25
7	0.35	20.40	9.03
8	0.59	21.37	9.93
9	0.80	21.45	10.81
10	3.97	20.49	11.73
11	4.70	25.49	15.02
12	2.35	26.02	15.75

The performance of the SERI (Bird) Model-II is indicated in tables 3.7 and 3.8 by comparing the model predictions with the measured data from 18 stations. The percent errors as a function of k_t from table 3.8 are also indicated by triangles in figures 3.8 and 3.9. This new cloud model [SERI (Bird) model-II] improves accuracy. This work demonstrates, for a much larger data set than the earlier Aerospace set, that opaque cover is the routine cloud observation best able to predict the hourly variation in the global insolation. Opaque sky cover gives more accurate results than either total cloud cover or combinations of layered cloud data and cloud types.

Several groups have worked on related aspects of modeling the effects of cloud cover on global insolation. We are presenting these for further reading and references.

Suckling and Hay (1977) have developed a model using more detailed

Table 3.8
Error statistics for average clearness index values for the Bird clear-sky model combined with the opaque-cloud ASHRAE model results obtained from comparing measured data at 18 sites for 4 years

Average K_t	%MBE	Hourly %RMSE	Daily %RMSE
0.275	6.49	43.63	26.41
0.325	2.86	37.98	20.51
0.375	2.48	32.71	18.27
0.425	1.26	29.81	16.16
0.475	0.07	27.26	14.52
0.525	−1.33	24.88	12.94
0.575	0.00	20.94	10.96
0.625	2.08	17.05	8.55
0.675	0.90	13.41	6.53
0.725	0.33	9.42	4.37

cloud information that corrects for the effects of multiple reflections between more than one cloud layer and for the characteristics of different cloud types.

Atwater and Ball (1981) and Ball, Atwater, and Thoren (1981) have evaluated how various methods of cloud analysis affect the accuracy of estimated insolation. They have also reviewed and summarized the literature on the effective transmittance of various cloud types.

Stochastic means were also explored to include the temporal correlation of both hourly and daily global insolation values by Biga and Rosa (1981), Mustacchi, Cena, and Rocchi (1979), Brinkworth (1977), and Goh and Tan (1977). These authors have shown for the climates they studied that a significant temporal correlation of global insolation values exist for up to 2 days.

3.3.2.3 SOLMET Global Insolation Data Accuracy Based on the results in the preceding section, mean daily global insolation values extracted from the SOLMET data base for the rehabilitated stations are probably accurate to within ±7.5%. The ERSATZ values are probably accurate to within ±10%. The procedures used to derive all of these global values, however, destroyed any evidence for long-term trends in clear sky insolation. Roosen and Angione's (1984) study of Smithsonian data back to the early years of the century indicate no long-term trends in atmospheric opacity. Effects of large volcanic eruptions are noted, but in the absence of recent volcanic events, the clear atmospheric transmittance away from urban areas appears to be stable. In an area of significant industrial development, we expect long-term changes. Furthermore, the uncertainty of individual hourly values is certainly greater.

The numerous modeling and interpolation assumptions required to produce the data values in the SOLMET/ERSATZ data base were done to produce the best estimate of mean values. Little consideration was given to preserving the statistical distribution of values or the temporal correlation of values. For some solar energy applications, the statistically infrequent may be the most important. For example, the probability of a sequence of hours with unusually low insolation may be the significant driving factor in determining the size of a backup energy system. Yet it is precisely this type of excursion that tends to be damped by the procedures used in rehabilitating insolation data in the SOLMET/ERSATZ data base. These effects need further study.

3.4 Direct Insolation Models

Suitable serial direct insolation data sets did not exist in the early 1970s to support the detailed studies of solar energy systems. The need for such data sets led to the development of models for estimating direct insolation, usually from global-horizontal insolation values. From a theoretical perspective, direct insolation models are somewhat simpler since they can ignore the complicated scattering processes giving rise to diffuse insolation. In fact, a direct insolation model is usually part of most clear-sky global insolation models. Many statistical correlation global insolation models also divide insolation into direct and diffuse components and, thus, implicitly contain a direct insolation component, although they are not usually expressed in this way. In this section, attention is restricted to describing those models used to prepare the existing, widely used insolation data bases (e.g., SOLMET/ERSATZ) for the United States.

3.4.1 Historical Models

The solar energy system mission analyses conducted by the Aerospace Corporation (Leonard and Randall, 1974) beginning in 1973 were primarily concerned with concentrating thermal systems and required direct insolation data. Computing the insolation from atmospheric data required meteorological parameter sets that were unavailable for all the locations of interest, so a statistical correlation procedure was adopted as suggested by the work of Bennett (1969) and Liu and Jordan (1960). Since the available regression relations were based on data from climates significantly different from the arid Southwest and only implicitly related direct and global insolation, a new regression relationship was developed. Data from Albuquerque and some data from Blue Hill, MA, were used to develop the correlation (Leonard and Randall, 1974). Hourly average direct insolation data were not available, so they used the published instantaneous direct insolation values taken at specific solar zenith angles as a basis for interpolating intermediate time values. Subsequent experience, however, demonstrated this was a poor choice. The resulting regression relation between hourly percent of possible global insolation ($K_t \times 100$) and direct insolation was then used with the global insolation values to generate complete data bases containing hourly surface meteorological data, global insolation, and direct insolation for 2 years (1962 and 1963). These data bases were prepared first for the southern California area (Leonard and Randall, 1974), then for the southwestern United States (Hall et al., 1974), and finally for 34 locations throughout the United States (Randall, 1976). These data bases were widely used at the Aerospace Corporation and elsewhere during the early 1970s.

Sandia National Laboratories-Albuquerque carried out similar mission analysis studies of solar photovoltaic conversion systems beginning shortly after the initiation of the Aerospace studies. Again, a need was identified for a standard set of realistic direct insolation data to support system comparisons. Sandia (Boes, 1975) also adopted a statistical regression approach. They were able to obtain strip chart recordings of direct insolation observations from Omaha, NE, Blue Hill, and Albuquerque. These were digitized and summed to provide better hourly direct insolation values on which to base their regression relationships. The resulting regression relationship was applied to hourly global data from NWS for 1958–1962 to obtain a data base for 26 U.S. locations (Boes et al., 1976). In addition to its use for specific system studies, this data base was also used to estimate insolation on surfaces oriented in various directions for the same 26 locations (Boes et al., 1977).

Watt (1978) also estimated mean monthly direct insolation values as part of a general evaluation of the solar radiation energy resource. Watt's model was based on an empirical modeling of the various physical processes resulting in the atmospheric attenuation of the incident direct solar radiation. As input to this model, he used climatic means of the relevant atmospheric properties, such as turbidity and water vapor. The results of this clear atmospheric modeling were then modified for the cloud cover statistics of the specific site. The thrust of Watt's model was to estimate climatic mean values, so he did not try to obtain hourly values for energy system simulations.

3.4.2 SOLMET Direct Insolation Modeling

3.4.2.1 Estimation Procedure Description During 1976–1977, several considerations led to a reexamination of the direct insolation models. There was a desire to prepare a data base covering a longer time period than Sandia's 5 years. The imminent completion of the National Oceanic and Atmospheric Administration (NOAA) global insolation rehabilitation project meant that for the first time an improved global insolation data base, covering a significant time period, would be available as a source of estimation. At the Aerospace Corporation, researchers were increasingly concerned that although the regression procedures everyone had used up to that time predicted the mean insolation values, these procedures did not correctly reflect the statistical variation of actual insolation values about the mean. These statistical variations were potentially significant, and detailed analyses were required to answer energy system questions, such as the amount of conventional backup capacity required. In addition, a comparison of the Sandia direct-normal estimates with the earlier Aerospace values indicated that the Aerospace values were 10–30% higher than Sandia values for similar locations.

To continue the solar photovoltaic work at Sandia and the solar thermal conversion work at Aerospace, researchers needed improved values of direct insolation as a standard data base to compare systems. Therefore, it seemed reasonable to develop a single estimation procedure and standard data base. While these improved direct insolation estimation procedures were being developed, NOAA and ERDA (now DOE) decided to include these estimates of direct insolation in their rehabilitated solar data sets, which became SOLMET. Randall and Whitson (1977) developed the algorithm and its application to provide direct insolation estimates for the 26 fundamental SOLMET stations. For general use, they provided a slightly improved version of the algorithm's computer code to SERI (Randall and Biddle, 1981).

As part of the preparation of the *Solar Radiation Energy Resource Atlas of the United States* (SERI, 1981), EG&C applied a modified version of the mean value portion of the 1977 Aerospace algorithm to estimate monthly mean values of direct insolation for the ERSATZ data stations. These mean values are the source for the *SERI Direct Normal Solar Radiation Data Manual* (Knapp and Stoffel, 1982). Note that the SOLMET-type data tapes for the ERSATZ stations do not contain hourly values of the direct insolation.

The improved direct insolation estimation procedure, developed by Randall and Whitson (1977), began by deriving a new regression relationship between observed hourly global and direct insolation measurements. These measurements were made with carefully calibrated instruments at four locations having widely diverse climates: Livermore, CA, Raleigh, NC, Ft. Hood, TX, and Maynard, MA. Some parts of the analysis also utilized some carefully reviewed data from Albuquerque for 1961–1964. Figure 3.18 shows the direct insolation as a function of global insolation for Maynard with both insolation values expressed as a percentage of the extraterrestrial insolation. This figure is typical of the relationships observed for all data locations.

Two observations from the figure furnished important guidelines regarding direct insolation simulation. First, higher direct insolation values tend to correspond to a higher percentage of possible global insolation. Second, for any given value of global insolation, a wide range of observed direct insolation values remains. This distribution is a real characteristic of hourly insolation values and is not attributable to errors in the measurements. An adequate data base for stimulating solar energy system simulations should reflect both of these characteristics. We summarize Aerospace's algorithm here; for details of the rather complicated procedures actually used, see Randall and Whitson (1977).

Aerospace researchers used an empirical approach to obtain the trend characteristics of the direct-global relationship rather than force the regression to fit a single functional form throughout its range. They segregated the direct-global insolation data pairs, based on the global insolation values, into 10 bins, each 10% wide. They then computed the mean value of all the direct insolation values associated with the global insolation in each bin. These values were then connected by straight lines to create a continuous mean value estimation algorithm. The relationship between direct and global insolation derived in this way was quite similar for all of the recent data sets considered. Figure 3.19 shows the mean relationship

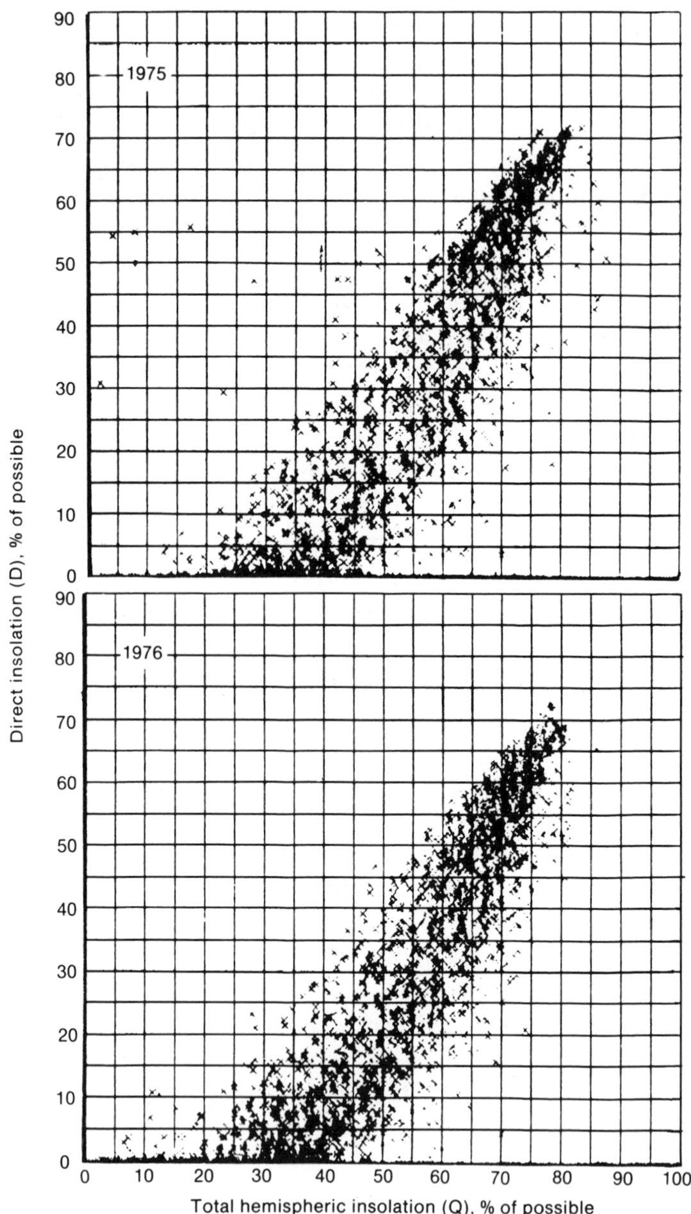

Figure 3.18
Scatter plot of observed hourly Maynard percentage of possible direct insolation $D(\%)$ as a function of observed percentage of possible total-hemispheric insolation $Q(\%)$. Source: Randall and Whitson (1977).

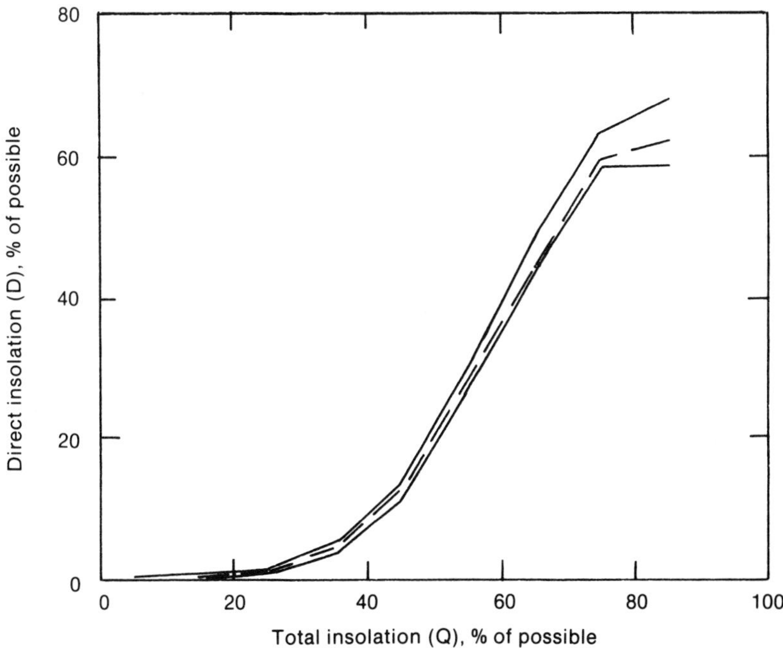

Figure 3.19
Extreme values of station mean direct insolation *D* as a function of total-hemispheric insolation *Q*. Computations were made for 10% intervals in *Q*. The broken line is the mean of the five station means. Source: Randall and Whitson (1977).

derived from all of the sites as a dashed line; the solid lines bound the values derived from each of the data sets individually. The site independence of the relationship between global and direct insolation is consistent with the regression results of other investigators going back to Liu and Jordan (1960). The data from all the sites, therefore, were combined into a single mean value estimating relationship. Although the difference among sites was small, the difference between months was significant, so a separate trend curve was developed from the data for each month.

The distribution of values about the mean was obtained using a set of empirically derived cumulative frequency distribution curves (ogives) describing the deviation of the observed direct insolation value from the mean. The solid line in figure 3.20 shows the mean value relationship between direct and global insolation derived from all the data sets using the previously described procedure. The actual parameter shown for direct

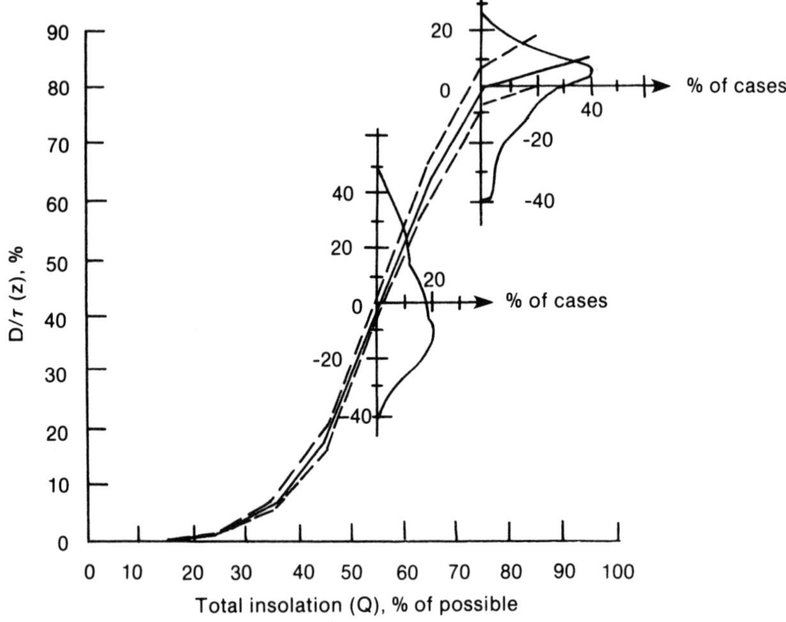

Figure 3.20
Composite plot D/τ with example distributions of deviations from D/τ for Maynard data ($\tau \equiv$ clear sky transmittance). Source: Randall and Whitson (1977).

insolation in figure 3.20 is the percentage of possible direct insolation divided by an empirical clear-sky transmittance value (Randall and Whitson, 1977). Also shown is the distribution of the Maynard direct insolation values about the mean for two of the global insolation bins previously described. These statistical distributions are clearly different from each other and not Gaussian, so a separate distribution was empirically determined for each percentage of possible global insolation bins in each data set. Insolation values were correlated sequentially to some extent by defining the K_t bin on the basis of not only the current hour but also the previous hour. The first and last daylight hours of the day were also considered separately. Figure 3.21 shows the ogives from a single global insolation bin for the four primary data locations. The ogives for the same percentage of possible conditions from all of the locations in the empirical data base were combined to make a final set for use with the estimation algorithm.

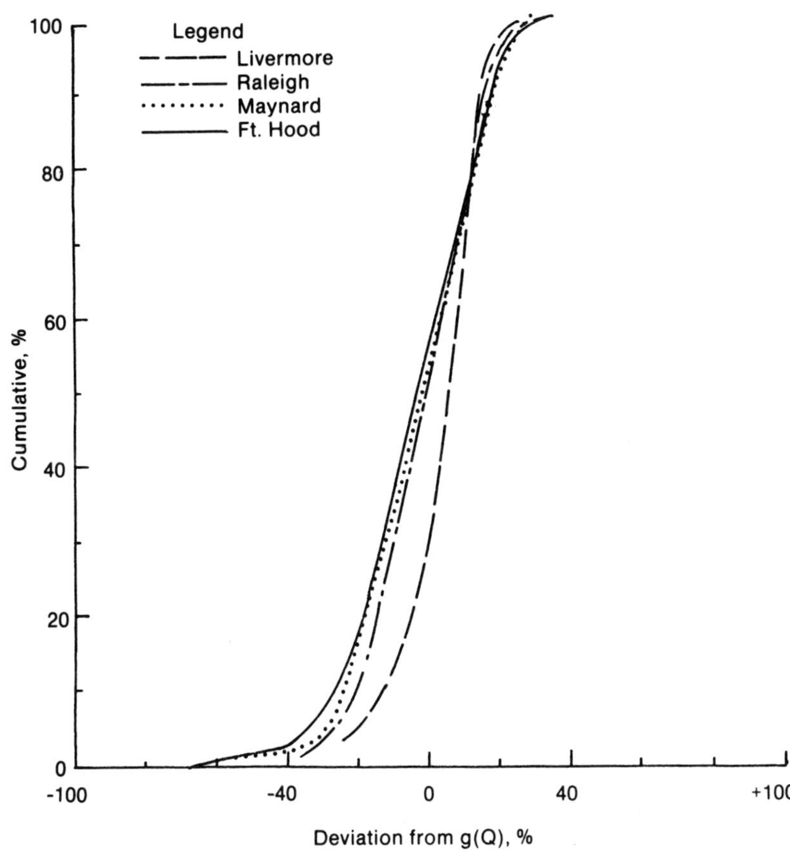

Figure 3.21
Cumulative frequency distributions (ogives) of deviations from $g(Q)$ for $60\% < Q < 70\%$ and midday hours. Source: Randall and Whitson (1977).

Aerospace researchers used the algorithm to estimate direct insolation by first computing the percentage of the possible global insolation value ($K_t \times 100$) for the hour. From this, they used a trend curve (e.g., figure 3.19) to estimate a mean direct insolation value for the hour. They then used appropriate ogives (e.g., figure 3.21) and a random-number generator to enter the observed distribution of the direct insolation about the mean value. Finally, they applied several constraints to assure that the estimated direct insolation value was physically realistic and consistent with the observed global insolation for the hour. The algorithm was used to estimate hourly direct insolation values for all of the 26 SOLMET stations. Subsequently, they supplied to SERI a slightly modified version of the computer code implementing this algorithm, known as the Aerospace Direct Insolation Prediction Algorithm (ADIPA) (Randall and Biddle, 1981).

The original SOLMET-type data base for the ERSATZ stations does not contain direct insolation. EG&G, however used a simplified version of the direct insolation algorithm to estimate direct insolation for the ERSATZ stations in preparing the *Atlas* (SERI, 1981). In addition, they extracted for each of the ERSATZ stations a typical meteorological year (TMY) that included direct insolation estimated using this simplified model; this is called the ERSATZ TMY (ETMY) model. The ETMY model did not use the statistical simulation procedures contained in the full ADIPA, and therefore the code ran much faster. The table that related the transmitted direct insolation to the percentage of possible global insolation (table 3.3 of Randall and Whitson, 1977) was changed to produce larger direct insolation values for global values between 75% and 95%. This compensated for a bias in the mean value curves that was subsequently corrected in the full ADIPA algorithm by the statistical application of the ogives but would have remained in the ETMY implementation without this compensation. Finally, the ADIPA used true solar time, and the ETMY algorithm used local standard time.

3.4.2.2 Evaluation of the SOLMET/ERSATZ Direct Insolation Data
Internal Estimates of Accuracy. In developing the 1977 Aerospace algorithm, researchers tested it by estimating the direct insolation values in the data base from which the parameters were derived. This provided the most optimistic estimate of accuracy. Annual mean daily direct insolation values agreed to within 4%. They estimated the mean daily direct insolation values for 66% of the months to within 6% of the observed values. The algorithm tended slightly to underestimate in the winter and overestimate

in the summer, but the test data set contained enough contrary examples to preclude adjusting the algorithm. The mean direct insolation was also compared as a function of global insolation value with the estimated statistical distributions of estimated values using the observed distributions. In all cases the agreement was satisfactory (Randall and Whitson, 1977).

Comparison with Earlier Models. Randall and Whitson (1977) compared the Aerospace algorithm applied to the rehabilitated global insolation of the SOLMET data bases with the older direct insolation models, and in most cases they found it predicted lower direct insolation. The climatic estimates of Watt (1978) agreed best with the mean values obtained from the SOLMET tapes. The new algorithm and revised global insolation data predicted significantly lower direct insolation for all the locations included in the earlier Aerospace insolation data bases (Leonard and Randall, 1974; Hall et al., 1974; Randall, 1976). Melton (1978) reported on the consequences of these changes for solar-thermal systems. The new estimated direct insolation values were also lower than the earlier Sandia estimates (Boes et al., 1976, 1977). As a consequence, Sandia issued an addendum to its original work so that others could continue to use their extensive tables and data sets.

A portion of the decrease in estimated direct insolation results from somewhat reduced average values of global insolation values, but the major difference is in estimating the mean value. Figure 3.22 compares the direct-global regression relationships used to prepare the hourly insolation data bases (discussed here) with the relationship developed by Liu and Jordan. The early Aerospace relationship clearly overestimates the direct insolation for low values of global insolation, probably because of how the direct insolation values were obtained for the original 1973 regression relationships. The Sandia regression relationship agreed well for low global insolation values, but it gives high direct insolation values for clear conditions.

Comparison with Independently Observed Values. Two levels of comparison are possible. The first level is to obtain an independent set of simultaneous hourly observations of hourly global and direct insolation. The global values are then used to drive the estimating algorithm, and the resulting estimated direct values are compared with the observed values. The second level is to compare mean estimated values from the SOLMET data base with recent observations. This latter comparison is necessary because the algorithms have inadequacies and the actual weather deviates from the

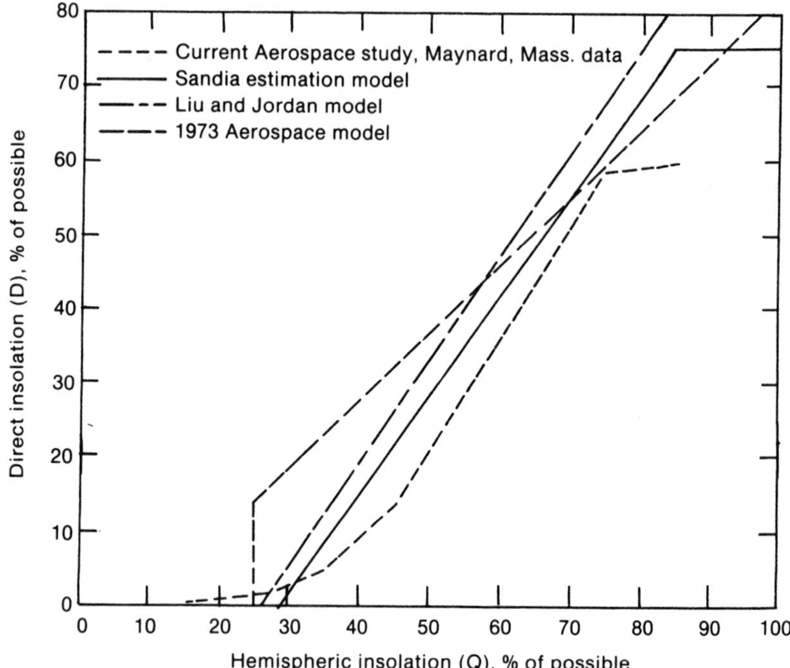

Figure 3.22
Comparison of direct insolation estimation regression relations. The Sandia estimation model is from Boes et al. (1976); the Liu and Jordan model is from Liu and Jordan (1960); and the Aerospace model is from Leonard and Randall (1974). Source: Randall and Whitson (1977).

climatic means reflected in the SOLMET summaries. Only the first method really tests the algorithm alone, but because the algorithm is complex and requires a great amount of data, this test is applied infrequently.

Mujahid and Turner (1980) compared observations for one year at Blytheville, AR, using the Aerospace-developed SOLMET direct insolation algorithm. They found that the algorithm overestimated the annual mean daily direct insolation by 9%. The overestimates were significantly higher in the winter than in the summer. They also compared the algorithm with 1 year of observations at Houston, TX, and found an annual underestimate of about 5%. They observed, however, the same relative error trend: Winter values were overestimated and summer values underestimated. These observations are consistent with the tendencies observed as a part of the original study.

Hall et al. (1980) have extensively compared the predictions of the 1977 Aerospace algorithm with independent data sets. These data came from three sources: Lawrence Berkeley Laboratories (LBL), WEST Associates, and John Hay of the University of British Columbia. Martin and Berdahl (1984b) from LBL measured direct insolation with an active cavity radiometer as part of their circumsolar radiation measurements program. Hall used LBL data from Albuquerque, Barstow and China Lake, CA, Argonne, IL, and Ft. Hood, TX. The WEST Associates measured direct insolation at several southwestern United States locations (Yinger, 1982). Hall used these data for Barstow during 1976 and 1977. Hay provided observations from Toronto and Vancouver, Canada. Hall implemented the 1977 Aerospace algorithm on Sandia computers and used it to predict the direct insolation from the observed hourly global insolation in each of these data sets. He then compared the predicted hourly direct insolation with the observed value. The monthly median values of the resulting hourly percentage differences are presented in table 3.9. These results suggest that the algorithm tends to underestimate direct insolation. If the Toronto results are ignored, there is additional support for the suggestion that the winter values are underestimated.

Hall also considered the differences between estimated and observed daily direct insolation. For each month of available data from each station, he computed the median daily direct insolation value. For the 112 station-months considered, he finds that 82 (73%) have monthly median difference values of absolute magnitude of 15% or less, 66 station-months (59%) have differences of magnitude of 10% or less, and only 36 station-months (32%) have differences of magnitude of less than 5%. When the station-months with differences greater than 5% are considered, most (80%) are underestimates of direct insolation.

Hall's findings that 50% of the months have differences of 10% or less are qualitatively consistent with the results of the tests applied as part of developing the original algorithm, which indicated that 66% of the months were within 6%. Although this consistency is reassuring, Hall's analyses are also limited by the data available to him. Some of the strongest indications of understimation come from the LBL data for Barstow. The WEST Associates observations made from the same rooftop in Barstow do not indicate such great differences. Before drawing definitive conclusions, these differences in the data need to be resolved. The data available for Hall's analysis are also not quite uniformly distributed throughout the year; the data consider a few more winter months than summer months.

Table 3.9
Median percent difference between hourly Aerospace estimated direct normal insolation and hourly observed direct insolation

			Barstow					
	Albuquerque	Argonne	West	LBL	China Lake	Ft. Hood	Toronto	Vancouver
1976								
Jan	—	—	−0.2	—	—	—	—	—
Feb	—	—	−5.2	—	—	—	—	—
Mar	—	—	−6.2	—	—	—	—	—
Apr	—	—	−5.8	—	—	—	—	—
May	—	—	−7.1	—	—	—		—
Jun	−6.0	—	−11.0	—	—	—	—	—
Jul	−11.4	—	−8.8	—	—	—	—	—
Aug	−9.4	—	−13.0	—	−24.0	−14.4	—	—
Sep	—	—	−1.6	—	—	—	—	—
Oct	−19.8	—	−9.6	—	−16.5	—	—	—
Nov	−22.1	—	−14.5	—	−12.6	−0.15	+23.8	—
Dec	−11.7	—	−7.5	—	−6.9	+0.1	+26.3	—
1977								
Jan	−13.5	—	−1.5	—	0.3	3.6	+45.7	—
Feb	−15.2	—	−10.4	—	−8.5	—	+25.4	—
Mar	—	—	−4.9	—	—	−0.2	+0.9	—
Apr	−0.1	—	−9.8	—	—	—	+0.8	—
May	−11.9	—	−7.8	—	—	—	−6.4	—
Jun	—	—	−17.2	—	—	—	−3.8	—
Jul	−8.6	—	−14.0	—	—	—	+1.9	—

Aug	—	—	−7.7	−19.4	—	—	+4.7	−14.5
Sep	—	−5.4	−12.1	−19.8	—	—	−0.5	+0.3
Oct	−10.0	−15.7	−8.9	−18.8	—	—	−0.9	+5.8
Nov	−16.0	−2.2	−5.6	−16.4	—	—	+7.6	+0.3
Dec	−20.8	−0.6	−2.8	−16.5	—	—	+3.6	+0.4
1978								
Jan	—	—	—	—	—	—	+54.4	+6.3
Feb	—	—	—	—	—	—	+2.5	+2.5
Mar	—	—	—	—	—	—	+5.5	+2.9
Apr	—	+6.8	—	—	—	—	—	+15.6
May	—	—	—	−16.7	—	—	—	−0.9
Jun	—	—	—	−19.3	—	—	—	+4.0
Jul	—	—	—	—	—	—	—	−2.3

Source: Hall et al. (1980).

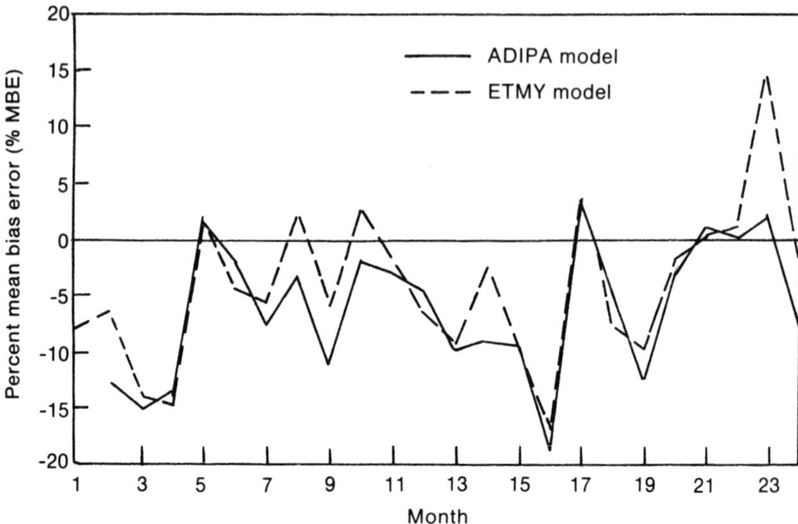

Figure 3.23
The percent mean bias errors (model-measured) of the ADIPA and ETMY models for West Los Angeles, CA.

SERI (Bird) researchers recently compared the direct insolation estimated using the ADIPA and the simplified ETMY algorithm with measured direct insolation from several locations: West Los Angeles, Palm Springs, and Barstow, CA; Alamosa, CO; Atlanta, GA; and Albany, NY. All comparisons were for the 24 months from January 1979 through December 1980, except for the Atlanta data, which covered the 22 months from August 1980 through May 1981. September and October 1979 data were missing from the Alamosa set, causing the ADIPA to produce values for only the first 9 months. For each hour, they compared the observed direct insolation with that estimated by the ADIPA and ETMY algorithm. For each station per month, they computed the %MBE for each model. The results are shown in figures 3.23–3.28. Clearly, both models underestimate the direct insolation. For the 126 station-months for which ADIPA comparisons are available, the average %MBE is −5.0% and the median is −5.5%. The %MBE of 68% of the 126 ADIPA comparisons are between −11.4% and +1.2%. For the 140 station-months for which ETMY comparisons are available, the average %MBE is −1.3% and the median is −3.3%. The %MBE of 68% of the 140 ETMY comparisons are between

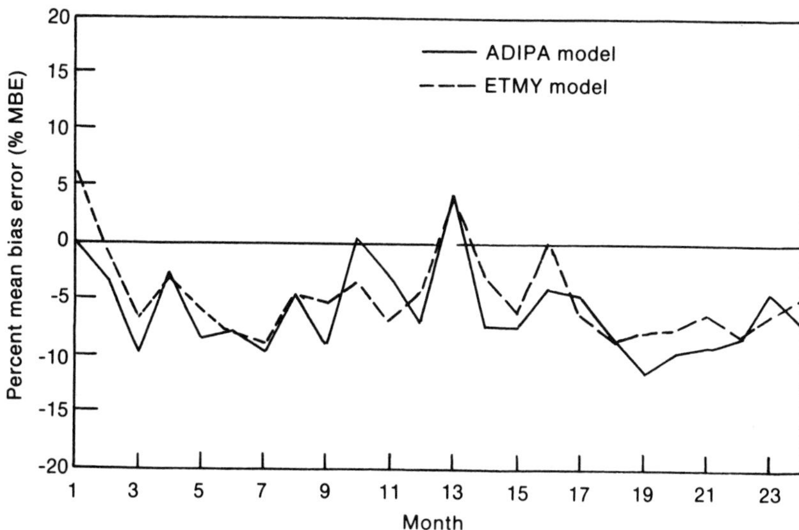

Figure 3.24
The percent mean bias errors of the ADIPA and ETMY models for Palm Springs, CA.

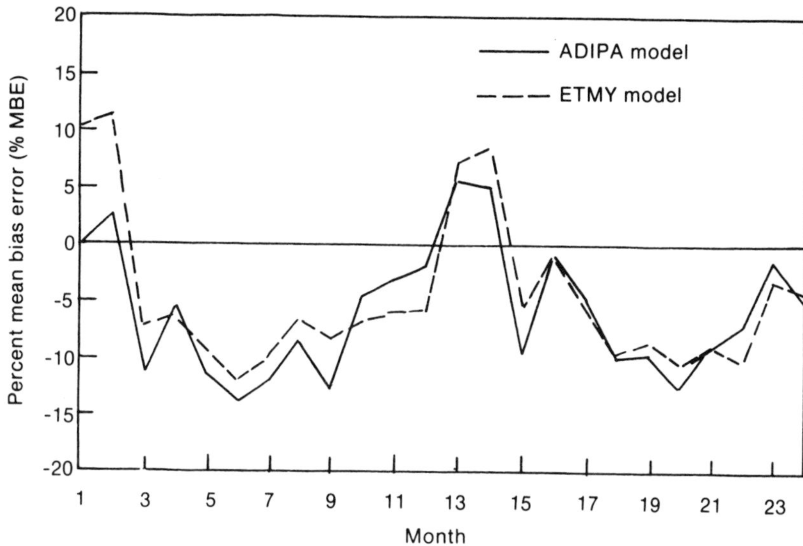

Figure 3.25
The percent mean bias errors of the ADIPA and ETMY models for Barstow, CA.

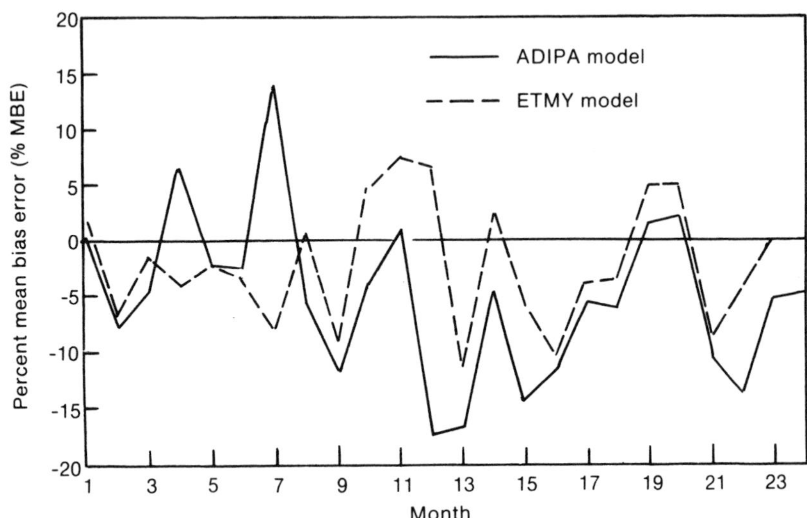

Figure 3.26
The percent mean bias errors of the ADIPA and ETMY models for Albany, NY.

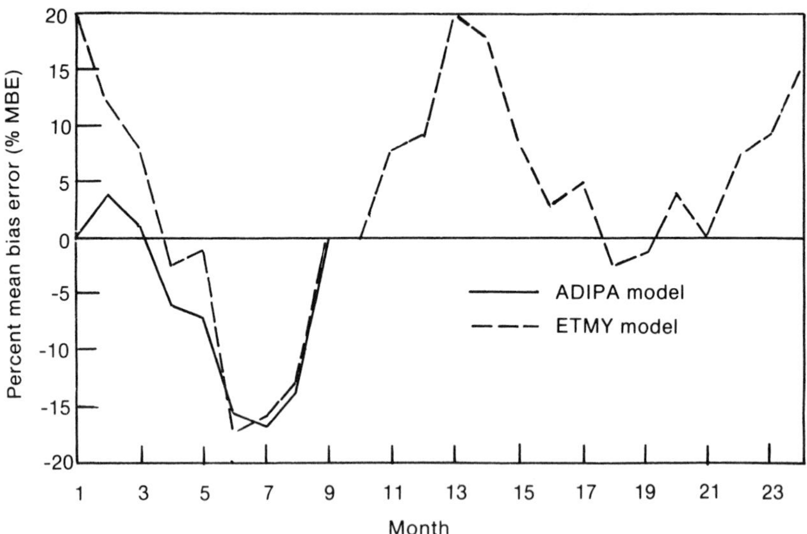

Figure 3.27
The percent mean bias errors of the ADIPA and ETMY models for Alamosa, CO.

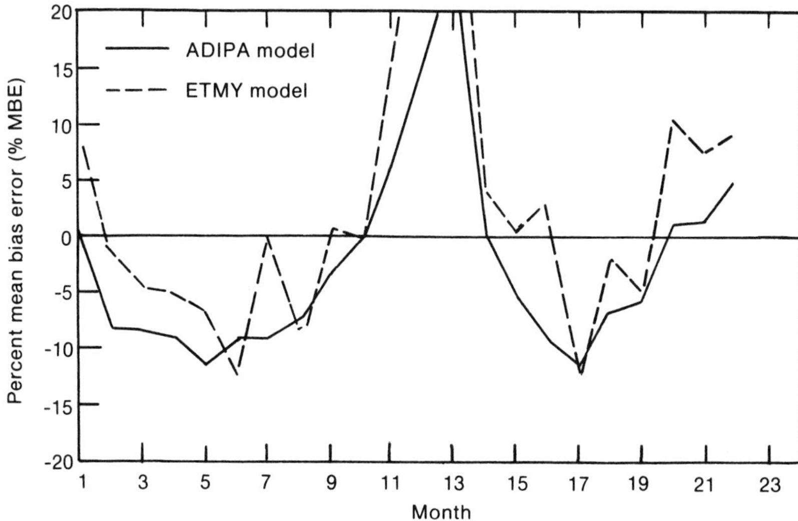

Figure 3.28
The percent mean bias errors of the ADIPA and ETMY models for Atlanta, GA.

−8.6% and +7.2%. The largest errors are during the summer at Atlanta and winter at Alamosa and are positive. A careful review of the Atlanta data suggests that the problem may be an error in the pyrheliometer tracking, causing an apparent reduction in the observed direct insolation, rather than a problem with the estimating algorithms. Pyrheliometer pointing errors are also suspected in the winter Alamosa data.

We can compare the monthly mean daily direct insolation values from the SOLMET direct data set with the observations from the new NOAA network for the 3 years (1978–1980) during which direct insolation measurements were made and published (NCC, 1980). Table 3.10 provides the observed monthly mean daily direct insolation (kJ/m^2) from the new network. Blank spaces indicate that insufficient observations were available to compute a monthly mean value for any year considered. The values in table 3.10 were obtained by taking the mean of all the reported monthly means. As a consequence, the values reported may be the average of from 1 to 3 monthly values. Estimated direct insolation values from the 18 rehabilitated SOLMET sites, which also have data from the new network, are compared in table 3.11. The table presents the percent difference of the

Table 3.10
Mean observed daily direct insolation (kJ/m^2), NOAA network 1977–1980

		Month											
Station name	Number	J	F	M	A	M	J	J	A	S	O	N	D
Albuquerque, NM	23050	14320	19886	21050	28323	26211	31343	28859	27943	25980	26034	14851	20842
Bismarck, ND	24011	9048	6397	13238	11480	22297	25018	25080	17522	21698	13171	9692	10831
Blue Hill, MA	14753												
Boise, ID	24131	7410	6329	15501	18296	22907	29052	31660	29912	21313	22235	10986	9581
Boulder, CO	94018	12322	17016	18435	19908	15551	25110	25450	23812	27498	20924	14580	15379
Brownsville, TX	12919	6815	8149	9154	10704	13016	20312	21826	14667	14117	12022	10425	8398
Burlington, VT	14742	7556	7170	12746	12957	15675	16890	16960	13770	16553	7201	6236	4577
Caribou, ME	14607	10945	11945	14857	9690	15670	18044	14018	15245	13960	6085	7321	6110
Columbia, MO	3945		9934	11063	10611	16283	19334	11261			14691	9392	12209
Desert Rock, NV	3161				30443		39223				29400		
Dodge City, KS	13985	5752	14952										
El Paso, TX	23044	15766	22534	24330	29605	30456	30348	27855	23937	21951	24652	19370	19325
Ely, NV	23154	11678	14281	16415	17207	29097	33039		34063	28150	22836	15775	17159
Fairbanks, AK	26411												
Fresno, CA	93193	4764							29672	23530	18978	13835	7963
Grand Junction, CO	23066	8667	12291	12338		14395						15791	13555
Great Falls, MT	24143	9829	8377	13981	16365	15586	22516	26854	20081	15977	16073	9654	6581
Guam, PI	41415	16553	14186	13722	15521	14367	13604			12176	8178	13356	14460
Honolulu, HI	22521												
Indianapolis, IN	93819												
Lake Charles, LA	3937	5758		6693	17744	10712	20770	9614	12889	9493	18400	15305	12407
Lander, WY	24021	7861	16467	19987	24144		29935	29621	27578	27667	22555	14187	12444

Las Vegas, NV	23169	12793	18912	22227	27431	30186	35938	33324	31293	29526	24833	20812	20129
Los Angeles, CA	23174	10919	15886	21045	19813	15607		24407	21526	16298	15186	16442	15136
Madison, WI	14837	7223		7575	8579				10824	17709	10366	6819	
Medford, OR	24225	8466	4235	12794	13665	19178	26075	33492	23211	22268	12908	5185	4198
Miami, FL	12839	12078		16653	14098	12567	13534	11650	11258	8720	10395	15673	10823
Midland, TX	23023		19870	21949	28609	22736		29530	24576	18688	20968	12165	17354
Montgomery, AL	13895	7472	13567	12428	17755	15025	15617	14242	14763	10588	19799	13026	13459
Nashville, TN	13897	5968	9141	11094	15829	13256	16567	12302	11728	11436	16627	10819	11576
Omaha, NE	94918		7145	15266	18155	16397			19648	19698	16601	13991	9536
Phoenix, AZ	23183	12756		21829	28830	28951	33453	28726	29455	28830	24112	15178	18621
Pittsburgh, PA	94823	2678	5062		9323	13784	16882	11868		12869	13168	5672	3806
Raleigh, NC	13722											4143	
Salt Lake City, UT	24127	6431	9060	14386	19927	22082	34361	32348	26283	27714	21462	11969	10612
San Juan, PR	11641			17892	13205	11611		18215	17236	15175			
Seattle-Tacoma, WA	24233					16634		17229	13126				
Sterling, VA	93734	7091	13231	13348	14061	16108	15674	13530	12524	13787	12868	11682	10719
Tallahassee, FL	93805	11875	12837	16112	15488						18106		13305

Table 3.11
Direct insolation percent error of SOLMET estimated insolation from NOAA measured insolation (1977–1980)

		Month											
Station name	Number	J	F	M	A	M	J	J	A	S	O	N	D
Albuquerque, NM	23050	42.0	12.8	15.8	28.2	15.8	1.8	−0.3	−0.3	4.2	−2.5	48.6	−4.7
Bismarck, ND	24011	3.0	−48.6	−13.1	−28.8	13.9	15.1	−0.5	−25.7	20.3	−12.3	2.0	40.6
Blue Hill, MA	14753												
Brownsville, TX	12919	−37.5	−33.8	−33.3	−28.4	−21.1	3.1	−2.7	−30.2	−18.0	−28.0	−21.1	−23.3
Caribou, ME	14607	−35.2	−9.6	−7.2	48.2	−7.2	−11.6	18.6	1.3	−11.5	48.5	−30.7	−14.0
Dodge City, KS	13985	177.9	21.1										
El Paso, TX	23044	−24.9	−4.7	−6.3	−0.4	−2.1	−4.3	−1.4	−12.5	−16.0	−3.8	−15.3	−5.2
Ely, NV	23154	43.7	36.1	38.6	44.2	−7.9	−11.3		−15.0	2.7	7.0	19.6	−7.7
Fresno, CA	93193	89.7							3.5	16.4	17.8	6.4	4.4
Great Falls, MT	24143	24.7	−28.0	−11.3	2.2	−18.8	0.4	−5.7	−17.7	−16.6	7.8	0.8	−4.8
Lake Charles, LA	3937	48.2		86.6	−27.8	44.5	−18.4	53.5	12.0	47.9	−16.3	−21.2	−22.8
Madison, WI	14837	20.1		76.8	61.1				64.6	−16.0	22.6	11.4	
Medford, OR	24225	−49.8	106.6	−7.7	20.4	7.9	−7.6	−10.0	13.4	−7.0	4.0	16.6	−15.1
Miami, FL	12839	−10.3		7.6	−14.7	−15.1	9.9	−11.1	−9.6	−23.6	−20.2	14.6	−22.5
Nashville, TN	13897	−25.0	−10.0	−1.9	11.0	−17.8	−4.1	−25.1	−28.4	−18.3	16.3	6.2	49.6
Omaha, NE	94918		−46.5	5.7	11.6	−12.9			−6.5	17.9	4.3	25.4	0.3
Phoenix, AZ	23183	42.2		8.4	0.3	10.5	−5.4	−2.5	−8.9	−8.8	0.3	34.0	−5.7
Seattle-Tacoma, WA	24233					−6.9		19.9	31.1				
Sterling, VA	93734	−16.2	29.0	13.0	4.2	6.0	−7.0	−11.2	−15.8	0.0	1.0	22.0	49.6
Mean percent difference		18.3	2.0	11.5	8.7	−0.7	−3.0	1.7	−2.6	−1.7	2.9	7.5	1.3
Mean percent difference (all cases together)				3.9									

monthly mean daily SOLMET estimates from the new observed values. Estimated mean direct insolation values obtained with the simplified algorithms for the 13 ERSATZ sites, which have direct insolation data from the new network, are compared in table 3.12.

Most of the differences presented in tables 3.11 and 3.12 are positive, indicating that estimates are greater than what was observed, in contrast to the results of the SERI study just presented. Furthermore, some of the overestimates are very great. We believe that these differences represent errors in the measurements rather than errors in the estimation algorithm. Measuring direct insolation requires the pyrheliometer to track the sun. Any tracking inaccuracy reduces the observed direct insolation. All pyrheliometers used in the network are mounted on solar trackers that cannot track actively; they have electrically driven polar axis clock drives that must be adjusted manually for changes in the solar declination and equation of time. These trackers must be adjusted daily during some seasons of the year. Furthermore, if the polar axis is not aligned correctly, the tracker will not provide the correct pointing all day, even if the declination and equation of time are set correctly at one time of day.

SOLMET/ERSATZ Direct Insolation Accuracy Summary. It is not possible to evaluate directly the accuracy of the SOLMET/ERSATZ direct insolation values empirically. The information in tables 3.11 and 3.12 should provide such a measure, but in reality it indicates the difficulties in routinely measuring direct insolation. We can indirectly evaluate the accuracy by considering separately the accuracy of the direct insolation estimating algorithm and the indications of the accuracy of the global insolation values from which the direct insolation is obtained.

The recent SERI study gives the best indication that the direct-insolation estimating algorithm is accurate. That study, summarized in the preceding section, shows that the monthly %MBE for the ADIPA used to provide the SOLMET direct insolation values is $-5.6 \pm 6\%$. The comparable value from the original study (Randall and Whitson, 1977) is $-0.9 \pm 6\%$. The comparison of recent global insolation observations with the SOLMET values presented in section 3.3.2.2 indicates that the SOLMET estimates are in the mean 3% higher than the recent observed values. In the SOLMET data base these two errors tend to compensate. In other words, an estimating procedure that gives values 5% too low is being applied to numbers that are 3% too high. Of course, this argument invoking means over numerous sites

Table 3.12
Percent difference of ERSATZ from observed direct insolation, NOAA network 1978–1980

		Month											
Station name	Number	J	F	M	A	M	J	J	A	S	O	N	D
Boise, ID	24131	2.6	98.7	7.1	17.9	15.5	−1.1	5.2	−3.6	18.7	−14.0	−1.1	−20.8
Burlington, VT	14742	−11.5	32.7	−6.0	11.1	9.1	9.2	15.1	29.0	−16.7	54.2	−12.7	−1.0
Columbia, MO	3945		24.9	18.8	42.5	11.9	6.2	95.1			3.3	21.0	−25.8
Grand Junction, CO	23066	94.2	58.5	73.5		95.1						19.8	25.2
Indianapolis, IN	93819												
Las Vegas, NV	23169	52.9	23.0	17.4	11.6	8.2	−5.8	−6.1	−5.2	−3.4	2.4	1.1	−6.7
Midland, TX	23023		7.7	10.0	−9.0	20.3		−8.1	4.4	19.3	6.3	72.7	11.1
Montgomery, AL	13895	35.6	−13.0	7.4	−10.4	12.3	7.6	6.5	9.1	33.0	−20.0	1.9	−18.6
Pittsburgh, PA	94823	83.5	21.5		7.6	−9.1	−19.1	13.9		−9.6	−18.5	0.5	−0.7
Raleigh, NC	13722											133.3	
Salt Lake City, UT	24127	82.3	78.1	39.1	15.2	29.1	−10.5	0.8	12.3	−4.0	2.8	24.6	7.8
San Juan, PR	11641			−9.4	18.5	17.4		−11.1	−3.3	−5.1			
Tallahassee, FL	93805	−10.5	−5.3	−14.6	1.8						−13.6		−17.1

does not necessarily apply to an individual site. Nevertheless, the monthly %MBE for the SOLMET direct insolation values appears to be no worse than the error in the estimating algorithm itself.

The errors in the ERSATZ global insolation data base and the ETMY direct insolation model are similar in magnitude and sign to the errors for the rehabilitated SOLMET data. Thus, the direct insolation values presented in the *Atlas* (SERI, 1981) and the *Direct Normal Solar Radiation Data Manual* (Knapp and Stoffel, 1982) may be accurate to within 10% or less. However, actual error depends on the specific site and time of year.

3.4.3 Related Modeling

We discuss several algorithms here that provide either an alternative to estimating direct insolation or information on a closely related parameter. We did not attempt to review the entire literature in these cases, but rather to indicate complementary approaches.

3.4.3.1 Diffuse Insolation Models The relationship among the global, direct, and diffuse insolation given in equation (3.1) demonstrate that an algorithm relating any two of these quantities is an algorithm for finding the third. In particular, several published models relating global to diffuse insolation exist that are implicitly direct insolation models. Indeed, the classic Liu and Jordan (1960) results are expressed in terms of diffuse-global insolation relationships. Relationships of hourly diffuse and global insolation values are provided in the work of Orgill and Hollands (1977), Bruno (1978), and Bugler (1977). Spencer (1983) compared several of these hourly estimation methods with Australian observed insolation data, concluding that Orgill and Hollands' relationship has more of the diffuse insolation estimates close to the observed values. The relationship between daily values of global and diffuse insolation was considered by Choudhury (1963), Stanhill (1966), Tuller (1976), Ruth and Chant (1976), and Collares-Pereira and Rabl (1979). Several of these authors also provide correlations of monthly values. Erbs, Klein, and Duffie (1982) have reviewed these earlier approaches and present additional analysis of recent data. The work of Collares-Pereira and Rabl and Erbs, Klein, and Duffie are based in part on the same experimental data base as the algorithm of Randall and Whitson (1977), used in preparing the SOLMET data sets.

The correlations of diffuse with global-horizontal insolation are usually expressed as a relationship between the diffuse fraction of the observed

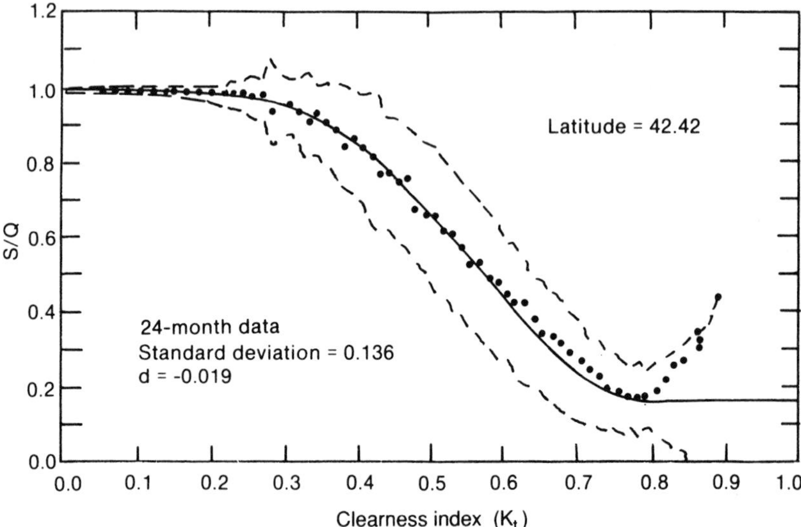

Figure 3.29
Diffuse insolation model of Erbs et al. compared with observed data from Maynard, MA (where S is the measured diffuse-horizontal insolation and Q is the measured global-horizontal insolation). Source: Erbs, Klein, and Duffie (1982).

global insolation and the percentage of possible global insolation K_t [equation (3.15)]. Figure 3.29 shows Erbs et al.'s analysis of the Maynard, MA, data shown in figure 3.18. The curve is the mean hourly diffuse model derived by Erbs from all the data considered. That empirically derived curve can be expressed by the following mathematical relations (Erbs, Klein, and Duffie, 1982):

$$S/Q = \begin{cases} 1.0 = 0.09K_t & \text{for} \quad K_t \leq 0.22 \\ 0.9511 - 0.1604K_t + 4.388K_t^2 \\ \quad - 16.388K_t^3 + 12.336K_t^4 & \text{for} \quad 0.22 < K_t \leq 0.80 \\ 0.165 & \text{for} \quad K_t > 0.80. \end{cases} \tag{3.20}$$

3.4.3.2 Standard Pressure Model In general, direct insolation is greater at higher altitudes and for drier conditions. If there is no additional information, we can estimate direct insolation based only on the pressure altitude of the location, a judgment concerning whether the site is dry or moist, and relevant solar position information. Lowry (1980) provides an algorithm for estimating the direct insolation from this limited information.

3.5 Models for Estimating Insolation on Flat, Tilted Collectors

Typically, available insolation data consist of either global-horizontal or direct-normal insolation or both. Most of these data, especially the historical data, are only global-horizontal insolation. To be used for solar energy applications, such data often must be converted to yield estimates of the insolation available to flat, tilted solar collectors.

Several investigators have developed algorithms for converting global-horizontal, diffuse-horizontal, and direct-normal insolation to apply to tilted surfaces. This section briefly reviews the more well-known or current ones. Special emphasis is given to those that apply to converting the hourly SOLMET/ERSATZ data (chapter 2).

All of the models are made up of three fundamental components: the direct insolation component, the diffuse-sky insolation component, and the ground-reflected insolation component, given as

$$Q_t = D_t + S_t + \mathrm{GR}_t, \qquad (3.21)$$

where

Q_t = total global insolation on the tilted surface,

D_t = direct insolation on the tilted surface,

S_t = diffuse insolation from the sky on the tilted surface, and

GR_t = ground-reflected insolation component.

3.5.1 Models for Converting Monthly Global-Horizontal Insolation

Monthly average daily global-horizontal insolation data are quite common, especially internationally and historically. Liu and Jordan (1961) developed one of the first methods; the University of Wisconsin (Klein, Beckman, and Duffie, 1978) also did extensive work in this area. This section summarizes such work and their results.

The average daily global insolation on a tilted surface $\bar{Q}_t$ can be related to the average daily global insolation on a horizontal surface $\bar{Q}$ by

$$\bar{Q}_t = \bar{R}\bar{Q} = \bar{R}\bar{K}_t\bar{Q}_x, \qquad (3.22)$$

where $\bar{R}$ is defined as the ratio of the daily average insolation on a tilted surface to that on a horizontal surface for each month. The terms K_t and Q_x represent monthly average daily quantities. Liu and Jordan (1961) proposed, assuming the diffuse skylight and ground-reflected insolation are equally distributed from all directions, that Q_t can be given by

$$\bar{Q}_t = (\bar{Q} - \bar{S})\bar{R}_b + \bar{S}(1 + \cos\beta)/2 + \bar{Q}\rho(1 - \cos\beta)/2, \tag{3.23}$$

where

$\bar{S}$ = monthly average daily diffuse insolation,

$\bar{R}_b$ = ratio of the average direct normal insolation on the tilted surface to that on a horizontal surface for each month,

β = tilt of the surface from the horizontal, and

ρ = ground reflectance (albedo).

If we only have the monthly average daily insolation data available, then we must somehow derive S, R_b, and ρ.

Liu and Jordan (1961) and Page (1961) showed the ratio $\bar{S}/\bar{Q}$ to be a function of $\bar{K}_t$, which can be calculated from

$$\bar{K}_t = \frac{\bar{Q}}{\bar{Q}_x}. \tag{3.24}$$

Several investigators (see Klein, Beckman, and Duffie, 1978) calculated the values for the monthly $\bar{Q}_x$. Liu and Jordan originally determined that

$$\bar{S}/\bar{Q}_x = 1.390 - 4.027\bar{K}_t + 5.531\bar{K}_t^2 - 3.108\bar{K}_t^3. \tag{3.25}$$

However, Page (1961) determined that

$$\bar{S}/\bar{Q}_x = 1.00 - 1.13\bar{K}_t. \tag{3.26}$$

Klein, Beckman, and Duffie (1978) consider Page's algorithm to be the better of the two because it agrees more closely with several other more recent algorithms (Choudhury, 1963; Stanhill, 1966) based on better experimental data.

The term $\bar{R}_b$ is a function of the actual atmospheric transmittance of the direct normal insolation, which is a variable that depends on atmospheric water vapor, cloudiness, and aerosols, as well as geometric factors dependent on the location of the sun in the sky. Liu and Jordan (1961), Page (1961), and Klein, Beckman, and Duffie (1978) have calculated $\bar{R}_b$ values for various site latitudes, months, collector tilt angles, and orientations. Use Klein, Beckman, and Duffie's publication for tabulating and formulating $\bar{R}_b$.

The ground reflectivity ρ can vary from 0.10 (e.g., plowed fields) to 0.80 (e.g., fresh snow cover). If such information is not available, assume an average value of 0.20.

The paragraphs, equations, and references presented here allow us to convert monthly average daily global-horizontal insolation data to flat, tilted surfaces. The University of Wisconsin investigators (Klein, Beckman, and Duffie, 1978), funded by DOE, have performed such conversions for global horizontal insolation data for 261 North American cities, including the SOLMET/ERSATZ stations (see chapter 2). They produced monthly average daily insolation for the following flat, tilted collector orientations:

Tilt angle (degrees)	Azimuth of collector surface (degrees)
20	0 (due south),
30	15 (E or W of south),
40	30 and 45 (E or W of south)
50	
60	
70	
80	
90	

Along with the global tilted surface insolation values, data are also given for monthly average daily global-horizontal insolation, K_t, monthly average temperature, and monthly average heating degree-days. This collection of data is particularly suited for solar thermal heating applications.

3.5.2 Hourly Average Direct-Normal, Global-Horizontal, and Diffuse-Horizontal Insolation Conversion Models

With the advent of the SOLMET/ERSATZ data base for hourly average direct-normal, global-horizontal, and diffuse-horizontal insolation, the need for an algorithm for converting these data to a tilted surface was highlighted. As a consequence, several investigators have developed a variety of models. This section reviews some of the more commonly known and used algorithms starting from equation (3.21).

Converting the direct normal insolation to a tilted surface is relatively straightforward:

$$D_t = D\cos\theta \tag{3.27}$$

where θ is the incidence angle of the direct solar beam to the tilted surface. We can calculate this angle from the known position of the sun and the elevation and azimuth of the normal to the tilted surface using a well-known formula given in many solar energy engineering texts (e.g., Sayigh, 1977). All of the models discussed in the following subsections calculate the direct-beam insolation component in this manner.

Most of the models estimate the ground-reflected insolation component on the assumption that the ground reflects insolation isotropically:

$$\mathrm{GR}_t = \tfrac{1}{2}[Q\rho(1 - \cos\beta)]. \tag{3.28}$$

Because of the uncertainties concerning ρ, this formula is usually considered sufficient for the intended purpose.

The models reviewed in the following subsections represent various methods for converting the diffuse-horizontal insolation (from the sky dome) S to a selected tilted surface S_t. Keep in mind that the complete algorithm also includes the D_t and GR_t terms.

3.5.2.1 Liu and Jordan Algorithm The diffuse sky term of this algorithm (Liu and Jordan, 1961) is often referred to as the isotropic algorithm and is expressed as

$$S_t = \tfrac{1}{2}S(1 + \cos\beta). \tag{3.29}$$

This expression assumes uniform radiance from each portion of the sky, i.e., isotropic diffuse radiation from all directions above the surface, resulting in no dependence on the azimuth angle ϕ of the sun.

3.5.2.2 Temps and Coulson Algorithm Temps and Coulson (1977) added a degree of complexity to the isotropic diffuse sky term:

$$S_t = \tfrac{1}{2}S(1 + \cos\beta)M_1 M_2, \tag{3.30}$$

where

$$M_1 = 1 + \sin^3(\beta/2), \tag{3.31}$$

$$M_2 = 1 + \cos^2\theta \sin^3 z, \tag{3.32}$$

and

θ = angle of incidence of the direct beam on the surface and

z = solar zenith angle.

The terms M_1 and M_2 are anisotropic modifiers to account for horizon brightening and for the brightening around the solar disk (circumsolar radiation), respectively. The term M_2 is a potential source of error since the measurement of the direct normal beam D encompasses a large part of the circumsolar radiation. Note also that M_2 cannot account for the increase in circumsolar radiation with the onset of cloud cover, which eventually leads to an anisotropic sky under overcast conditions. For this reason the Temps and Coulson algorithm is applicable only to cloudless conditions. Under overcast skies there is horizon "darkening"—another reason why this algorithm is inappropriate for other than clear sky conditions.

3.5.2.3 Klucher Algorithm A refinement of the diffuse sky term provided by Klucher (1979) adds a cloudiness function F to the Temps and Coulson algorithm; i.e.,

$$S_t = \tfrac{1}{2}S(1 + \cos\beta)M_3 M_4, \tag{3.33}$$

where

$$M_3 = 1 + F\sin^3(\beta/2), \tag{3.34}$$

$$M_4 = 1 - F\cos^2\theta \sin^3 z, \tag{3.35}$$

$$F = 1 + (S/Q)^2. \tag{3.36}$$

Examining the cloudiness function F reveals that under overcast skies ($S = Q$) the Klucher diffuse sky term reduces to the isotropic term of Liu and Jordan, which was the intent.

3.5.2.4 Hay Algorithm The diffuse sky term described by Hay and Davies (1978) also provides for anisotropy by considering both circumsolar and isotropic terms. The diffuse sky term in the Hay algorithm is expressed as

$$S_t = S[(D\cos\theta/D_x\cos z) + \tfrac{1}{2}(1 + \cos\beta)(1 - D/D_x)]. \tag{3.37}$$

The first term inside the brackets represents circumsolar radiation whose intensity is a function of atmospheric transmissivity D/D_x, the angle of incidence of the direct beam, and the solar zenith angle. The inverse relation with $\cos z$ represents an increase in circumsolar radiation with an increase in air mass (increasing solar zenith angle). The second term represents the isotropic radiation from the sky that increases with decreasing transmissivity—i.e., with increasing turbidity or cloud cover. Note that the Hay algorithm also simplifies to the isotropic case for overcast conditions.

3.5.2.5 Perez Algorithm This algorithm, like the Klucher and the Temps and Coulson algorlithms, accounts for horizon brightening as well as circumsolar radiation (Perez, Stewart, and Scott, 1983). This is done by describing the sky dome as isotropic except for a circular region of variable size around the sun and a horizontal band of variable height at the horizon. The equation for the diffuse sky term of the Perez algorithm is

$$S_t = S\left[\frac{\frac{1}{2}(1+\cos\beta) + 2(F_1-1)(1-\cos\alpha)Z_c}{1+2(F_1-1)(1-\cos\alpha)Z_h}\right.$$

$$\left.\cdots\frac{+\,2(F_2-1)\xi/\pi\sin\xi'}{+\,\frac{1}{2}(1+\cos 2\xi)(F_2-1)}\right], \tag{3.38}$$

where

F_1 = ratio of the radiance in the circumsolar region to isotropic sky radiance,

F_2 = ratio of the radiance in the horizon band to the isotropic sky,

α = half-angle width of the circumsolar region,

ξ = angular width of the horizon band,

ξ' = altitude angle of the apex of the horizon band with respect to the tilted surface,

$$Z_c = \begin{cases} \cos\theta, & 0 < \theta < (\pi/2 - \alpha) \\ \dfrac{(\chi_c \sin\chi_c)}{\alpha}, & (\pi/2 - \alpha) < \theta < (\pi/2 + \alpha) \\ 0, & \theta > (\pi/2 + \alpha), \end{cases}$$

$\chi_c = (\pi/2 + \alpha - \theta)/2$,

θ = angle of incidence of circumsolar radiation on a tilted surface,

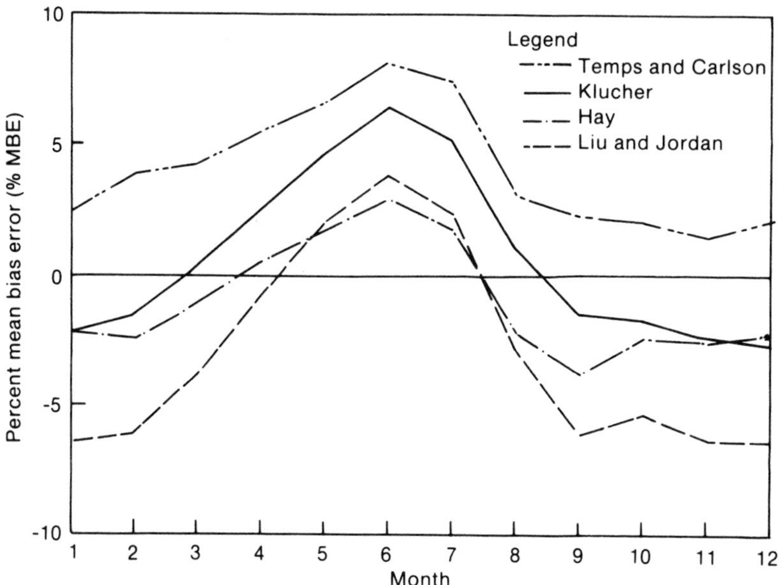

Figure 3.30
Evaluation of four insolation-tilted surface algorithms for Atlanta, GA.

$$Z_h = \begin{cases} \cos Z, & 0 < Z \leq (\pi/2 - \alpha) \\ \dfrac{(\chi_h \sin \chi_h)}{\alpha}, & Z > (\pi/2 - \alpha), \end{cases}$$

χ_h $= (\pi/2 + \alpha - Z)/2$, and

Z = angle of incidence of circumsolar radiation on a horizontal surface.

The terms F_1 and F_2 are empirically derived from measured data at the location of interest. Values for F_1 and F_2 are available for Albany, NY, and San Antonio, TX (Perez, Stewart, and Scott, 1983).

3.5.3 Evaluation of Algorithms

SERI researchers evaluated the hourly algorithms discussed in section 3.5.2. Part of this evaluation was done for fixed south-facing surfaces at four sites. The results of an analysis of 22 months of data at Atlanta and 38 months at Albany are presented in figures 3.30 and 3.31 for a tilt angle equal to the latitude of each site. This evaluation covered the Liu and Jordan, Temps and Coulson, Hay, and Klucher algorithms.

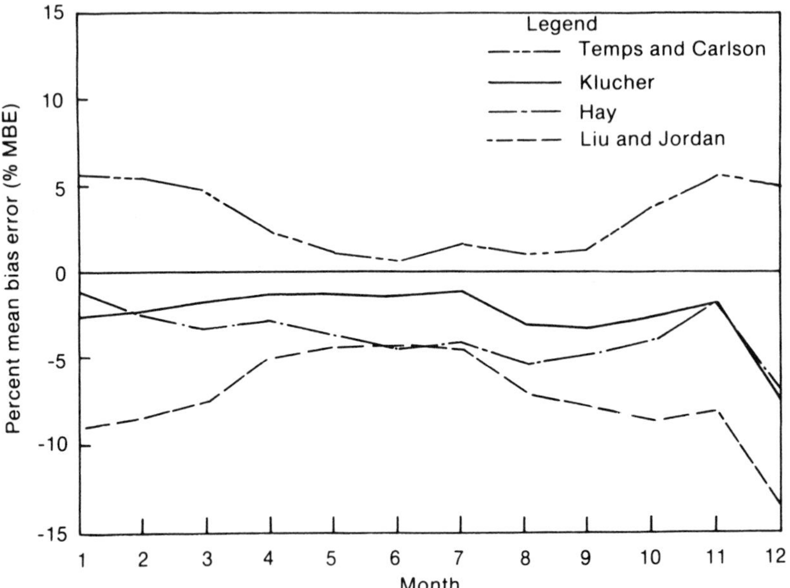

Figure 3.31
Evaluation of four insolation-tilted surface algorithms for Albany, NY.

A comparison of the Liu and Jordan, Hay, and Klucher algorithms for three south-facing tilt angles was performed by Ma and Iqbal (1983) at Woodbridge, Ontario, Canada. Their results are similar to SERI's with the additional conclusion that the errors are slightly larger at 90° tilt angles. Smietana et al. (1984) performed a limited study that resulted in similar conclusions.

The Perez (Perez, Stewart, and Scott, 1983) anisotropic algorithm can be more accommodating to changing atmospheric conditions. Stewart and Perez (1984) compared their algorithm and the algorithms of Liu and Jordan, Hay, and Klucher with several months of data at Albany and San Antonio. They found that their algorithm was consistently more accurate than the others for surfaces facing north, east, south, and west. One significant result from their study was that the Hay and Klucher algorithms give very different results on 90° (vertical) north-facing surfaces. At Albany, where the K_t values are relatively small, the Hay model underestimated by 3.8% and the Klucher model overestimated by 39%. At San Antonio, with

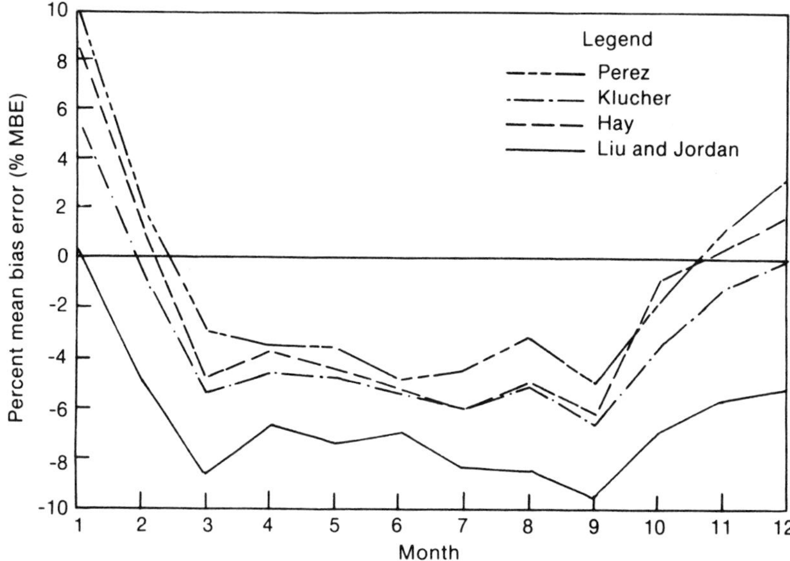

Figure 3.32
A comparison of the Liu and Jordan, Hay, Klucher, and Perez algorithms for a two-axes tracking surface at Phoenix, AZ.

higher K_t values, the Hay model underpredicted by 40% and the Klucher model overpredicted by 9.3%. Note that for 90° north-facing surfaces, the insolation is almost entirely diffuse radiation, which accounts for the relatively large errors.

SERI researches compared measured and algorithm-generated data at two sites for a two-axes tracking flat plate (a flat surface normal to the sun). They used the Liu and Jordan, Hay, Klucher, and Perez algorithms in the comparisons to gather 1 year of data at a site near Phoenix, AZ, and 7 months at Goodnoe Hills, WA. Comparison results near Phoenix are shown in figure 3.32, and the Goodnoe Hills results are provided in figure 3.33. The Perez algorithm is the most accurate of all the algorithms for most of the months shown. The other algorithms perform very much like they did for fixed, south-facing surfaces. Note that November and December at Goodnoe Hills were very foggy, which could lead to measurement errors affecting the comparison with all the models. Based on these and other comparisons, SERI researchers recommend the Perez algorithm. However, depending on the application, the Hay algorithm may also be quite adequate.

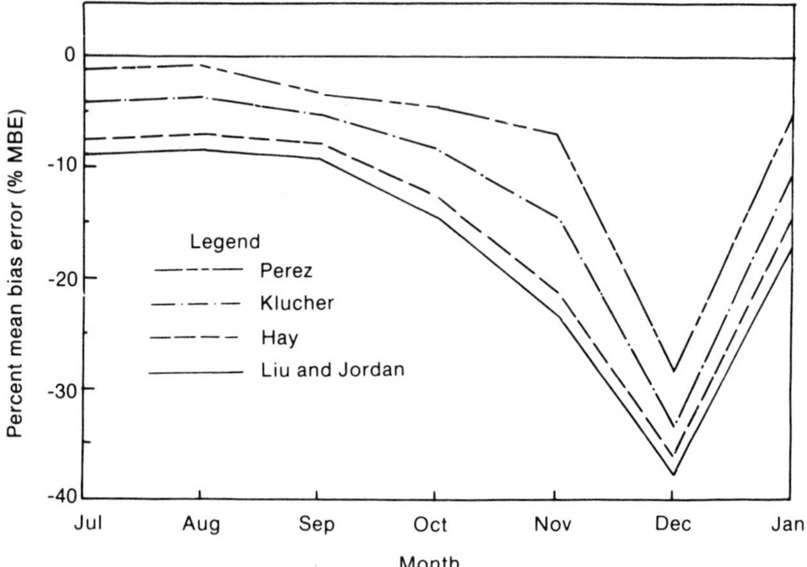

Figure 3.33
A comparison of the Liu and Jordan, Hay, Klucher, and Perez algorithms for a two-axes tracking surface at Goodnoe Hills, WA.

3.6 Recommendations

We recommend that the direct-insolation estimating algorithm be evaluated. Hall et al. (1980) made the most comprehensive effort to do this, although they were limited by the amount of simultaneous hourly global and direct insolation observations available at that time. Even some of the data available had unexplained inconsistencies. The recommended study must be based on carefully qualified data from a wide range of climates. Such data should be available now as a product of network observations from 1978 through 1980. The results of this evaluation will either validate the existing algorithm as described in Randall and Whitson (1977) and Randall and Biddle (1981) or, more likely, indicate improvements to it. The resulting validated algorithm can then be used to model direct insolation.

We need advanced algorithms that can simulate more accurately the true statistical characteristics of the insolation resources. Almost all of the algorithms developed to date have concentrated on accurately estimating the average value of the insolation parameter. The small increase in uncertainty of the ERSATZ global insolation values compared with the rehabilitated data (table 3.10) suggests that mean value algorithms for global insolation may be about as good as weather variations allow. Although the mean values are important, the nature of the distribution about the mean is also important. In particular, the serial correlation of the insolation is important in defining the amount of energy storage and backup capacity required. The direct insolation model used in generating the SOLMET data sets did attempt to simulate the distribution of insolation values, but had a very artificial way of including the serial correlation of insolation values.

Additional accurate insolation data must be obtained for an extended period of time for a climatically diverse set of locations for evaluating and improving new and existing insolation modeling algorithms.

References

Allen, C. W., 1973. *Astrophysical Quantities*. London; The Athlone Press, University of London.

Atwater, M. A., and J. T. Ball, 1981. A surface solar radiation model for cloudy atmospheres. *Monthly Weather Review* 109:878–888.

Bahm, R. J., ed., 1981. Satellites and forecasting of solar radiation. In *American Section of ISES Workshop Proceedings*. February 2–5, 1981.

Ball, J. T., M. A. Atwater, and S. J. Thoren, 1981. Sensitivity of computed incoming solar radiation at the surface to cloud analyses. *Monthly Weather Review* 109:889–894.

Bennett, I. 1965. Monthly maps of mean daily insolation for the United States. *Solar Energy* 9:145–158.

Bennett, I., 1969. Correlation of daily insolation with daily total sky cover, opaque skycover and percentage of possible sunshine. *Solar Energy* 12:391–393.

Berdahl, P., and R. Fromberg, 1982. The thermal radiance of clear skies. *Solar Energy* 29:299–314.

Biga, A. J., and R. Rosa, 1981. Statistical behaviour of solar irradiation over consecutive days. *Solar Energy* 27:149–157.

Bird, R. E., and R. L. Hulstrom, 1980. *Direct Insolation Models*. SERI/TR-335-344. Golden, CO: Solar Energy Research Institute.

Bird, R. E., and R. L. Hulstrom, 1981. *A Simplified Clear Sky Model for Direct and Diffuse Insolation on Horizontal Surfaces*. SERI/TR-642-761. Golden, CO: Solar Energy Research Institute.

Boes, E. C., 1975. *Estimating the Direct Component of Solar Radiation*. SAND75-0565. Albuquerque, NM; Sandia National Laboratories.

Boes, E. C., 1981. Fundamentals of solar radiation, chapter 2, *Solar Energy Handbook*. New York: McGraw-Hill. Sandia Report SAND79-0490, December 1979, pp. 2-1–2-78.

Boes, E. C., H. E. Anderson, I. J. Hall, R. R. Prairie, and R. T. Stromberg, 1977. *Availability of Direct, Total, and Diffuse Solar Radiation to Fixed and Tracking Collectors in the USA*. SAND77-0885. Albuquerque, NM: Sandia National Laboratories.

Boes, E. C., I. J. Hall, R. R. Prairie, R. T. Stromberg, and H. E. Anderson, 1976. *Distribution of Direct and Total Solar Radiation Availabilities in the USA*. SAND76-0411. Albuquerque, NM: Sandia National Laboratories.

Brinkworth, B. J., 1977. Autocorrelation and stochastic modelling of insolation sequences. *Solar Energy* 19:343–347.

Bruno, R., 1978. A correction procedure for separating direct and diffuse insolation on a horizontal surface. *Solar Energy* 20:97–100.

Budyko, M. I., ed., 1963. *Atlas Teplovoga Balansa Zemnogo Shara* (*Atlas of the Heat Balance of the Earth*). Moscow: Glav. Geofiz. Obs. im A.I. Voeikova. Translated in 1964 by A. Donehoo, WB/T-106, Washington, DC: Weather Bureau, U.S. Department of Commerce.

Bugler J. W., 1977. The determination of hourly insolation on an inclined plane using a diffuse irradiance model based on hourly measured global horizontal insolation. *Solar Energy* 19:477–491.

Centeno, V. M., 1982. New formulae for the equivalent night sky emissivity. *Solar Energy* 28:489–498.

Choudhury, N. K. D., 1963. Solar radiation at New Delhi. *Solar Energy* 7:44–52.

Collares-Pereira, M., and A. Rabl, 1979. The average distribution of solar radiation—correlation between diffuse and hemispherical and between daily and hourly insolation values. *Solar Energy* 22:155–164.

Cotton, Gerald F., 1979. ARL models of global solar radiation. *SOLMET Volume 2: Hourly Solar Radiation—Surface Meteorological Observations*. TD-9724, Frank T. Quinlan, ed. Asheville, NC: National Oceanic and Atmospheric Administration, pp. 165–184.

Dave, J. V., P. Halpern, and N. Braslau, 1975. *Spectral Distribution of the Direct and Diffuse Solar Energy Received at Sea-Level of a Model Atmosphere*. G320-3332. Palo Alto, CA: IBM Palo Alto Scientific Research Center.

Davies, J. A., J. E. Hay, 1978. Calculations of solar radiation incident on a horizontal surface. In *Proc. First Canadian Solar Radiation Data Workshop*, pp. 32–58.

Elsasser, W. M., and M. F. Culbertson, 1960. Atmospheric radiation tables. *Meteorological Monographs* 4(23):1–43.

Erbs, D. G., S. A. Klein, and J. A. Duffie, 1982. Estimation of the diffuse radiation fraction for hourly, daily and monthly-average global radiation. *Solar Energy* 28:293–302.

Fröhlich, C., J. Geist, J. Kendall, Sr., and R. M. Marchgraber, 1973. The third international comparisions of pyrheliometers and a comparison of radiometric scales. *Solar Energy* 14:157–166.

Goh, T. N., and K. J. Tan, 1977. Stochastic modeling and forecasting of solar radiation data. *Solar Energy* 19:755–757.

Grether, D., D. Evans, A. Hunt, and M. Wahlig, 1981. *The Effect of Circumsolar Radiation on the Accuracy of Pyrheliometer Measurements of the Direct Solar Radiation*. LBL-12707. Presented at the Annual Meeting of the American Section of the International Solar Energy Society, May 26–30, 1981.

Hall, I. J., R. R. Prairie, H. E. Anderson, and E. C. Boes, 1980. *Solar Radiation Model Validation*. SAND80-1755. Albuquerque, NM: Sandia National Laboratories.

Hall, R. T., S. L. Leonard, C. M. Randall, and N. A. Fiamengo, 1974. *Solar Thermal Conversion Mission Analysis Volume III: Southwestern United States Insolation Climatology*. ATR-74(7417-16)-2. Los Angeles, CA: Aerospace Corporation.

Hanson, K. J., 1979. Method of determining standard-year clear-sky, solar-noon irradiance values for 26 station over the contiguous United States. *SOLMET Volume 2: Hourly Solar Radiation—Surface Meteorological Observations*. TD-9724. Frank T. Quinlan, ed. Asheville, NC: National Oceanic and Atmospheric Administration, pp. 71–118.

Hay, J. E., 1976. A revised method for determining the direct and diffuse components of the total short-wave radiation. *Atmosphere* 14:278–287.

Hay, J. E., and J. A. Davies, 1978. Calculation of the solar radiation incident on an inclined surface. *Proceedings of the First Canadian Solar Radiation Data Workshop*, pp. 59–72.

Hoyt, D. V., 1978. A model for the calculation of solar global insolation. *Solar Energy* 21:27–35.

Hoyt, D. V., 1979. Theoretical calculations of the true solar noon atmospheric transmission. *SOLMET Volume 2: Hourly Solar Radiation—Surface Meteorological Observations*. TD-9724. Frank T. Quinlan, ed. Asheville, NC: National Oceanic and Atmospheric Administration, pp. 119–163.

Iqbal, M., 1983. *An Introduction to Solar Radiation*. New York: Academic Press.

Jordan, R. C., and B. Y. H. Liu, eds., 1977. *Applications of Solar Energy for Heating and Cooling of Buildings*. ASHRAE GRP 170. New York: American Society of Heating, Refrigerating, and Air Conditioning Engineers.

Kimura, K., and D. G. Stephenson, 1969. Solar radiation on cloudy days. *ASHRAE Transactions* 75:227–234.

Klein, S. A., W. A. Beckman, and J. A. Duffie, 1978. *Monthly Average Solar Radiation on Inclined Surfaces for 261 North American Cities*. Revised, EES Report no. 44–2, Madison WI: Engineering Experiment Station, University of Wisconsin.

Klucher, T. M., 1979. Evaluation of models to predict insolation on tilted surfaces. *Solar Energy* 23:111–114.

Knapp, C. L., and T. L. Stoffel, 1982. *Direct Normal Solar Radiation Data Manual*. SERI/SP-281-1658. Golden, CO: Solar Energy Research Institute.

Kneizys, F. X., et al., 1983. *Atmospheric Transmittance/Radiance: Computer Code LOWTRAN6*. AFGL-TR-83-0187 (Environmental Research Paper No. 846). Hanscom Air Force Base, MA: Air Force Geophysics Laboratory.

Kneizys, F. X., et al., in press. *Users Guide to LOWTRAN7*. AFGL-TR-88-0177. Hanscom Air Force Base, MA: Air Force Geophysics Laboratory.

Leonard, S. L., and C. M. Randall, 1974. *Solar Thermal Conversion Mission Analysis Volume III: Southern California Insolation Climatology*. ATR-74(7417-05)-1. Los Angeles, CA: Aerospace Corporation.

Liu, B. Y. H., and R. C. Jordan, 1960. The interrelationship and characteristic distribution of direct, diffuse and total solar radiation. *Solar Energy* 4:1–19.

Liu, B. Y. H., and R. C. Jordan, 1961. Daily insolation on surfaces tilted toward the equator. *ASHRAE Journal* 3(10):53–59.

Liu, B. Y. H., and R. C. Jordan, 1963. The long-term average performance of flat-plate solar-energy collectors. *Solar Energy* 7:53–74.

Löf, G. O. G., J. A. Duffie, and C. O. Smith, 1966. World distribution of solar energy. *Solar Energy* 10:27–37. Also, Report 21, Madison WI: Engineering Experiment Station, University of Wisconson.

Lowry, W. P., 1980. Clear-sky direct-beam solar radiation versus altitude: a proposal for standard soundings. *J. of Applied Meteorology* 19:1323–1327.

Ma, C. C. Y., and M. Iqbal, 1983. Statistical comparisons of models for estimating solar radiation on inclined surfaces. *Solar Energy* 31:313–317.

Martin, M., and P. Berdahl, 1984a. Summary of results for the spectral and angular sky radiation measurement program. *Solar Energy* 33:241–252.

Martin, M., and P. Berdahl, 1984b. Characteristics of infrared sky radiation in the United States. *Solar Energy* 33:321–336.

McClatchey, R. A., et al., 1972. *Optical Properties of the Atmosphere* (3rd ed.). AFCRL-72-0497 (Enviromental Research Paper No. 411). Bedford, MA: Air Force Cambridge Research Laboratory.

Melton, W. C., 1978. *Performance, Value, and Cost of Solar Thermal Electric Central Reciever Plants outside of the Southwest*. ATR-78(7689-04)-1. Los Angeles, CA: Aerospace Corporation.

Mujahid, A., and W. D. Turner, 1980. Solar radiation modeling and comparisons with current solar radiation models. In *Proc. 1980 Solar Energy Conference*, Phoenix, AZ.

Mustacchi, C., V. Cena, and M. Rocchi, 1979. Stochastic simulation of hourly global radiation sequences. *Solar Energy* 23:47–51.

National Climatic Center (NCC), 1980. *National Climatic Center Monthly Summary Solar Radiation Data—1977–1980 (Monthly)*. Asheville, NC: NCC.

Orgill, J. F., and K. G. T. Hollands, 1977. Correlation equation for hourly diffuse radiation on a horizontal surface. *Solar Energy* 19:357–359.

Page, J. K., 1961. The estimation of monthly mean values of daily total short-wave radiation on vertical and inclined surfaces from sunshine records for latitudes 40°N–40°S. Paper No. 35/5/98. In *Proc. UN Conference on New Sources of Energy*. Also, Sheffield, United, Kingdom: Dept. of Building Science, Faculty of Architectural Studies, University of Sheffield.

Paltridge, G. W., and C. M. R. Platt, 1976. Radiative processes in meteorology and climatology, chapter 5, *Developments in Atmospheric Science*. New York: Elsevier Scientific Publishing Company.

Perez, R., R. Stewart, and J. T. Scott, 1983. An anisotropic model of diffuse solar radiation with application. SUNY 870. Albany, NY: State University of New York.

Quinlan, F. T., ed., 1977. *SOLMET Volume 1: Hourly Solar Radiation—Surface Meteorological Observations*. Asheville, NC: National Oceanic and Atmospheric Administration.

Quinlan, F. T., ed., 1979. *SOLMET Volume 2: Hourly Solar Radiation—Surface Meteorological Observations*. Asheville, NC: National Oceanic and Atmospheric Administration.

Randall, C. M., 1976. *Insolation Data Base Available from The Aerospace Corporation*. ATR-76(7523-11)-9. Los Angeles, CA: Aerospace Corporation.

Randall, C. M., and J. M. Biddle, 1981. *Hourly Estimates of Direct Insolation: Computer Code ADIPA User's Guide*. ATR-81(7878)-1. Los Angeles, CA: Aerospace Corporation.

Randall, C. M., and M. E. Whitson, Jr., 1977. *Hourly Insolation and Meteorological Data Bases Including Improved Direct Insolation Estimates*. ATR-78(7592)-1. Los Angeles, CA: Aerospace Corporation. Also, SAND78-7047. Albuquerque, NM: Sandia National Laboratories.

Randall, C. M., B. R. Johnson, and M. E. Whitson, Jr., 1980. *Measurements of Typical Insolation Variation at Daggett, California*. ATR-80(7747)-1 (2 volumes). Los Angeles, CA: Aerospace Corporation.

Randall, C. M., M. E. Whitson, Jr., and B. R. Johnson, 1978. *Measurements of Insolation Variation over a Solar Collector Field*. ATR-79(7747)-2. Los Angeles, CA: Aerospace Corporation.

Rapp, D., 1979. Critique on the solar rehabilitation procedures used in SOLMET-2. *Energy Conversion* 19:101–110.

Roosen, R. G., and R. J. Angione, 1984. Atmospheric transmission and climate: results from Smithsonian measurements. *Bulletin American Meteorological Society* 65:950–967.

Rothman, L. S., et al., 1982. AFGL atmospheric absorption line parameter compilation: 1982 edition. *Applied Optics* 22:2247–2256.

Ruth, D. W., and R. E. Chant, 1976. The relationship of diffuse radiation to total radiation in Canada. *Solar Energy* 18:153–154.

Sayigh, A. A. M., ed., 1977. *Solar Energy Engineering*. New York: Academic Press.

Smietana, P. J., Jr., R. G. Flocchini, R. L. Kennedy, and J. L. Hatfield, 1984. A new look at the correlation of K_d and K_t ratios and at global solar radiation tilt models using one-minute measurements. *Solar Energy* 32:99–107.

Solar Energy Research Institute (SERI), 1981. *Solar Radiation Energy Resource Atlas of the United States*. SERI/SP-642-1037. Golden, CO: SERI.

Solar Energy Research Institute (SERI), 1984. *Insolation Assessment Studies Progress Report: FY 1982/83*. SERI/PR-215-2214. Golden, CO: SERI.

Spencer, J. W., 1971. Fourier series representation of the position of the sun. *Search* 2:172.

Spencer, J. W., 1982. A comparison of methods for estimating hourly diffuse solar radiation from global solar radiation. *Solar Energy* 29:19–32.

Stanhill, G., 1966. Diffuse sky and cloud radiation in Israel. *Solar Energy* 10:96–101.

Stewart, R., and R. Perez, 1984. Validation of an anisotropic model estimating insolation on tilted surfaces. *Progress in Solar Energy*, Volume VII. Boulder, CO: American Solar Energy Society.

Stine, W. B., and R. W. Harrigan, 1985. *Solar Energy Fundamentals and Design*. New York: John Wiley & Sons.

Suckling, P. W., and J. E. Hay, 1977. A cloud layer-sunshine model for estimating direct, diffuse and total solar radiation. *Atmosphere* 15:194–207.

Temps, R. C., and K. L. Coulson, 1977. Solar radiation incident upon slopes of different orientations. *Solar Energy* 19:179–184.

Thekaekara, M. P., and A. J. Drummond, 1971. Standard values for the solar constant and its spectral components. *Nature Physical Science* 229:6–9.

Tuller, S. E., 1976. The relationship between diffuse, total and extraterrestrial solar radiation. *Solar Energy* 18:259–263.

Turner, W. D., and A. Mujahid, 1984. The estimation of hourly global solar radiation using a cloud cover model developed at Blytheville, Arkansas. J. Climate and Applied Meteorology 23:781–786.

van de Hulst, H. C., 1981. *Light Scattering by Small Particles*. New York: Dover Books.

Vittitoe, C. N., and F. Biggs, 1981. Six-Gaussian representation of the angular-brightness distribution for solar radiation. *Solar Energy* 27:469–490.

Walraven, B. R., 1978. Calculating the position of the sun. *Solar Energy* 20:393–397, 22:195.

Watt, A. D., 1978. *On the Nature and Distribution of Solar Radiation*. HCP/T2552-01. Washington, DC: U.S. Department of Energy.

Watt, A. D., 1980. *Circumsolar Radiation*. SAND 80-7009. Albuquerque, NM: Sandia National Laboratories.

White, O. R., ed., 1977. *The Solar Output and Its Variation*. Boulder, CO: Colorado Associated University Press.

Willson, R. C., 1984. Measurements of solar total irradiance and its variability. *Space Science Reviews* 38:203–242.

Willson, R. C., H. Hudson, and M. Woodard, 1984. The inconstant solar constant. *Sky and Telescope* 67:501–503.

Yinger, R. J., 1982. *Solar Energy Measurements at Selected Sites throughout the Southwest During 1980*. Report 82-RD-4. Rosemead, CA: Southern California Edison Company.

4 Solar Radiation Monitoring Networks

Kirby Hanson and Thomas Stoffel

4.1 Introduction

In this chapter we provide a historical perspective on the solar radiation monitoring networks that have operated in the United States from the turn of the century to the present. We also briefly review broadband radiometer development and how this instrumentation is important in monitoring network design and data interpretation. We did not attempt to compile or review all of the measuring networks or sites where solar radiation measurements have been taken in the United States. Rather, we focus attention on the networks funded by the federal government (Department of Commerce and Department of Energy) and electrical power utilities. For a catalog of solar radiation and meteorological data collection activities in the United States, we refer the reader to the Solar Energy Research Institute (SERI) publication *Solar Radiation Data Directory* (SERI, 1983).

Solar radiation data from monitoring networks can be used in a variety of applications:

- solar energy conversion technologies (photovoltaics, active and passive solar thermal systems, and biofuels),
- meteorology (radiative transfer, atmospheric physics, etc.),
- climatology (greenhouse effect, nuclear winter, effects of urbanization, etc.),
- agriculture (evapotranspiration, Photosynthesis, etc.),
- medicine (skin cancer research), and
- materials (polymer degradation, reflective coatings, etc.).

An individual needing solar radiation resource data for a location in the United States is often frustrated by the current lack of information. Conflicts between desired information and available data can arise from many areas:

- type of solar measurements (global horizontal, global on a slope, direct normal, diffuse, etc.),
- data accuracy (most data are model estimates based on cloud observations),
- data frequency (minute, hour, day, month, season, annual, etc.),
- period of record (when do you have enough data to represent all the design criteria?),

- collateral data (wind, temperature, pressure, cloud cover, etc.), and
- storage media (printed matter, magnetic tape, diskette, etc.).

This is in spite of the measurements made by the National Solar Radiation Network of the National Oceanic and Atomospheric Administration (NOAA) and its predecessor agencies (the Weather Bureau and the Environmental Science and Services Administration) for more than 80 years at over 250 locations.

4.2 Historical Perspective

4.2.1 Early Instrumentation Developments

Samuel P. Langley (1834–1906) of the Smithsonian Institution, Washington, DC, became interested in the sun as early as 1873 and spent the next three decades developing instruments and collecting measurements to study the spectrum of the sun's direct component (Coulson, 1975). During the same period, Knut Angstrom (1857–1910), Sweden, developed the electrical compensation pyrheliometer (1893), which provided a means for measuring the integrated solar spectrum to relatively high precision and also for standardizing these measurements to electrical standards. By the turn of the century interest in solar measurements spawned by Langley's work and improvement in pyrheliometry led the U.S. Weather Bureau on a course that eventually established solar measurements within the their network of meteorological stations. In 1902, the Bureau Chief documented this new direction in his annual report (Weather Bureau, 1902):

> SOLAR HEAT AND ATMOSPHERIC ABSORPTION
>
> In July, 1901, the Bureau received three copies of Angstrom's Electric Compensation Pyrheliometer, which instrument is intended to measure in calories the amount of heat received by radiation from any distant source, including, of course, the sun. *It is intended to use these three instruments in carrying out researches on the amount of solar heat and of atmospheric absorption and allied questions.* One of them is kept as a standard at the Weather Bureau and may be used in Washington; the others are now located, respectively, in Baltimore, in care of Prof. J. S. Ames, and the other in Providence, R.I., in care of Prof. Carl Barus. Numerous investigations must be carried on by these physicists as preliminary to the main object of our research. Articles published in the *Monthly Weather Review* by prof. C. F. Marvin and Prof. F. W. Very and by Mr. C. G. Abbott and Prof. S. P. Langley have given a general idea of the scope that the investigation must take.

Thus began the first measurements of solar radiation by the U.S. Weather Bureau.

During the following year, the Bureau Chief sent H. Kimball to Asheville, NC, where he obtained pyrheliometer measurements from the fall of 1902 to the spring of 1903. The Bureau Chief reported in 1903 that the observations had been carried out "with faithfulness ..." and also made a point (which could have been prophetic for all those who would subsequently collect solar data) that "the two memoirs containing the work ... are recommended for publication in a separate bulletin, as they are too voluminous for the *Monthly Weather Review*" (Weather Bureau, 1903). No doubt this was a point of some sensitivity since the Bureau Chief was also the editor of the *Review*. Kimball later established the solar radiation network within the Weather Bureau.

In his annual report of 1902, the chief took the opportunity to indicate the applications for which these measuremnents were intended (Weather Bureau, 1903):

PYRHELIOMETRIC MEASUREMENTS

The solar radiation, as measured ... is that which is actually received by every object at the earth's surface; the datum is directly applicable to problems in agriculture as well as to meteorology.... The actual amount of solar energy received at the outer limit of the atmosphere can also be obtained from this observed datum after it has been learned how to interpret it by the study of the bolometric work that is carried on professionally by Professor Langley.

Note that in parallel with collecting pyrheliometric measurements at Asheville in 1902 and 1903, Kimball also obtained measurements of atmospheric polarization to try to establish a connection between atmospheric absorption and polarization.

The history of solar measurement activities in the Weather Bureau indicates that a principal purpose for these early measurements was to establish a better understanding of the physical processes of atmospheric attenuation of solar radiation, i.e., the scattering and absorption components. The Weather Bureau also saw applications of this information for solar energy, meteorology, climatology, physical science, and agriculture.

4.2.2 Evolution of Monitoring Stations (1909–1975)

By July 1909, the Weather Bureau had begun routinely collecting measurements of "total solar and sky radiation on a horizontal surface" (now usually termed global radiation) at Washington, DC (Kimball, 1915). Similar measurements were initiated at Madison, WI, in April 1911 and at Lincoln, NE, in June 1915. At all 3 of these sites, the global radiation measurements were obtained initially with Callendar pyranometers (then termed pyrheliometers), and 10-day average solar radiation values were published periodically in the *Monthly Weather Review* in various notes by Kimball from the beginning of the record for each station until March 1915. The following month, the *Review* began regular monthly publication of global solar radiation for these stations.

Routine measurements of the *direct* solar component were started in July 1910 at Madison, Lincoln, and Santa Fe, NM. During January 1916, they were also begun at Washington, DC. The instrument used for measurements at these locations was the Marvin pyrheliometer (Kimball, 1919). The pyrheliometer was mounted in a tracker for continuous alignment with the solar disk. As with the pyranometer data (global) before April 1915, the pyrheliometer (direct) data were published in notes by Kimball in the *Review*. Beginning in April 1915, the *Review* began publishing solar radiation data regularly.

By 1925, Chicago, IL, and New York, NY, were added to the 3-station pyranometer network, and by 1930, an additional 5 stations were added. The stations of the pyrheliometer network remained the same during the 1920s, as did publication of the data in the *Review*.

By 1935, the pyranometer network had increased to a total of 18 measurement locations in the United States, 8 being at Weather Bureau sites and 10 being at "cooperator" sites not operated by the Weather Bureau. The Callendar pyranometer was still used at only 3 sites in 1935; the relatively new thermoelectric pyranometer, based on the design and development work of Kimball and Hobbs (1923) but produced by the Eppley Laboratory, was used at the other 15 sites (Hand, 1935).

The 1940s saw continued slow growth (figure 4.1) of the pyranometer network in the United States to 28 stations by 1947. The majority of these stations were cooperators, typically at a college or university, and the number of Weather Bureau stations remained small. During the late 1940s, increased interest in solar radiation research brought about a major decision in the Weather Bureau to expand its pyranometer network. Rapid growth in 1950 and 1951 brought the number of pyranometer stations to 78 in 1951, mainly by additional measurements at Weather Bureau stations. In addition to this expansion, the quality of the radiation data base

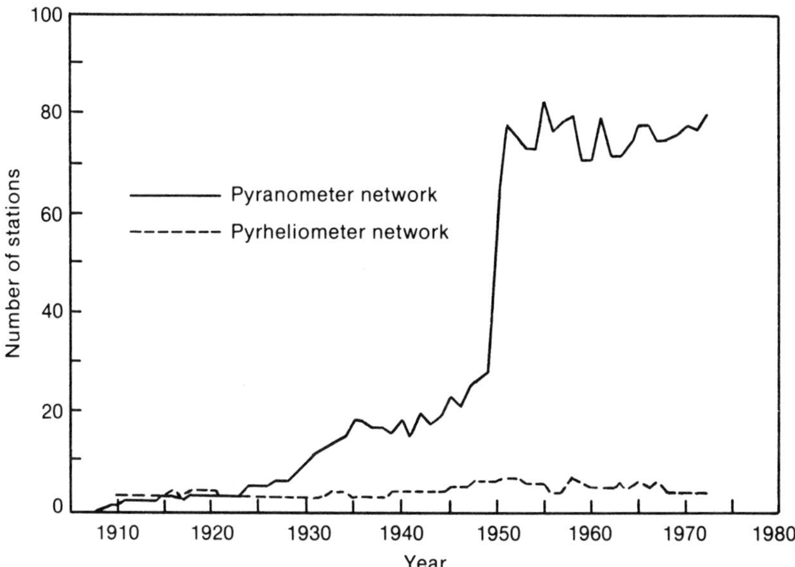

Figure 4.1
The number of stations in the national solar radiation network 1908–1972.

was significantly improved. Integration recorders were added at some stations, a data quality control effort was established, and measurement accuracy was improved through calibration work and studies of the characteristics of the Eppley pyranometer, the radiometer being used exclusively in the network at that time (MacDonald, 1951; MacDonald and Foster, 1954; Fritz, 1984).

Why did the Weather Bureau expand the pyranometer network significantly at that time? Fritz (1984), who was responsible for the Weather Bureau solar radiation group, cited several reasons for this expansion:

- Some thought was given then to the idea that solar radiation information might be used for monthly and seasonal climate prediction.
- Climatological information on solar radiation over the United States needed to be bolstered. This information was somewhat lacking because of the previous low density of stations and limitations in data quality at some cooperator stations.
- The solar energy community needed long-term average monthly solar radiation maps.

Figure 4.2
The Eppley Model 50 pyranometer.

• Finally, many people had inquired about various kinds of solar radiation and solar energy information over the United States. These inquiries ranged from "Do you refrigerate railroad cars in Arizona?" to "Where would you find the minimum ultraviolet radiation?"

Thus, the size of the Weather Bureau pyranometer network dramatically increased in 1950 and 1951. As with the initial establishment of selected solar radiation measurements in 1902, the network expansion in 1950 to 1951 was intended to serve a variety of needs.

During the 20-year period following the expansion, the pyranometer network remained at about the same number of stations (70–80). Almost exclusively, the instrument used in the network during that period was the Eppley (180°) pyranometer (figure 4.2).

The pyrheliometer network continued at about the same number of stations (4–7). Eppley normal incidence pyrheliometers (NIP) were used in the network during this period, although the bureau standards for calibration included Smithsonian-scale and Ångstrom-scale pyrheliometers.

Publication of the NOAA pyranometer and pyrheliometer network measurements ended in September 1972 without explanation, although the observations continued at network stations and the data continued to be archived at NOAA's National Climatic Data Center (NCDC), NOAA, at Asheville (NOAA, 1972).

By 1973, the pyranometer network had grown to 90 sites. Sixty were National Weather Service (NWS) sites and 30 were cooperators. Eppley Laboratory's newly designed Model II pyranometer was being used at about one-quarter of the NWS sites in 1973. The remaining sites continued to use the Eppley (180°) pyranometer (Jessup, 1973).

From the beginning of routine pyranometer measurements in 1909 until the end of 1976, 150 stations obtained pyranometer data, which are now available at NCDC. Carter, Wells, and Williams (1976) summarized the various times and locations for the large pyranometer data base. Of the 150 measurement locations, 57% were at NWS locations and the remaining were cooperator locations. The map in figure 4.3 gives these locations in the contiguous United States (Carter, Wells, and Williams, 1976).

4.2.3 Measurement Errors

Although errors of about ±5% might be expected from operating the Eppley (180°) pyranometer commonly used in the network from 1950 to 1970, subsequent studies made it clear that errors of at least 10% are possible for some conditions (Drummond, 1965; Latimer, 1971; Flowers, 1973). In addition, a study of the calibration history of network instruments showed that it was not possible to evaluate the significance of network errors from pre- and postcalibration information, primarily because of the lack of an adequate number of postcalibrations. Apparently, too many of the instruments were damaged or destroyed before postcalibrations could be obtained (Hanson, 1979).

Although several error sources could have affected the data during that period, the most significant error resulted from the particular blackening used on the detector surface of some Eppley (180°) pyranometers during their manufacture. In 1956, the Eppley Laboratory changed the coating of the blackened (hot) receiver ring of the pyranometer detector from lampblack to Parson's optical black lacquer. Most of these particular instruments manufactured after 1956 received the latter coating. By the early 1960s, some of the instruments with the black lacquer had been calibrated against Weather Bureau lampblack standards and were used in the national network.

Figure 4.3
Location of solar radiation stations with data archived at the National Climatic Data Center, Asheville, NC. Numbers indicate more than 1 station in area.

From the vantage point of hindsight, we now know that this created two kinds of problems. First, for field instruments calibrated under tungsten light against lampblack standards, their subsequent radiation measurements in the field were about 5% to 10% lower than if they had been calibrated under natural light. Thus, for several instruments in the field during that period, a bias error of −5% to −10% was introduced. The second problem was that long field exposure caused an optical whitening (degradation) of the blackened surface. This also resulted in field radiation measurements that were too low. Unfortunately, the degree of degradation varied with environmental conditions, defying simple schemes to correct for the degradation (Flowers, 1973). In short, the accuracy limitations resulting from the initial 1923 instrument design, the two problems that arose from using Parson's optical black lacquer, and normal measurement problems associated with solar data collection all combined to provide errors of 20% or more in the network data.

4.2.4 National Data Base Rehabilitation

NOAA considered two important questions concerning the solar radiation network data base during the mid-1970s. Was it necessary, for solar energy needs, to correct the bulk of the data collected in the network from 1953 to 1975? Was it feasible to determine corrections that would enhance the accuracy of the measurements? A solar energy workshop was held in 1974 that dealt with these questions (Turner, 1974). Workshop participants decided that previously collected data were too erroneous for immediate use in solar energy applications and that an attempt should be made to "... rehabilitate the pyranometer data for the U.S. network of stations to at least a 5 percent accuracy for all possible stations for a period of 10 years or longer where possible. [In addition, it was recommended that] the resulting rehabilitated data should be presented in a form useful for solar energy applications ... [by] including meteorological data together with the [solar] radiation information. ..."

These recommendations set in motion efforts by the Department of Energy (DOE) and NOAA to improve and expand the U.S. solar radiation data base for the period 1953–1975.

There were three major parts to this NOAA effort (Quinlan, 1979):

- The *hourly* global-horizontal solar data for 26 observing stations from 1953 to 1975 were adjusted to provide more representative values and were combined with simultaneously observed meteorological data to make them more useful to the solar community. The resulting data base, termed SOLMET, has been available on request from NCDC since 1978 (NOAA, 1978; NOAA, 1979a). Similarly, the *daily* solar data for 27 observing stations for the period 1953–1975 were adjusted and combined with simultaneously observed meteorological data. The resulting data base, termed SOLDAY, has been available from NCDC since 1979 (NOAA, 1979b). A map of station locations is given in figure 4.4.

• For each of the 26 SOLMET locations, a typical mteorological year was assembled that comprised specific calendar months selected from the entire recorded span of historical data (generally 1953–1975). The selection of months was based on a statistical analysis that ensured that the selected month was the most representative or typical of that station (Quinlan, 1979). The data set is available from NCDC.

• Solar radiation values were modeled for two groups of network stations based on solar zenith angle and meteorological information. For the first group of stations (26 SOLMET stations), solar radiation values were modeled to fill in missing hourly observations at these stations from approximately 1952 to 1976. For the second group of stations (222 meteorological stations), all solar radiation values were modeled from the available surface weather observations. The resulting solar data base (approximately 1952–1976) is entirely based on modeled values since there were no solar observations at these stations. Cotton (1979) presents a complete description of this modeling effort and the data limitations associated with it. The 222-station network is indicated as "Derived Data Stations" in figure 4.5, also known as the ERSATZ data base (see chapter 2).

4.2.5 DOE-Funded Research

When NOAA was rehabilitating the historical data and redesigning the Solar Monitoring Network, DOE provided additional research support leading to an improved National Solar Radiation Data Base. Specifically, they funded the Aerospace Corporation to develop a means of estimating the hourly direct-normal solar component using the rehabilitated global-horizontal data from the 26 SOLMET stations (Randall and Biddle, 1981). Funding was also provided to reinstrument the solar monitoring network in 1976 (see section 4.3.1).

Once the historical data set was assembled, in the form of computer tapes, the data needed to be analyzed to make it available to a wide

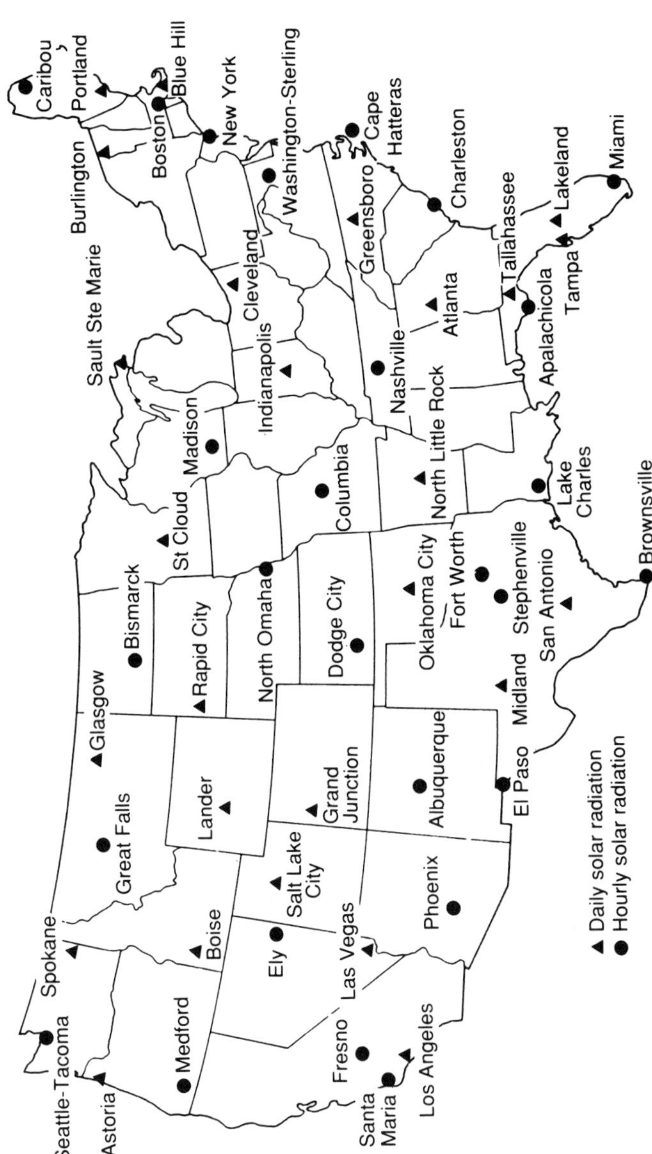

Figure 4.4
Solar radiation data rehabilitation stations.

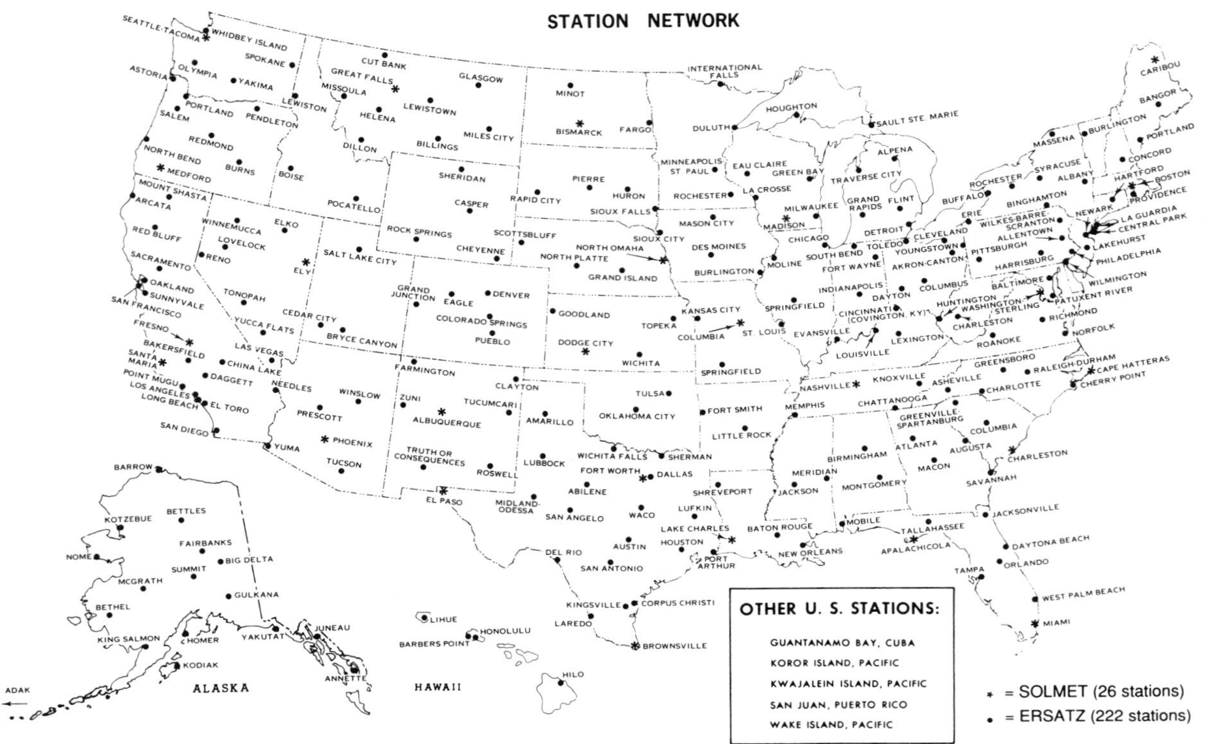

Figure 4.5
The 222 stations with derived solar radiation information.

audience. DOE supported the preparation of a national solar atlas (SERI, 1981a) and two other publications containing highly condensed data summaries (SERI, 1980, 1982a) (see chapter 2).

4.3 U.S. National Solar Monitoring Network

4.3.1 Deployment and Operations (1977–1980)

In March 1977, NOAA (1977) announced the inauguration of a new national solar radiation network:

> A nationwide solar radiation network designed to help solar technology is being inaugurated by NOAA in cooperation with the Energy Research and Development Administration's Division of Solar Energy.
>
> Solar radiation measurements will be taken at 35 National Weather Service stations, and will give scientists and engineers information on how much solar energy is available at each site, on the average, for use in home heating or cooling, agricultural drying, heating water, generation of electricity, and many other purposes.
>
> A new improved pyranometer will be used, measuring with great accuracy solar energy falling on a surface. All of the pyranometers in the network have been calibrated by NOAA's Air Resources Laboratories at Boulder, Colo., against the international standard of the World Meteorological Organization, and will be periodically recalibrated.
>
> In solar-energy technology, solar radiation is not the only weather information needed. Other data such as temperature and wind must be known to determine local energy needs and to design collectors of optimum size for home heating and cooling. NOAA's Environmental Data Service at Asheville, N.C., combines conventional weather data with solar data onto a single magnetic tape, called SOLMET, for archival and distribution to the public.
>
> The new network includes 10 special stations at which diffuse, or scattered, solar radiation also is to be monitored. This provides an indirect means of computing the direct solar beam—important in applications where solar radiation is to be focused on a small area to produce high temperatures, and useful for calculating the amount of solar energy falling on inclined surfaces.

The distribution of stations measuring global radiation in early 1978 is shown in figure 4.6. By late 1978, 3 stations were added (Puerto Rico, Guam, and Honolulu) to bring the network to 38 stations. Also in 1978, measurements of the direct component were added at all stations except Fairbanks, AK. A summary of the solar measurements by station, as of December 31, 1978, is given in table 4.1 (U.S. DOE, 1979).

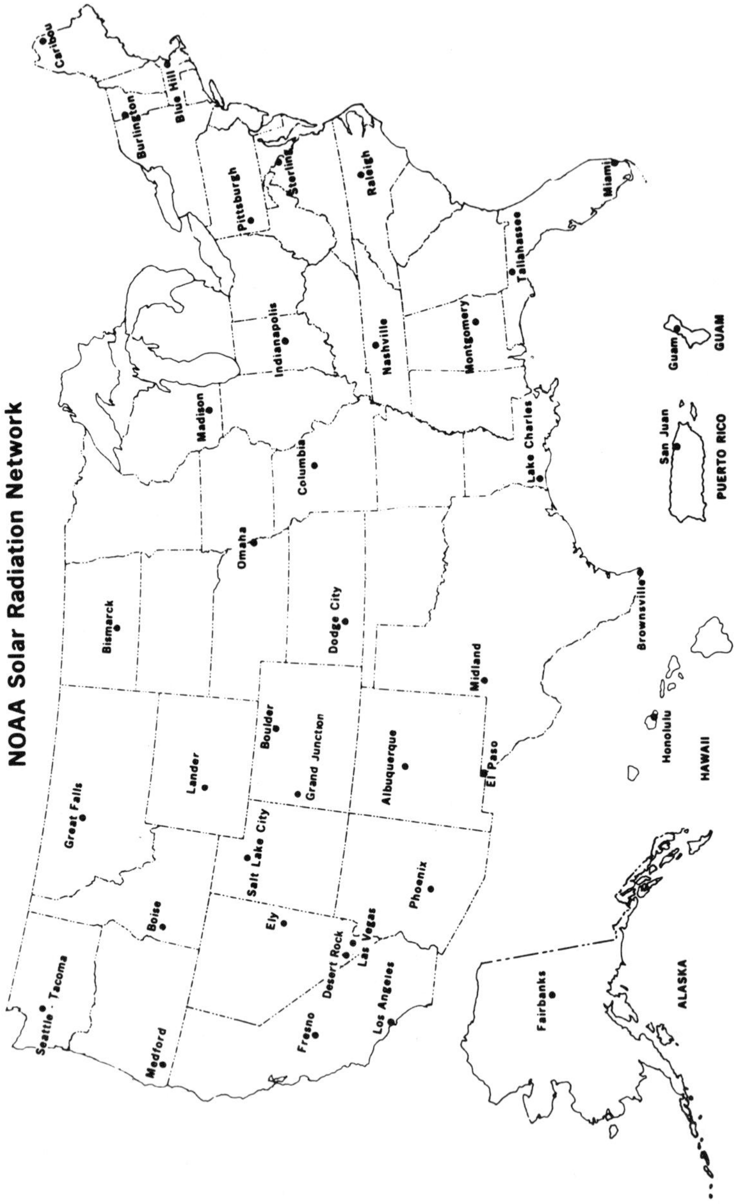

Figure 4.6
NOAA/DOE solar radiation 35-station network in early 1978.

Planning for the new national network dates back to 1975, when NWS decided to reduce the number of its stations observing global solar radiation from 60 (in 1974) to 17 by mid-1975. However, they also decided to use only improved pyranometers at these 17 continuing stations (NOAA 1975).

In 1976, NWS continued planning for this new network. One of the questions examined was whether these 17 stations would be adequate to "characterize the U.S. (solar) climate." To answer this question, they considered station densities based on various siting criteria. In the end, the one accepted was a World Meteorological Oganization (WMO) (1969) criterion of a 500-km spacing between network stations. This equated to 45 stations in the contiguous states, 2 in the Pacific, 1 in Puerto Rico, and 4 in Alaska for a total of 52 stations. Thus, a 52-station U.S. network was NWS's desired goal in 1976.

The specific plan for reaching that goal was indirect, however. According to Riches (1976), "The plan was to use the 35-station network [figure 4.6] ... as a basic network but add others such as NOAA sites, ERDA laboratories, and several high quality cooperatives. Thus a complete climatic network can be designed ... to show the overall solar climate of the United States."

Between late 1976 and early 1977, a 35-station network was equipped with WMO Class I pyranometers for global-horizontal measurements. With funds from DOE, NWS deployed pyrheliometers mounted in sun-following trackers from January to December 1978. They equipped 10 stations with shadow-band stands for diffuse component measurements. All radiometers used in the network were regularly calibrated (12–18-month cycle) by the NOAA Solar Radiation Facility in Boulder, CO, using an absolute cavity radiometer traceable to the World Radiometric Reference scale of measurement maintained by the World Radiation Center in Davos, Switzerland.

By January 1977, 27 stations were collecting global-horizontal insolation data. They began measuring the diffuse radiation component in January 1978 and discontinued it in April 1982.

Data were acquired on digital cassette tapes and printed paper tapes and transmitted regularly to NCDC in Asheville, NC, for processing. The cassette records were 1-min instantaneous values of the available radiation components. The printed paper tapes contained the integrated hourly values, which were used to fill missing cassette records and serve as a data quality control check. Analog strip-chart recorders were added to all

Table 4.1
Solar radiation network as of December 31, 1978 (after U.S. DOE, 1979)

Station	Station number	Latitude (N) (degrees, minutes)	Longitude (W) (degrees, minutes)	Sensor height (m)	Sensor height above subjacent ground (m)	Parameter		
						Global	Diffuse	Direct
Eastern Region								
Blue Hill, MA	509	42.13	71.07	199.6	6.1	X	X	X
Burlington, VT	617	44.28	73.09	111.9	10.4	X		X
Caribou, ME	712	46.52	68.01	195.0	17.0	X		X
Pittsburgh, PA	520	40.30	80.13	371.3	20.4	X		X
Raleigh, NC	306	35.52	78.47	137.1	4.6	X		X
Sterling, VA	403	38.59	77.28	86.9	1.5	X	X	X
Southern Region								
Albuquerque, NM	365	35.02	106.37	1623.2	5.2	X	X	X
Brownsville, TX	250	25.54	97.26	11.7	5.9	X	X	X
El Paso, TX	270	31.48	106.24	1206.1	11.9	X		X
Lake Charles, LA	240	30.07	93.13	18.9	14.0	X		X
Miami, FL	202	25.49	80.17	7.9	5.8	X		X
Midland, TX	265	31.57	102.11	872.3	1.5	X		X
Montgomery, AL	226	32.18	86.24	68.2	9.8	X		X
Nashville, TN	327	36.07	86.41	185.6	9.1	X		X
Tallahassee, FL	214	30.23	84.22	17.7	1.5	X	X	X
Puerto Rico (PR)	78526	18.26	66.00	24.9	21.0	X		X
Central Region								
Bismarck, ND	764	46.46	100.46	511.1	1.7	X	X	X

Boulder, CO	001	40.00	105.02	1634.0	24.0	X	X	X
Columbia, MO	445	38.49	92.13	277.3	6.7	X		X
Dodge City, KS	451	37.46	99.58	795.2	8.2	X		X
Grand Junction, CO	476	39.07	108.32	1472.8	2.4	X		X
Indianapolis, IN	438	39.44	86.16	244.1	2.7	X		
Lander, WY	576	42.49	108.44	1699.2	5.5	X		X
Madison, WI	641	43.08	89.20	270.5	9.0	X		X
Omaha, NE	553	41.22	96.01	404.2	5.2	X		X
Western Region								
Boise, ID	681	43.34	116.13	872.6	1.8	X		X
Ely, NV	486	39.17	114.51	1912.0	6.1	X	X	X
Fresno, CA	589	36.46	119.43	102.4	3.4	X		X
Great Falls, MT	775	47.29	111.22	1118.3	1.8	X		X
Las Vegas, NV	386	36.05	115.10	670.0	11.0	X		X
Los Angeles, CA	295	33.56	118.24	36.6	7.0	X	X	X
Medford, OR	597	42.22	122.52	412.3	12.4	X		X
Phoenix, AZ	278	33.26	112.01	339.2	1.8	X		X
Salt Lake City, UT	572	40.46	111.58	1288.1	1.5	X		X
Seattle, WA	793	47.27	122.18	142.6	5.0	X	X	X
Alaska Region								
Fairbanks, AK	261	64.49	147.52	142.7	9.8	X		
Pacific Region								
Guam	91217	13.33	1445.50E	111.7	1.6	X		X
Honolulu, HI	91182	21.20	157.56	3.7	1.6	X		X

stations in June 1977, when formatting problems were encountered on the cassette tapes. Because of the poor quality cassette data, most of the data processed in 1977 were extracted from the printed tapes. A small amount was derived from the strip-chart records. By April 1978, the cassette problem was corrected.

At the network's January 1977 inception, there were 34 stations within the conterminous United States, providing a mean distance of 310 statute miles between stations and 1 station at Fairbanks. All but 2 of the 35 stations were collocated with NWS offices, so hourly meteorological data complemented the solar radiation monitoring. The exceptions were Boulder and Blue Hill, MA. Each station, except for Boulder, had at least a 20-year published climatological record.

Interagency memoranda of understanding between DOE and NOAA provided for the continuity of responsibilities needed for operating and managing NOAA's solar radiation network. DOE provided fiscal support through fiscal year 1980 with the understanding that NOAA would seek funds through their budget process for fiscal year 1981 and beyond. During this period, despite repeated attempts, NOAA had no operational budget base to support the network's total funding needs.

By mid-1978, all of the stations in the network were monitoring direct normal irradiance with the exception of Indianapolis, IN. During October–December 1978, DOE supported the addition of Honolulu, HI, Guam, and San Juan, PR, to the network. These sites, initially equipped to monitor global horizontal radiation, joined the pyrheliometric network in January 1979.

Data in the form of digital cassette tapes, printed paper tapes, and strip-chart records were processed by NOAA during this period. Hourly records of the solar data were constructed from these sources and reviewed for accuracy. The quality control techniques applied to the data consisted of graphical comparisons with the extraterrestrial radiation (ETR), clear-sky model maximum value, and the consistencies expected between the global horizontal, diffuse, and direct normal solar components at any given time of day. Funding reductions caused NOAA to stop publishing data and quality control processing after December 1980.

4.3.2 Operations 1981–1985

During this period, budget restrictions resulted in limited support to the network. Without maintenance support, except for radiometer recalibrations, the stations continued collecting solar data until a piece of equipment failed involving the printed paper tape and the cassette recorder portions of the data acquisition systems. The data successfully recorded on either media was routinely sent to NCDC for transfer to magnetic tapes without the benefit of the previous quality control processing.

A few changes were made to the network in 1982. The solar monitoring equipment at Blue Hill, MA, was moved to a station in Concord, NH, to avoid local obstructions created by TV antenna towers. The measurements of diffuse irradiance at ten stations were replaced with the old style (Model 50) Eppley pyranometers for assessing the rehabilitation methods used on the historical data base.

By 1985, data recovery had reached an unacceptable level, and the network was shut down until adequate funding could be secured for new data acquisition equipment and data processing.

4.3.3 Current Status

With funding made available from the Air Resources Laboratory of NOAA in 1986, the network is upgrading data acquisition equipment and receiving automatic solar trackers for 31 stations. The network will resume operation after plans are completed for data processing (transmittals, quality control methods, publication, access, etc.) sometime in 1988.

4.4 Other Solar Monitoring Networks

Since the turn of the century, solar radiation measurements have been obtained at more than a thousand locations (other than the DOE/NOAA network) in the United States. Some potential users of these measurements were either unaware of these measurements or unable to obtain the data. Recognizing this problem, DOE funded three efforts from 1975 to 1985 to identify and catalog sources of solar radiation measurements. The resulting data directories and a brief summary of selected measurement programs are presented in this section.

4.4.1 Data Directories

To catalog solar radiation measurement sites within the United States, DOE provided funding to prepare and publish three directories (Carter, Wells, and Williams, 1976; SERI, 1983, 1985). These publications identify more than 40 national or regional solar measurement networks encompassing over 300 observing stations. Each directory summarizes aspects of the data collection that could be important to a potential user:

- station location,
- measurement parameters,
- instrumentation,
- period of record, and
- contact address.

Taken together, these publications are quite voluminous and are not easy to summarize in this chapter. We encourage the interested reader to consult them.

4.4.2 Summary of Selected Solar Monitoring Networks

Although network design studies for the United States often produced rather uniformly spaced stations, the climatic variables (which influence solar resource variations) are *not* uniformly spaced over such a network. Balling and Vojtesak (1983) reminded us of the United States's spatially nonuniform solar climate in their work in which they determined 18 solar climatic areas in the United States. Two of the areas are so large that together they cover about 76% of the United States (figure 4.7).

As a result of this information mismatch, DOE, state and local governments, and industry have conducted site-specific or regional solar monitoring programs to obtain the necessary data. Some of these networks are described in this chapter. More information about these and other networks can be found in the directories (SERI, 1983, 1985).

4.4.2.1 Solar Energy Meteorological Research and Training Sites With funding from DOE in 1975, eight universities began a research-quality solar monitoring network representing a range of climatic regions in the United States (figure 4.8). High resolution (1-min) data from a comprehensive collection of radiometers and meteorological instruments created this network data set that spans 1979–1983. Summary reports submitted by the individual unversities described the instrumentation, data processing, and research projects (SERI, 1981a, 1982b).

4.4.2.2 Northeast Utilities Environmental Data Acquisition Network A 4-station mesoscale network was started in 1973 by the Northeast Utilities

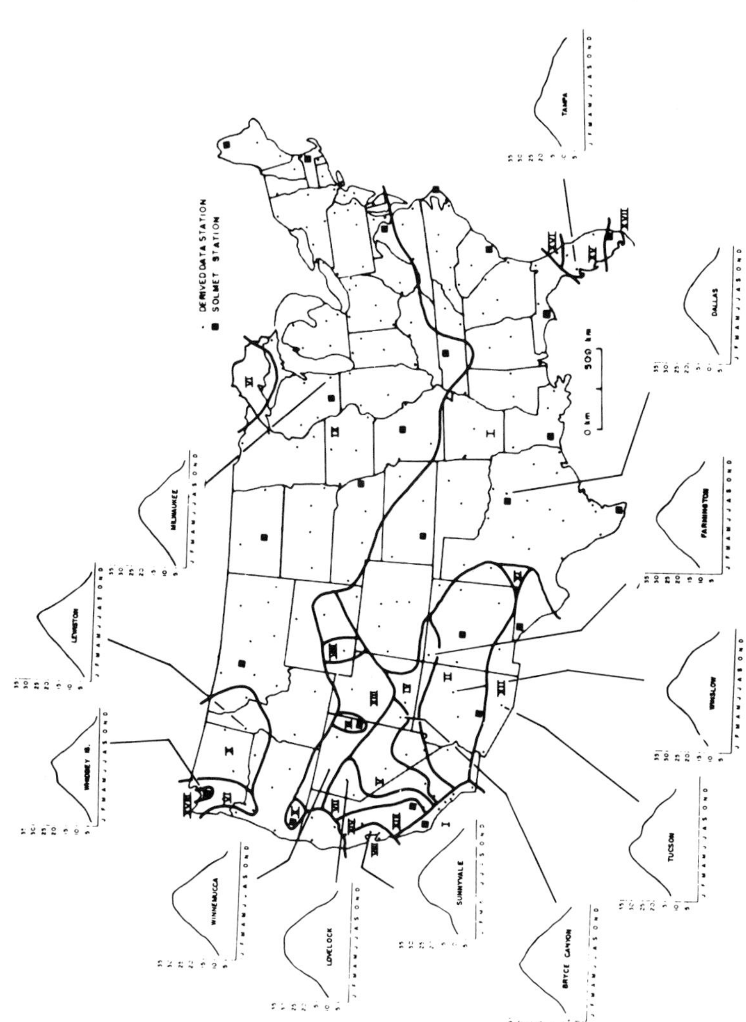

Figure 4.7
Solar climates of the United States.

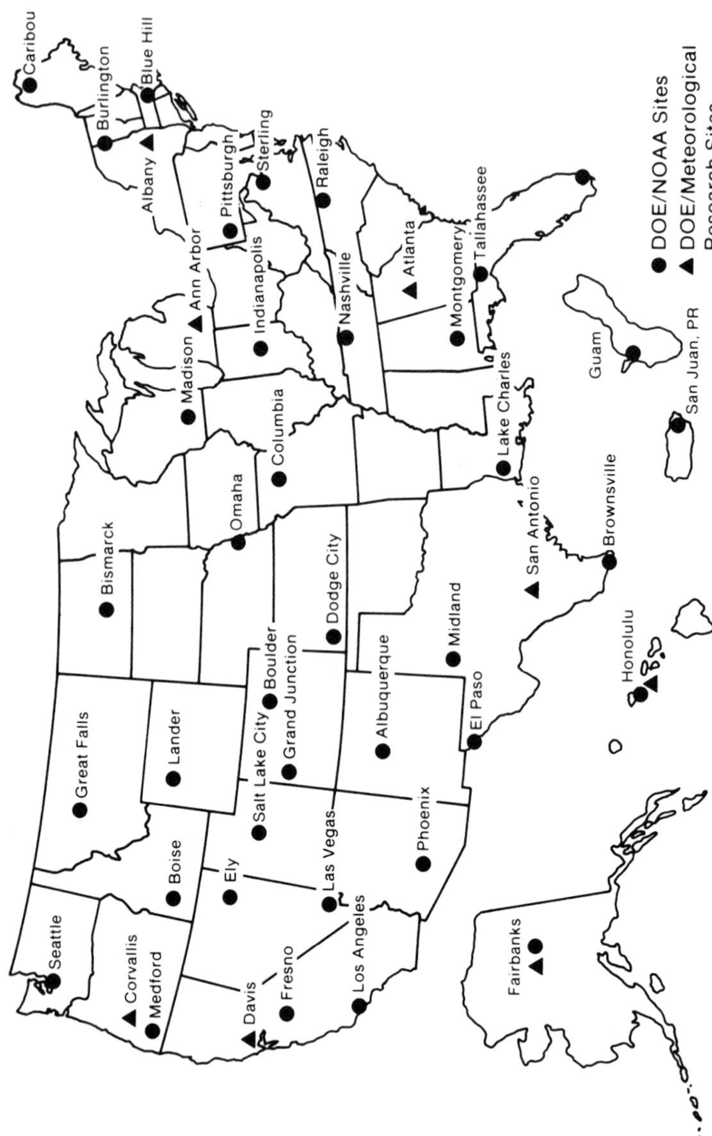

Figure 4.8
NOAA Solar Radiation Monitoring Network and DOE Solar Energy Meteorological Research Training Sites observing stations.

Service Company (figure 4.9). All stations are located at utility power plants except one, which is at a community-sponsored science center. Global-horizontal and direct-normal solar radiation data are available as 15-min averages for 1974–1981.

4.4.2.3 Western Energy Supply and Transmission (WEST) Associates WEST Associates began a solar monitoring network in 1976 to evaluate the solar resources available in the Southwest (figure 4.10). Data from 50 stations through 1980 were archived on magnetic tapes. These data consist of 15-min averages of global radiation and dry-bulb temperature.

4.4.2.4 University of Oregon Solar Radiation Monitoring Laboratory An 11-station network in the Pacific Northwest began collecting solar radiation data in 1975 (figure 4.11). Global-horizontal radiation and air temperature are measured at all sites. Some sites monitor direct-normal radiation. Edited data are available as hourly totals in a computerized format (University of Oregon, 1986).

4.4.2.5 Historically Black Colleges and Universities SERI established this 6-station solar monitoring network with funds from DOE in 1985 (Stoffel, 1987). The sites were selected to complement the planned NOAA network (figure 4.12). All sites monitor global- and diffuse-horizontal irradiance. The Bluefield, WV, station is also equipped to measure the direct-normal irradiance component. The averaged data are recorded on cassette tapes and printer listings every 5 min. The data are sent weekly (listings) and monthly (cassettes) to SERI for analysis and processing.

4.5 Summary

Measurements of the solar radiation resources are essential to assess accurately the design and performance of a solar energy conversion system. Since the early 1950s, the federal government has attempted to maintain a form of the national solar monitoring network with the regular meteorological observing stations providing climate and weather forecasting information. These efforts to provide national solar radiation data have fluctuated with the varying federal budgets and spending priorities. To date, the national solar monitoring network has experienced three distinctly different periods of performance. Unsatisfactory measured data from 1952 to 1975 for 26 stations had to be rehabilitated and were then

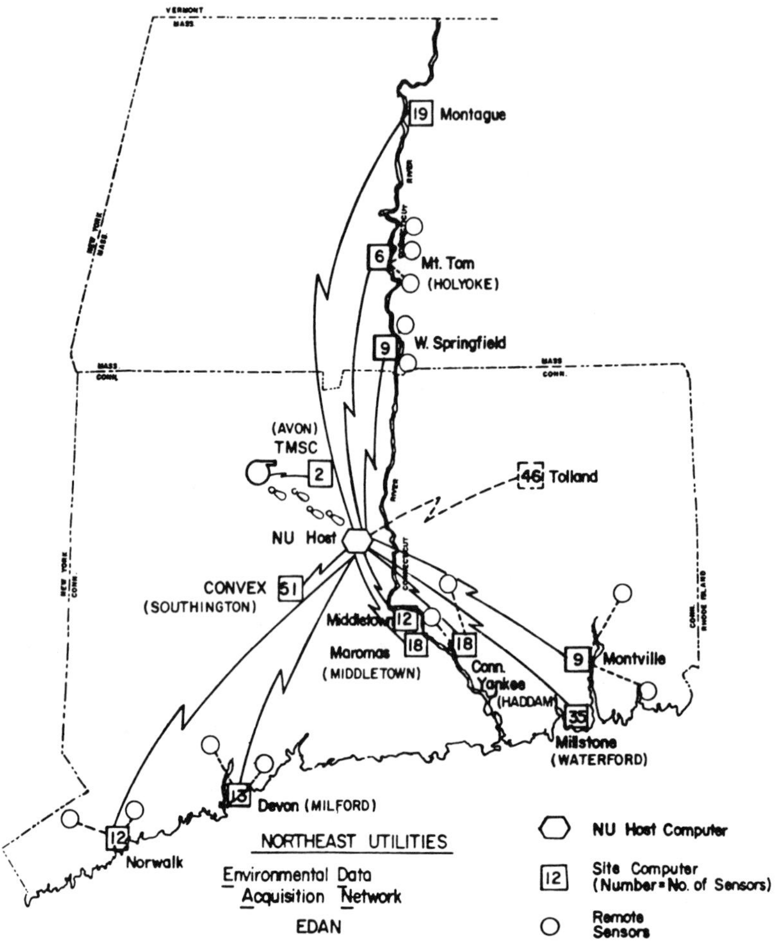

Figure 4.9
The Northeast Utilities Network in Connecticut and Massachusetts.

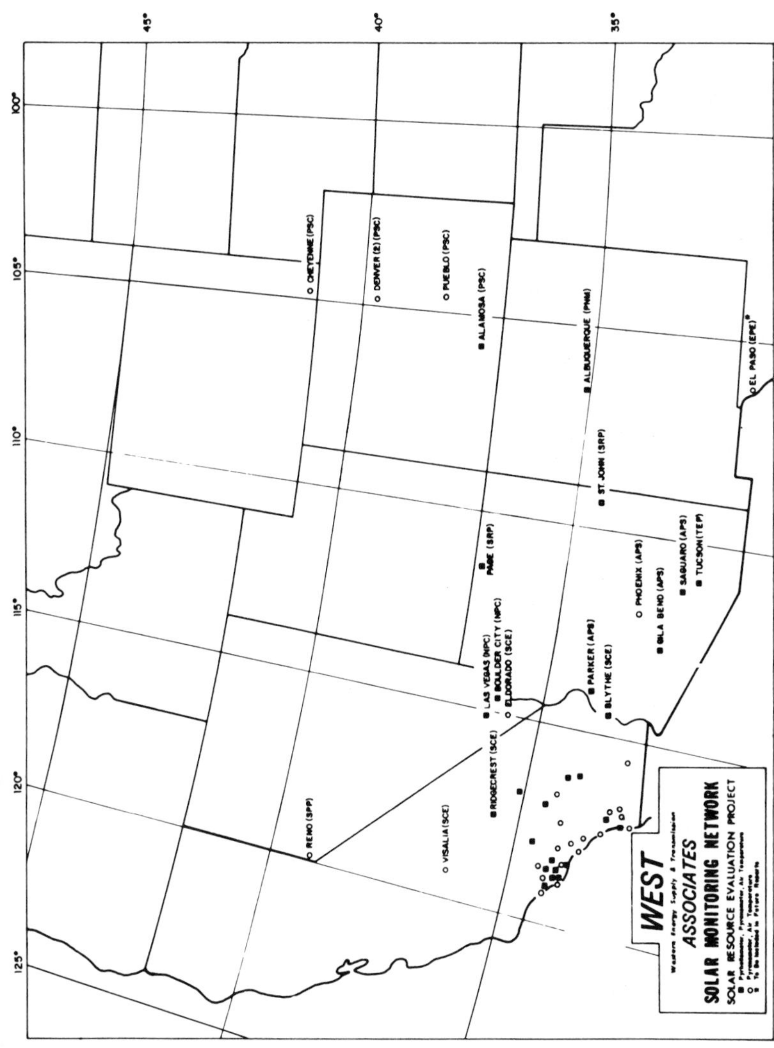

Figure 4.10
The WEST Associates Network in the southwestern United States.

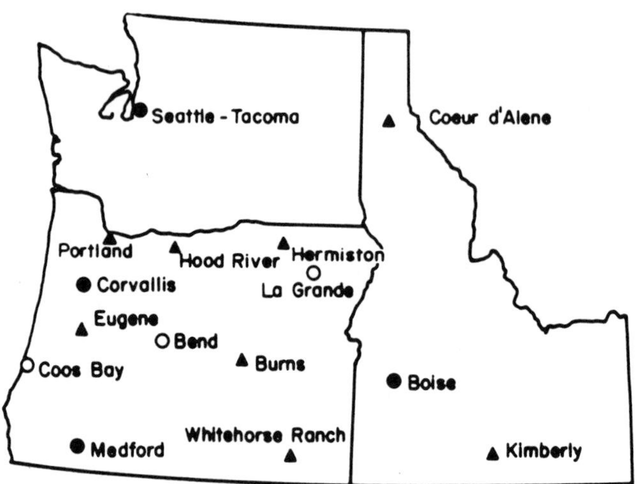

Figure 4.11
Pacific Northwest Solar Monitoring Network. Open circles indicate global horizontal measurements. Solid triangles indicate sites monitoring global and direct solar components. Closed circles show the locations of NOAA sites.

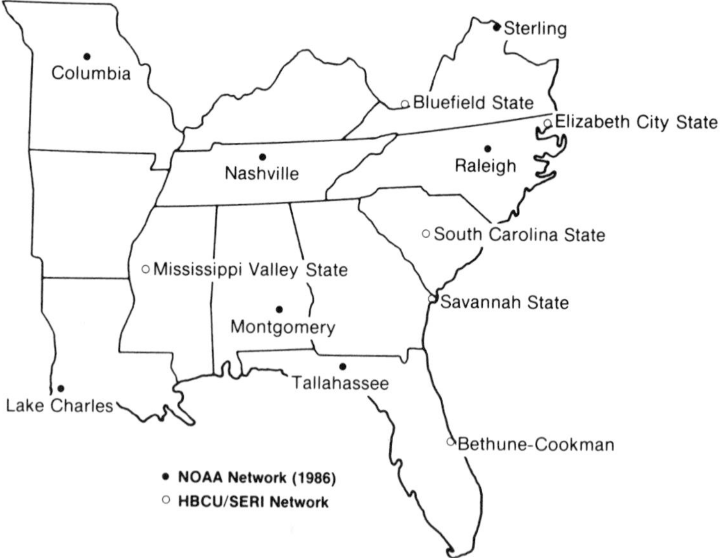

Figure 4.12
Map of the Historically Black Colleges and Universities Solar Radiation Network and NOAA stations in the Southeast (1986).

made available in a standard archive format. With the renewed interest in solar radiation after the 1973 oil embargo, a 38-station NOAA/DOE solar monitoring network upgraded and provided valuable data from 1977 until funding was reduced in 1981. Due to reduced federal budgets, the network again deteriorated from 1981 to 1985, resulting in less than satisfactory data. Current funding levels permit a 31-station network to resume operation with improved data acquisition systems and automatic solar trackers for pyrheliometer measurements sometime in 1988.

The spatial and temporal variablilities of the solar radiation resources usually require the user of such data to seek site-specific measurements. Since 1973, DOE, state and local governments, and industry have established more than 40 solar monitoring networks. Directories are available that describe the key aspects of these networks. This chapter briefly described five of these monitoring networks.

References

Balling, Robert C., and M. J. Vojtesak, 1983. Solar climates of the United States based on long-term monthly averaged daily insolation values. *Solar Energy* 31(3):283–291.

Carter, E., R. E. Wells, and B. B. Williams, 1976. Solar radiation observation stations with complete listing of data archived by the National Climatic Center, Asheville, N.C., and initial listing of data not currently archived. Huntsville, AL: Center for Environmental Studies, University of Alabama.

Cotton, Gerald F., 1979. ARL models of global solar radiation. *SOLMET Volume 2: Hourly Solar Radiation—Surface Meteorological Observations*. TD-9724. Asheville, NC: Air Resources Laboratory, pp. 165–184.

Coulson, Kinsell L., 1975. *Solar and Terrestrial Radiation*. New York: Academic Press.

Drummond, A. J., 1965. Techniques for the measurement of solar and terrestrial radiation fluxes in plant biological research: a review with special reference to arid zones. In *Proc. Montpiller Symp.*, UNESCO.

Flowers, Edwin C., 1973. The "so-called" Parson's black problem with old-style Eppley pyranometers. In *Proc. of 1973 Solar Energy Data Workshop* . NSF-RA-N-062. Washington, DC: Superintendent of Documents, U.S. Government Printing Office, pp. 28–30.

Fritz, S., 1984. Private communication, Asheville, NC: National Oceanic and Atmospheric Administration.

Hand, Irving F., 1935. Solar radiation measurements during January 1935. *Monthly Weather Review* 63(1):25.

Hanson, Kirby J., 1979. Method of determining standard-year, clear-sky, solar-noon irradiance (SYI) values for 26 stations over the contiguous United States. *SOLMET Volume 2: Hourly Solar Radiation—Surface Meteorological Data*. TD-9724. Asheville, NC: National Climatic Data Center, pp. 71–118.

Jessup, Edward, 1973. A brief history of the solar radiation program. In *Proc. of 1973 Solar Energy Data Workshop*. NSF-RA-N-062. Washington, DC: Superintendent of Documents, U.S. Government Printing Office, pp. 13–18.

Kimball, Herbert H., 1915. The total radiation received on a horizontal surface from the sun and sky at Washington, D.C. *Monthly Weather Review* 43(3): 104.

Kimball, Herbert H., 1919. Variations in the total and luminous solar radiation with geographical position in the United States. *Monthly Weather Review* 47(11): 769–793.

Kimball, Herbert H., and H. E. Hobbs, 1923. A new form of the thermoelectric pyrheliometer. *Monthly Weather Review* 51: 239–242.

Latimer, J., 1971. Radiation measurements. *International Field Year for the Great Lakes*, Tech Manual Series, No. 2, Ottawa, Canada: The Secretariat, Canadian National Committee for the International Hydrological Decade.

MacDonald, Torrence H., 1951. Some characteristics of the Eppley pyrheliometer. *Monthly Weather Review* 79(8): 153–159.

MacDonald, Torrence H., and N. B. Foster, 1954. Pyrheliometer calibration program of the U.S. Weather Bureau. *Monthly Weather Review* 82(8): 219–227.

National Oceanic and Atmospheric Administration (NOAA), 1972. *Climatological Data—National Summary*. Jan. 1950–Aug. 1972. Asheville, NC: U.S. Department of Commerce.

National Oceanic and Atmospheric Administration (NOAA), 1975. *Program Development Plan: Improvement of Solar Radiation Data*. Prepared for the National Science Foundation, Washington, DC, p. 33.

National Oceanic and Atmospheric Administration (NOAA), 1977. Collection of solar data set by NOAA. *NOAA NEWS* 2(6). Washington, DC: U.S. Department of Commerce.

National Oceanic and Atmospheric Administration (NOAA), 1978. Hourly solar radiation-surface meteorological observations. *SOLMET Volume 1: Hourly solar Radiation—Surface Meterological Observations*. Asheville, NC: National Climatic Data Center.

National Oceanic and Atmoshpheric Administration (NOAA), 1979a. *SOLMET Volume 2: Hourly Solar Radiation—Surface Meteorological Observations*. TD-9724. Asheville, NC: National Climatic Data Center.

National Oceanic and Atmospheric Administration (NOAA), 1979b. *SOLDAY User's Manual: Daily Solar Radiation-Surface Meteorological Data*. TD-9739. Asheville, NC: National Climatic Data Center.

Quinlan, Frank T., 1979. Availability of solar radiation data in the United States. *ASHRAE Transactions* 85(1): 736–741.

Randall, C. M., and J. M. Biddle, 1981. *Hourly Estimates of Direct Insolation: Computer Code ADIPA User's Guide*. ATR-81(7878)-1. El Segundo, CA: Aerospace Corporation.

Riches, Michael R., 1976. *Selection of the NOAA 35 Station Solar Radiation Network*. Unpublished report—copies available from the author, U.S. DOE, Washington, DC.

Solar Energy Research Institute (SERI), 1980. *Insolation Data Manual*. SERI/SP-755-789. Golden, CO: SERI.

Solar Energy Research Institute (SERI), 1981a. *Solar Radiation Energy Resource Atlas of the United States*. SERI/SP-642-1037. Golden, Co: SERI.

Solar Energy Research Institute (SERI), 1981b. *Solar Energy Meteorological Research and Training Site Program, First Annual Report*. SERI/SP-642-947. Golden, CO: SERI.

Solar Energy Research Institute (SERI), 1982a. *Direct Normal Solar Radiation Data Manual*. SERI/SP-281-1658. Golden, CO: SERI.

Solar Energy Research Institute (SERI), 1982b. *Solar Energy Meteorological Research and Training Site Program, Second Annual Report*. SERI/SP-290-1478. Golden, CO: Solar Energy Research Institute.

Solar Energy Research Institute (SERI), 1983. *Solar Radiation Data Directory*. SERI/SP-281-1973. Golden, CO: Solar Energy Research Institute.

Solar Energy Research Institute (SERI), 1985. *The Directory of Mesoscale Solar Radiation Information Sources*. SERI/SP-271-2715. Golden, CO: Solar Energy Research Institute.

Stoffel, T. L., 1987. *HBCU Solar Radiation Network Annual Report: FY 1986*. SERI/PR-215-3098. Golden, CO: Solar Energy Research Institute.

Turner, Charles, ed., 1974. *Solar Energy Data Workshop*. NSF-RA-N-062. Washington, DC: Superintendent of Documents, U.S. Government Printing Office, pp. 218.

University of Oregon, 1986. *Pacific Northwest Solar Radiation Data*. DOE/BP-24574-2. Eugene, OR: Solar Energy Center, Department of Physics.

U.S. Department of Energy (U.S. DOE), 1979. *NOAA's Role in the Solar Energy Program; 1978 Annual Report to the Department of Energy*. DOE/ET-0110. Washington, DC: U.S. Department of Energy, Assistant Secretary for Energy Technology, Division of Distributed Solar Technology, p. 10.

Weather Bureau, 1902. *Report of the Chief of the Weather Bureau*. Washington, DC: Government Printing Office, p. XVII.

Weather Bureau, 1903. *Report of the Chief of the Weather Bureau*. Washington, DC: Government Printing Office, p. XIX.

World Meteorological Organization, 1969. Radiation climatology: suggestions for increasing the number of actinometric stations. *Commission for Climatology*, Fifth Session, Geneva (Doc. 30, Appendix A, p. 2).

5 Solar Radiation Instrumentation

Gene Zerlaut

5.1 Introduction

The need for improving the state of the art of solar radiation measurements has received much attention in the past 4 years. Researchers and users have voiced concern about the quality of data from United States and various worldwide solar radiation networks and, more recently, about the reliability of calibration of pyranometers used for the accurate determination of solar irradiance on surfaces tilted from the horizontal. These topics have been addressed through a series of national and international seminars, workshops, and round-robin instrument comparisons. As a result, intercomparison protocols and procedures have been established, and consensus-developed instrument calibration and calibration-transfer standards have been promulgated in the United States and in the international community.

These improvements in the state of the art of solar radiation measurements are associated primarily with calibration and instrumentation techniques—although at least one new high-class thermopile pyranometer and two new self-calibrating pyranometers represent recent developments in the field.

5.2 Problems in Solar Radiation Measurements

5.2.1 Measurement Problems in Resource Assessment

The problems that plagued the original U.S. network operated by the National Weather Service encompassed a range of instrument problems: poor calibration techniques, poor traceability, a lack of temperature compensation in the instruments deployed, and generally unsatisfactory maintenance (Lufkin, 1980; Hanson, 1974). Although calibration procedures and traceability were improved in the late 1950s when the United States adopted the International Pyrheliometric Scale in 1956 (IPS 1956) in support of the International Geophysical Year, further errors in the U.S. network data in the 1960s and into the early 1970s were caused by the now well-known Parson's Black deterioration in Eppley Model 50 pyranometers (Flowers, 1974). The quality of data from those pyranometers that were neither replaced nor recalibrated steadily worsened, further reducing confidence in the network data.

The status of solar radiometry improved significantly when the United States adopted first the IPS 1956 and then the Absolute Radiometric Scale traceable to the World Radiation Reference (WRR) maintained for the World Meteorological Organization (WMO) by the Physikalisch-Meteorologisches Observatorium (PMO), Davos, Switzerland. The bases for these solar radiation scales and their significance are discussed in some detail in section 5.5.

The National Oceanic and Atmospheric Administration (NOAA) announced in early 1977 that its then new solar radiation network would be calibrated and maintained to the absolute scale defined by the absolute cavity radiometers compared in Davos at the 1975 International Pyrheliometric Conference (IPC IV). (This and subsequent regional and international comparisons are discussed in detail in sections 5.5 and 5.6.)

Although many of the instrumental problems of the old NOAA network have been solved by converting to Class I instruments (Lufkin, 1980) as defined by WMO,[1] the Solar Radiation Facility (SRF) of NOAA has observed deterioration in the sensitivity of certain types of so-called WMO Class I pyranometers as a function of environmental exposure (Flowers, 1982). It was previously thought that Class I-type instruments (hereafter referred to as First Class instruments) suffered significant exposure-related degradation only under the high-ultraviolet and high-temperature conditions of the southwestern desert, as reported by this author (Zerlaut, 1981a). More recently, the National Bureau of Standards (NBS) has reported aging-related decreases in instrument sensitivity. In one case, the instrument constant of one Eppley Model PSP pyranometer had decreased by more than 7% in 5 years (Reed, 1984).

A remaining problem inherent in many pyranometers used in meteorology and resource assessment, one that can be solved by preselecting pyranometers (to obtain a truly First Class instrument in terms of the 1983 WMO classification), is the severity of deviations from the cosine law. This deviation is variously described as cosine error or cosine correction—in other words, the ratio of the instrument response at a given angle of incidence (with respect to the receiver) to the product of the response at normal incidence and the cosine of the angle of incidence. Incident angle effects have often been denoted by the cosine response measured in two orthogonal planes, one of which passes through the electrical connector (north) of the pyranometer, and by a family of 360° azimuth response curves for a number of selected incident angles for a horizontally mounted pyranometer (figures 5.1 and 5.2). This definition may be challenged because it does not describe the actual range and path of incident angles of pyranometers in use at 0° horizontal and at south-facing tilts from the horizontal.

Problems occur because pyranometers exhibiting deviations from true cosine response are difficult to calibrate accurately to account for the range of solar altitudes encountered throughout the year. To measure accurately the irradiance required in performance evaluation of solar devices, a number of instrument constants should be used, preferably as file information in the computers used. However, this procedure would be far too costly and cumbersome for daily continuous measurements of irradiance in support of meteorological and resource assessment applications.

Flowers (1981) calibrates the pyranometers by obtaining the regression expression from several days of continuous ratioing to the NOAA/SRF reference pyranometer(s). At DSET Laboratories, Inc., we adjust the instrument constant of meteorological instruments to a value that provides the same integrated response (total integrated radiation) over a 3–5-day period as obtained with our reference instrument (which has a flat cosine response) with the test and reference instruments at the same tilt. This requires a reference instrument of small cosine, azimuth, and tilt sensitivity (without which precise characterization of the reference instrument is required).

The advantage of assigning an instrument constant that results in a total integral equal to that of the reference instrument is seasonal accuracy. However, for optimal accuracy the instrument must be calibrated about four times a year. Furthermore, unless a variety of sky conditions is included, significant errors can result. In the NOAA regression technique, the reference is calibrated by the shade method and the instrument constant assigned is for a value of 50° incident angle, virtually assuring that the calibrations are all referenced or normalized to 1000 W/m^2 (317 Btu/ft^2 h) (Flowers, 1981).

Finally, the most accurate approach to calibration is to characterize a pyranometer completely by measuring its deviation of response from a set of norms in terms of cosine and azimuth response, temperature effects, tilt effects, and linearity. As the cost of interfacing networks with sophisticated

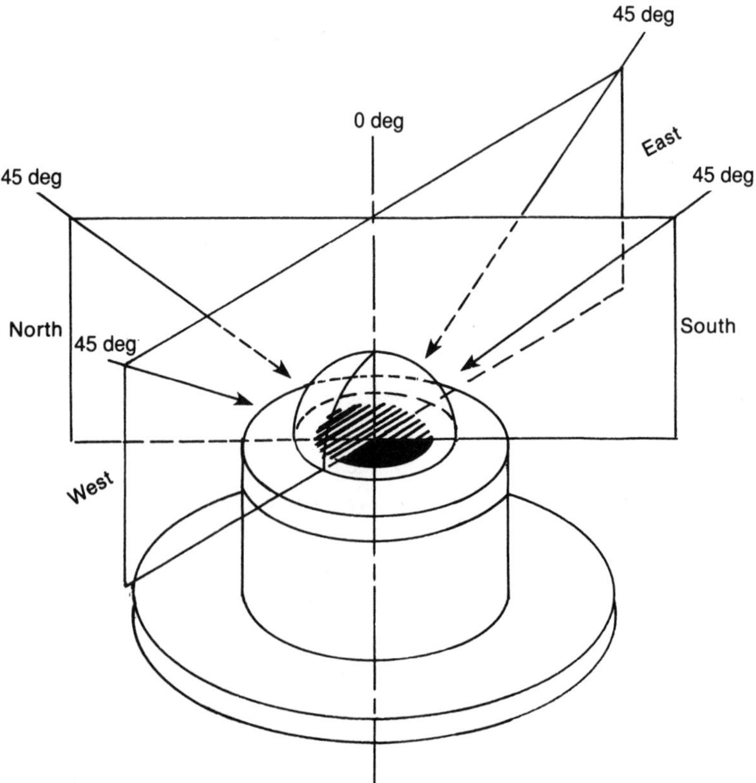

Figure 5.1
Schematic of solar cosine response angles to plane of a pyranometer receiver in two orthogonal planes passing through the pyranometer axis.

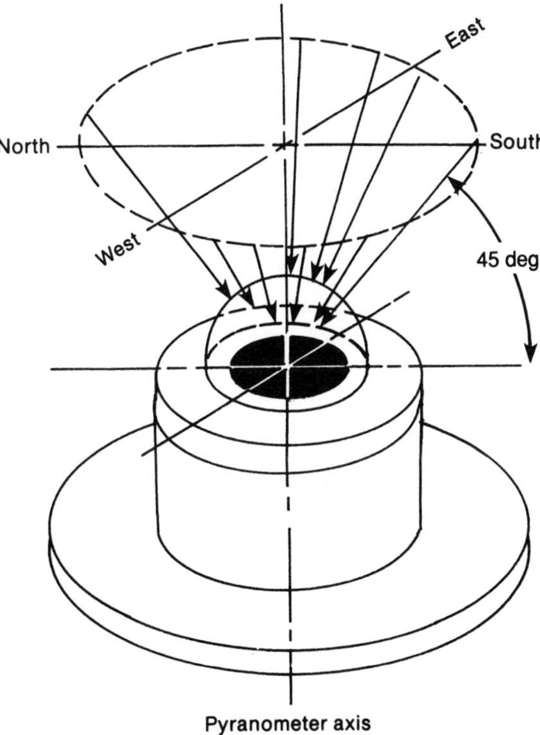

Figure 5.2
Schematic of solar azimuth response angles to the plane of a pyranometer receiver at a given incident angle (45°).

computers decreases, it may become possible to effect real-time compensation for deviation of each pyranometer from an arbitrarily chosen operational norm. Response deviations seem to be much less affected by aging than is the primary instrument constant selected as the reference point, and periodic checking of the primary instrument constant may be all that is always required.

5.2.2 Measurement Problems in Performance Testing of Solar Devices

Historically, the solar industry has expected, if not demanded, that thermal performance (efficiency) tests of solar thermal collectors tested to ASHRAE Standard 93–77 be accurate to within $\pm 2\%$. Complaints often arise when two test laboratories differ from each other by as little as 2% (or only 50% of the permissible spread within the significance of an allowable $\pm 2\%$).

Yet my coworkers and I have shown that the probable error contributed to the efficiency expression,

$$\dot{m} = \frac{\eta C_p \Delta T}{I_t} = F_R k_{\tau,\alpha} \alpha\tau, \tag{5.1}$$

by a presumably *well-calibrated* pyranometer can easily be $\pm 2\%$ alone. The possible error for a well-calibrated instrument can exceed $\pm 4\%$ (Zerlaut 1981c).[2] In equation (5.1), $\dot{m}$ is the mass flow rate, C_p is the specific heat, ΔT is the difference between inlet and outlet temperatures of the heat transfer fluid, I_t is the instantaneous solar irradiance, F_R is the plate efficiency factor, $k_{\tau,\alpha}$ is the incident angle modifier, α is the solar absorptance of the plate (receiver), and τ is the transmittance of the cover system.

These problems with accuracy first came to public light when the results of the rather poorly conceived, NBS-sponsored, Round Robin flat-plate solar collector test program were scrutinized (Streed et al., 1978). Two generically different collectors were tested by 21 laboratories (including 3 government laboratories) with the result that differences in the optical efficiencies exceeded an astounding 20%. Because the optical efficiency is the same as equation (5.1) for an inlet temperature approximating the ambient temperature, as defined by the Hottel-Whillier governing equation of the collector (ASHRAE, 1977), the ΔT expression in equation (5.1) is optimized. Thermocouple errors in the measurement of ΔT are extremely small—even for those inexperienced laboratories that participated in the Round Robins. Thus, most of the errors in the discrepancies obtained for the optical efficiency of the two Round Robin collectors resulted from the measurement of mass flow $\dot{m}$ and solar irradiance I_t. Given that most of the laboratories determined mass flow by a simple gravimetric measurement of flow for the specified interval, a significant proportion of errors in equation (5.1) must be caused by errors in the measurement of I_t, the solar irradiance. Streed et al. (1978) stated that a significant portion of the differences in optical efficiency could most likely be ascribed to problems in measuring solar irradiance.

Unfortunately, time has not solved the pyranometry problem inherent in the NBS Round Robin test sequence. A subsequent international Round Robin collector test program managed by the International Energy Agency (IEA) (Talarek, 1979) resulted in serious, but somewhat smaller, differences between laboratories. More recently, results of comparative collector performance tests performed in solar simulators have shown similar disparities in measuring the optical efficiency (Ley, 1979) of flat-plate solar collectors.

Subsequent investigations, coupled with an improved understanding of the problem, indicate that many errors in solar irradiance measurement resulted from one or more of the following: (1) use of WMO Second Class pyranometers that have subsequently been shown to exhibit tilt errors of as much as 10–12%, (2) use of First and Second Class pyranometers that had not been calibrated in some time, (3) use of First and Second Class pyranometers that exhibited significant deviations from the cosine law (which becomes increasingly important when testing collectors at a fixed tilt angle as opposed to altazimuth tracking of the sun), and (4) failure of most of the participating laboratories to use the temperature sensitivity correction factor that is furnished with the purchase of all Eppley pyranometers (and pyrheliometers). Indeed, the solar collector manufacturing and testing community has made poor progress in understanding these problems. Consequently, much still needs to be done to sell the notion that pyranometers used to measure "instantaneous irradiance" must first be preselected and then thoroughly characterized with respect to instrument constant and incident angle effects at the tilt angles used.

5.2.3. The Need for Calibration and Measurement Standards

Webster's *New Collegiate Dictionary* defines standard as "something set up and established by authority as a rule for the measure of quantity ... value, or quality." Within this context, standard methods of calibrating solar radiation instruments are urgently required. Experience shows that standard practices, or standard operating procedures, are also required if accurate and meaningful solar irradiance measurements are to be obtained.

Confidence in the calibration and maintenance of calibration of the U.S. network has largely been restored by the thoroughness of NOAA/SRF during the past several years. Nevertheless, standardized procedures are still required for meteorological applications so that pyranometer calibration techniques used by NOAA/SRF to support the U.S. network and those used by organizations to support certain regional and state networks yield equivalent and comparable results.

Perhaps even more critical than the need for standard calibration methods in meteorological applications is the need for standardized procedures for calibrating pyranometers that measure solar irradiance in support of solar device testing. Without either accurate calibrations or intercomparison procedures, agreement between laboratories on the thermal efficiency characteristics of solar collectors is fortuitous at best. Yet, as shown here, the calibration procedures required for meteorological applications and for laboratory testing of solar devices are necessarily somewhat different. Solar test laboratories around the world have too readily relied on meteorological calibrations; they have incorrectly and improperly trusted inappropriate, but otherwise quite correct, meteorlogical-type calibrations for testing the thermal efficiency of solar collectors. Calibration authorities, being largely devoted to meteorological applications, have been slow to accept the distinction between the calibration requirements for resource assessment and those for the precise determination of solar irradiance at various angles of incidence to the plane of a solar device or receiver—the latter requiring much greater accuracy in the sense of being a "physical optics" measurement.

Standardization efforts are also needed in the field use of pyrheliometers and pyranometers. Here the distinction between the requirements for resource assessment and precision performance testing diminishes. Both applications require daily maintenance such as cleaning and alignment checks (for pyrheliometers) and can tolerate only minimal interferences with the field of view. Alignment of the pyranometer's receiver with the spirit level (bubble level) is important to both applications, although it is particularly critical to the accurate assessment of instantaneous irradiance in performance testing.

Philosophically speaking, the voluntary consensus process traditionally has been the means by which U.S. industry and society develop and produce standards that form the basis of American commerce. This is the only process that can minimize the forces of economic chaos and government overregulation, thus permitting the exercise of a truly competitive marketplace. Without voluntary consensus standards promulgated largely by a healthy, self-policing industry, narrow segments of society can erect impediments to market penetration and government can react with so-called mandatory standards and costly regulations (Zerlaut and Garner, 1983). This scenario is no less true in the broad field of solar energy application, where incorrect assessment of solar irradiance leads directly to incorrect sizing calculations and concomitant errors in the computation of energy efficiency.

5.3 Instrumentation for Measurements of Total Solar Irradiance

5.3.1 The Solar Irradiance Measurements Problem

5.3.1.1 Natural Sunlight The sun radiates to earth as a 5,770K blackbody, except that the ultraviolet region below 400 nm is more closely represented by a 4,500K Planckian radiator. This is because the solar ultraviolet emanates from the outer surface of the sun, and its blackbody continum approximates that from a cooler radiator. The ultraviolet continum is further modified on the sun's surface as a result of Fraunhofer absorption by various metals, which imparts distinct character to the ultraviolet reaching the earth's atmosphere.

The solar radiation reaching the earth's chemosphere [110km (68.3mi)] extends in wavelength from about 200 nm to more than 2,500 nm. However, ultraviolet of 242.4-nm wavelength is absorbed by oxygen to produce ozone, which itself strongly absorbs ultraviolet in the entire wavelength region, 200–300 nm (the Hartley region). The spectral transmittance of ozone is shown in figure 5.3 for three values of *lm*, the product of ozone thickness in centimeters at normal termperature and pressure (NTP) and the optical mass for ozone (Iqbal, 1983). This is understandable when one considers that the measured absorption coefficient α for ozone is 10^{10} cm^{-1} at 295 nm (Inn and Tanaka, 1953). Ozone also exhibits moderate absorption in the 500–700-nm wavelength region (the Chappius band) to account for the flattening of the visible region of the extraterrestrial spectrum.

Ozone exists in greatest concentration above the stratosphere in a band about 20 km thick at altitudes between 20 and 30 km. It varies both diurnally and seasonally, generally being greatest in early morning and in late winter and early spring. Hence, the ultraviolet portion of terrestrial solar radiation varies in relation to ozone concentration and to surface atmospheric changes.

As sunlight passes through the chemosphere, its spectral distribution is further modified by water vapor absorption and absorption by mixed gases such as O_2, CO, CO_2, NO, and CH_4. Water vapor absorption occurs at wavelengths of 720, 810, 940, 1,100, 1,380, 1,870, 2,700, and 3,200 nm. As a result, the extraterrestrial solar spectral distribution shown in curve A of figure 5.4 is changed as depicted by curves B and C in that figure.

Although absorption mechanisms are largely responsible for the spectral character of terrestrial solar radiation in the ultraviolet and infrared spec-

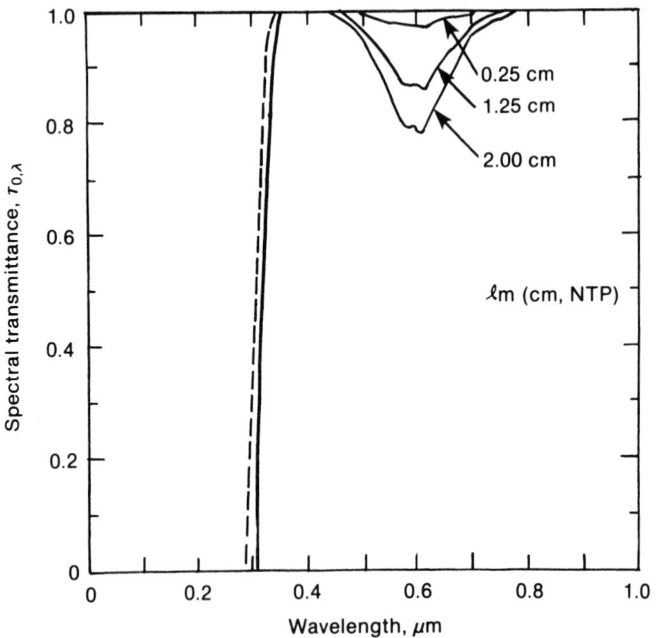

Figure 5.3
The spectral transmittance of ozone as a function of *m*. Source: Iqbal (1983).

tra (curves B and C in figure 5.4), atmospheric scattering accounts for the bulk of both diurnal and day-to-day variations in the spectral intensity of the direct beam (curve C in figure 5.4) and, to a lesser extent, the global normal irradiance component. Indeed, the difference between the two distributions (B and C) is known as the normal diffuse solar radiation and is a function of the total radiation (which is the difference between D and the direct beam normalized to 0° horizontal). Attenuation of sky radiation caused by absorption is low.

These processes may be summarized by the equation

$$[\delta = \sigma_r + \sigma_m + \alpha_g + \alpha_p]_\lambda, \tag{5.2}$$

where δ is the total spectral attenuation factor at any wavelength λ, σ_r is Rayleigh scattering, σ_m is Mie scattering, α_g is absorption by gases (carbon dioxide, oxygen, etc.), and α_p is absorption by pollutants and localized gaseous and aerosol contaminants. Rayleigh scattering affects particles that vary in size from molecules to Ångström size and gives the sky a blue cast. Mie scattering predominates for particles and droplets with diameters the size of the wavelength of light. Obviously, the radiation scattered out of the direct beam, except that lost to space, is added to the sky radiation component. These processes result in shift of the maximum in the solar spectral distribution toward longer wavelengths with increasing path length.

Scattering is both forward and backward, and the backward portion, which contributes to the earth's albedo, further attenuates the sky or diffuse portion of solar radiation. Under clear sky conditions, single-particle scattering phenomena predominate, causing the diffuse component to be concentrated in the circumsolar and near-circumsolar region surrounding the sun. However, under dense sky conditions (such as high haze), multiple-particle scattering is of increased importance, and the shorter solar wavelengths are preferentially scattered throughout the sky dome, resulting in increased sky brightness at the lower elevations.

Defined as the secant of the zenith angle of the sun, air mass varies from a high at sun-up (about 11 at a solar elevation of 5° at sea level) to a low at solar noon (about 1.05 at 34° north latitude at solar noon during the summer solstice). Figures 5.5 and 5.6 show the influence of varying air mass on the measured solar spectral energy distributions at New River, AZ, for direct beam and global solar irradiance, respectively; these spectra dramatically depict the shift in peak wavelengths to the infrared with increasing air mass.

5.3.1.2 Implications for Measuring Total Irradiance

Beam Irradiance. Beam irradiance, or direct solar radiation, is measured with a device called a pyrheliometer. Pyrheliometers consist of a sensor and reciver located at one end of an occluding tube and a limiting aperture, and usually a window, at the other end. The geometric relationship between the receiver and the limiting aperture defines the field of view of the instrument. Obviously, such a device requires a precise sun-tracking mount to follow the sun continuously when used in the field to make continuous measurements.

In our nearly 5 years of regular periodic spectroradiometric measurements of the sun, we have observed that changing air mass and changing atomspheric constituents (moisture, dust, pollution) have a far greater effect on the spectral character of the direct beam than they do on global horizontal, inclined global, or normal spectral energy distributions. Hence, although photovoltaic pyranometers (whose spectral responsivity resides

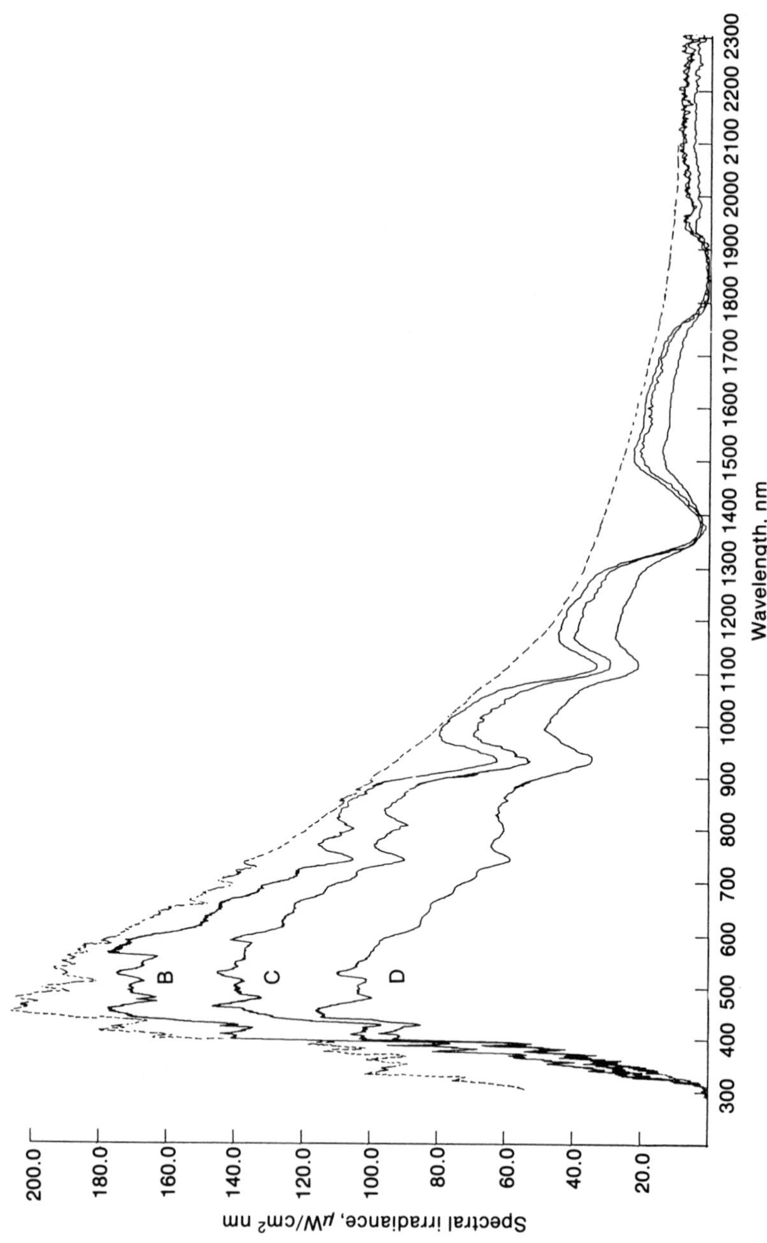
Spectral irradiance, μW/cm² nm
200.0
180.0
160.0
140.0
120.0
100.0
80.0
60.0
40.0
20.0
B
C
D
300 400 500 600 700 800 900 1000 1100 1200 1300 1400 1500 1600 1700 1800 1900 2000 2100 2200 2300
Wavelength, nm

W/m^2	Spectra	
1367	---	Labs & Neckel revised extraterrestrial spectrum (1970)
1055	B	Hemispherical irradiance at normal incidence
891	C	Direct (beam) irradiance for 6° f.o.v.
629	D	Hemispherical (global) irradiance at 0° horizontal

Figure 5.4
Terrestrial solar spectral energy distributions at AM1.37 compared to AMO (measured at New River, AZ on 6 October 1982).

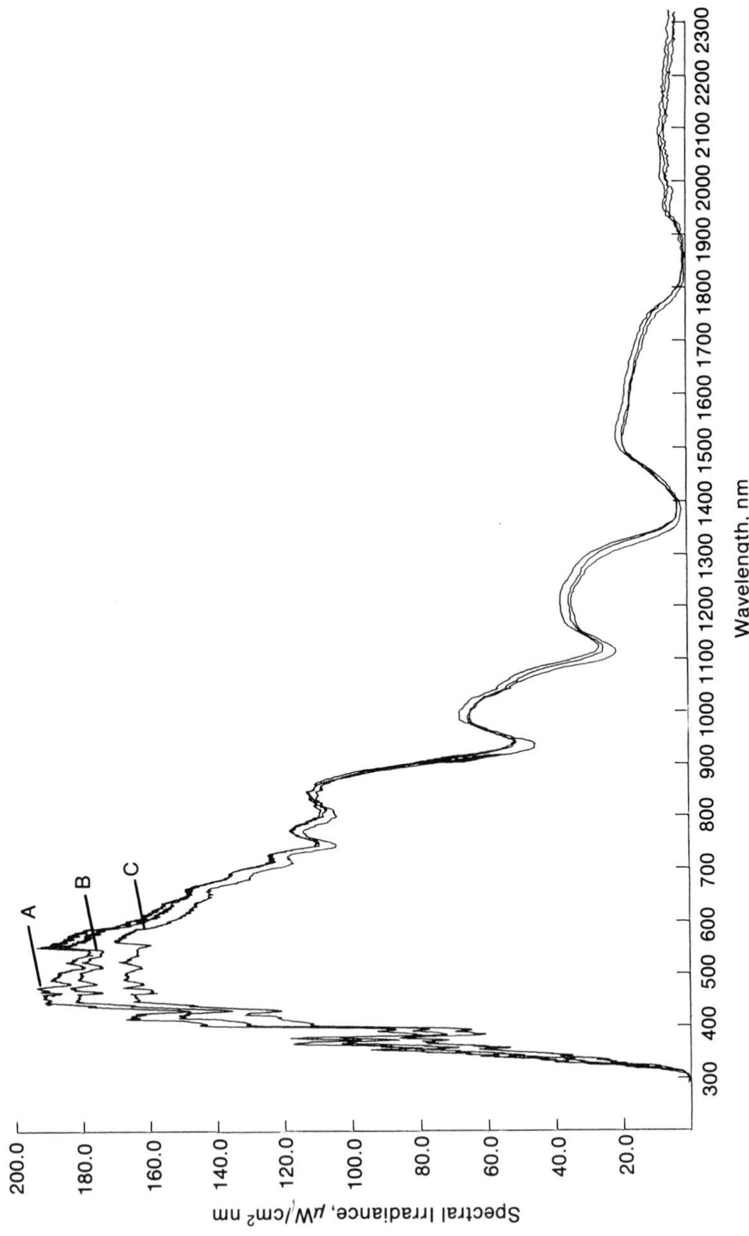

Spectral Irradiance, μW/cm² nm
200.0
180.0
160.0
140.0
120.0
100.0
80.0
60.0
40.0
20.0
A
B
C
300 400 500 600 700 800 900 1000 1100 1200 1300 1400 1500 1600 1700 1800 1900 2000 2100 2200 2300
Wavelength, nm

AM	Spectra	Integrated Irradiance
1.02	A	1075 W/m^2
1.21	B	1029 W/m^2
1.47	C	953 W/m^2

Figure 5.6
Hemispherical (global) solar spectral irradiance at normal incidence as a function of air mass at New River, AZ (June 1982).

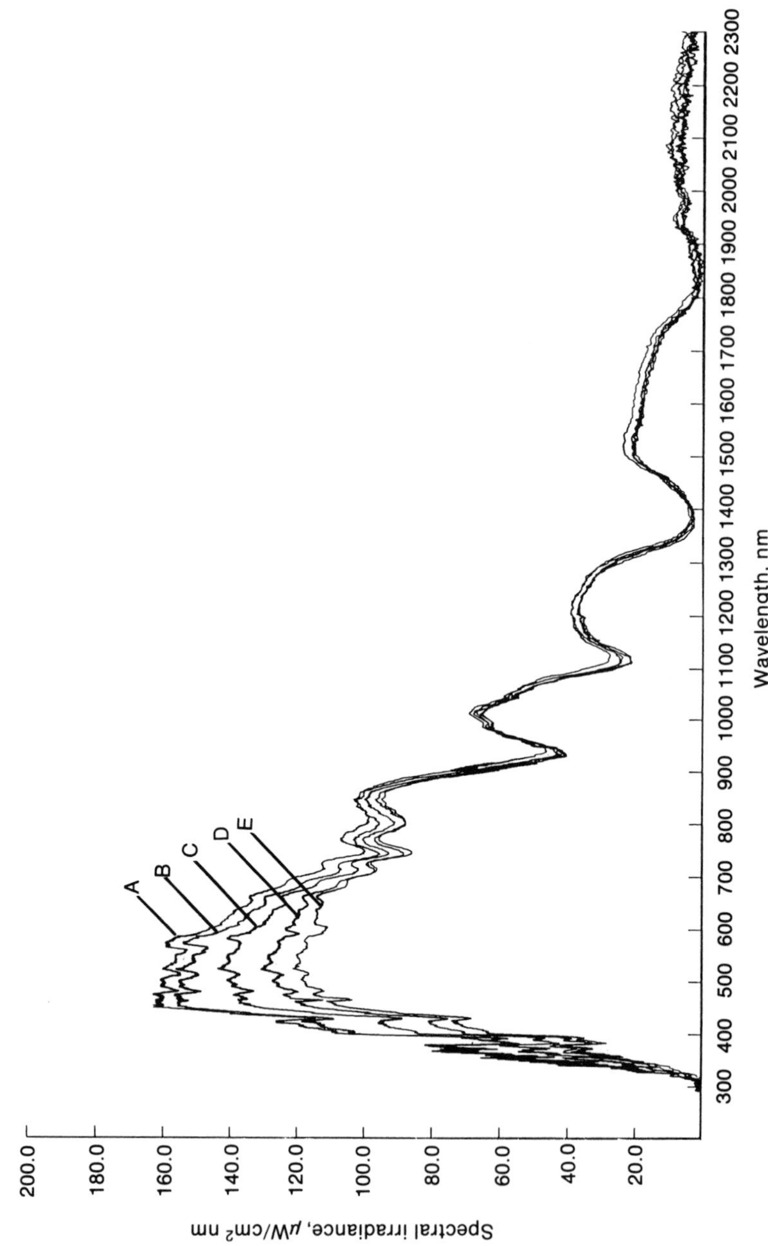
Spectral irradiance, μW/cm² nm
200.0
180.0
160.0
140.0
120.0
100.0
80.0
60.0
40.0
20.0
A
B
C
D
E
300 400 500 600 700 800 900 1000 1100 1200 1300 1400 1500 1600 1700 1800 1900 2000 2100 2200 2300
Wavelength, nm

AM	Spectra	Integrated Irradiance
1.02	A	961 W/m^2
1.20	B	918 W/m^2
1.31	C	860 W/m^2
1.66	D	829 W/m^2
2.17	E	816 W/m^2

Figure 5.5
Direct normal solar spectral irradiance as a function of air mass at New River, AZ (June 1982).

only in the 400–1,000-nm range) are useful for measuring global solar irradiance on horizontal and inclined surfaces, pyrheliometers based on photovoltaic sensors are subject to serious diurnal and seasonal errors in measuring total beam irradiance. These errors result largely from the changing ratio of visible to infrared radiation in the direct beam that is in direct proportion to changing air mass—the direct beam infrared is proportionately less diminished by increased air mass and increased dust and pollution than is the visible portion of the solar spectrum to which the photovoltaic device is predominately sensitive.

Although one can thus make a very strong case for thermopile pyrheliometers with very flat spectral response characteristics for determining total beam irradiance in support of solar thermal applications, they are not useful for measuring direct beam irradiance in support of concentrating photovoltaic applications. In other words, thermopile pyrheliometers with flat spectral response characteristics tend to overstate the beam irradiance when used to determine the electrical performance of concentrating photovoltaic devices, as well as for measuring the solar resource in support of sizing analyses for concentrating photovoltaic arrays. Zerlaut has shown that a standard normal incident pyrheliometer can overstate the beam energy available to a concentrating photovoltaic device by as much as 10% on a clear day in, for example, an urban desert environment such as Phoenix, AZ.

Thus, to assess properly the electrical performance of concentrating photovoltaic systems, one must use a pyrheliometer based on a photovoltaic sensor having the same spectral response characteristics as the photovoltaic device being evaluated.

The Eppley Model NIP (normal incidence pyrheliometer) is the most widely used pyrheliometer. Its field of view is 5°36′, which is adequate for most solar thermal concentrator device testing and resource assessment applications. However, its field of view encompasses a much greater portion of circumsolar radiation than is desirable for highly concentrating parabolic trough and dish solar collectors, and it is totally inadequate for assessing the performance characteristics of central receiver optical systems. Grether et al. (1980, 1981) have constructed a precision circumsolar telescope that mechanically scans through an arc of ±3° referenced to the center of the sun's disk. The brightness of the sun is recorded every 1.5′ of arc, and its value is related to calibration against an absolute cavity radiometer. Measurements have been made at four locations in the United

States. The authors concluded that haze and cirrus can affect the performance of highly concentrating solar systems; therefore, an assessment of circumsolar radiation is necessary to properly evaluate system performance. Figure 5.7 is a typical plot where the circumsolar ratio $(C/C + S)$ is the ratio between the circumsolar radiation defined within the $\pm 3°$ arc and the sum of the circumsolar plus true beam irradiance from the sun.

Global Irradiance. Global irradiance comprises the direct beam plus the sky, or diffuse, irradiance and is measured with a radiometric device known as a pyranometer. The most widely used pyranometer consists of a blackened receiver attached to the hot junction of a thermopile that is mounted in a thermally isolated case and enclosed in either one or two glass domes. The principal purpose of the glass domes is, first, to prelude large transient effects caused by convective cooling of the receiver and, second, to exclude long-wavelength sky and terrestrial radiation. Also, with two domes, the infrared exchanges are between the cold outer and the warm inner domes (and between the inner dome and receiver), thus minimizing direct infrared exchange between the outer dome and the hot receiver. The outer dome of several pyranometers is interchangeable with domes formed from various cut-off glass filters to provide broadband spectral information. The black receiver coatings currently in use are generally good diffuse receivers of solar radiation and are spectrally nonselective, and hence do not inherently exhibit significant angular dependence. Optical asperities in the glass domes themselves, though, coupled with poor optical design (receiver shape) and poor alignment over the receiver may cause the pyranometer to exhibit significant deviations from the cosine law.

The black coatings used on themopile receivers change with constant exposure, which may explain the propensity of some pyranometers to exhibit increased angular dependence with age. Furthermore, the instrument-to-instrument difference in angular dependence of thermopile receivers is ascribed in part to variations in thickness of the black receiver coating, resulting in differences in infrared reflectance (or, more properly, absorptance). Thus, at the off-angles associated with irradiance measurements early and late in the day when the sun is infrared-rich (compared to midday intensities), small differences in spectral response of the receiver resulting from variations in receiver thickness may cause response deviations that are mistaken for so-called cosine errors.

Nonetheless, we observe that pyranometers are inherently less sensitive than are pyrheliometers to changing spectral energy distributions of the

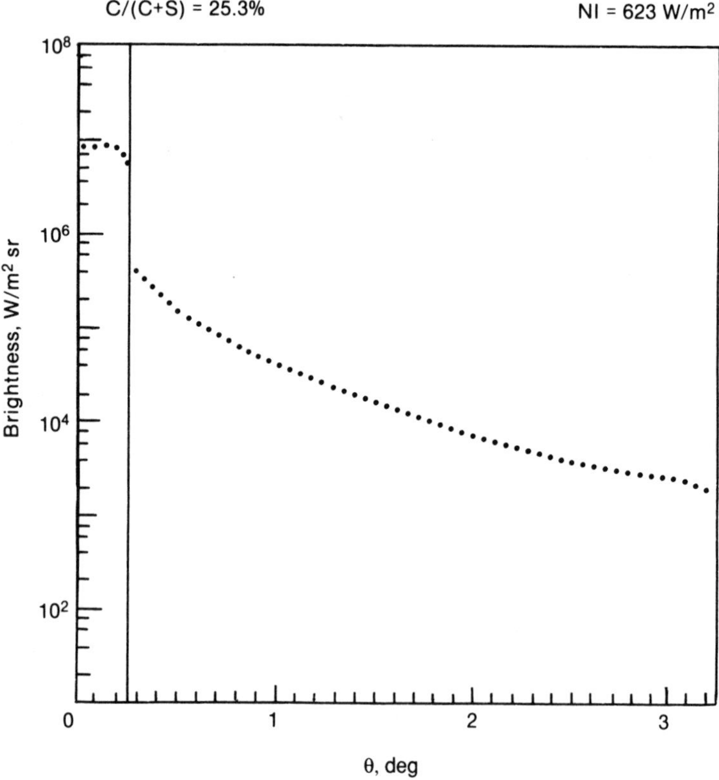

Figure 5.7
Circumsolar telescope scan for conditions of high altitude haze at Barstow, CA, on 6 May 1979. Source: Grether et al. (1981).

sun. As a result, pyranometers of the photovoltaic, or silicon solar cell, type have found increasing use in measuring global irradiance. Although "nude" photovoltaic sensors are sensitive to the incident angle of illumination, Teflon® and ground glass diffusers can be designed largely to preclude significant deviation from the cosine law. Their use in measuring hemispherical solar irradiance is based on the assumption that their spectral response distribution covers approximately 85% of the energy represented by AM1.5 solar spectral energy distribution, for example. Hence, photovoltaic sensors are said proportionately to measure diurnal air mass-induced changes to the solar spectrum and, when properly calibrated, to measure total irradiance with reasonable accuracy. However, cloud cover, haze, and pollution, which disproportionately affect the spectral response region of photovoltaic cells (that is, the infrared region is less sensitive to such atmospheric and weather-related conditions), will result in increased errors in the photovoltaic measurement of global irradiance. Likewise, solar cell pyranometers do not accurately measure infrared-rich, high air mass-solar energy distribution (early and late in the day). Finally, while photovoltaic pyranometers enjoy an advantage over some thermopile pyranometers in that their photoresponse is unaffected by tilts from the horizontal, changes in spectral irradiance as a function of the included sky dome can result in errors. Variations of 5% or more have been observed by Flowers (1984b) when the sensitivity of some photovoltaic receivers is determined first for horizontal exposure to a cloudless sky and then at tilt where the receiver sees the sun, a hazy sky, and reflected ground radiation.

5.3.2 Thermal Detectors

As discussed by Iqbal (1983), solar radiation detectors may be divided into four broad classes: thermomechanical, calorimetric, thermoelectric, or photoelectric. Because of their inaccuracies and imprecision, we are not discussing thermomechanical pyranographs that are based on a blackened bimetallic strip, the differential thermal expansion of which provides a measure of irradiance. Also, with a single exception, we are not dwelling on calorimetric sensors in which temperature rise is a measure of the incident radiation. We are discussing the single and double chamber water flow pyrheliometers developed by Abbot and Aldrich (1908, 1932) because, first, irradiance was calculated from first principles directly in calories, and, second, these pyrheliometers were among the first true absolute instruments, and were significant in developing SIS-13 and ultimately IPS 1956.

For a more complete historical perspective, see Anson (1980), Iqbal (1983), and the volume by Budde (1983) on optical radiation detectors.

5.3.2.1 Thermopile Detectors The most common type of thermal detector is the thermopile, in which the sensitive element consists of a series of thermocouple junctions, the emf or voltage of which depends on the difference in temperature between the hot and cold junctions. The additive voltage produced results from the Seebeck current generated when junctions of dissimilar metals are maintained at different temperatures. Figure 5.8 shows the three types of thermopile configurations—the so-called linear, radial, and Moll-type (a heat-sunk, linear thermopile).

The sensitive element of thermopile radiation detectors is usually a blackened, electrically insulated, thermally conductive metallic receiver attached to the thermopile's hot junction. The emf voltage of thermopile detectors is linear with flux intensity over a wide range. When properly chosen, the black receiver coatings are reasonably stable and spectrally nonselective. Additionally, well-designed thermopile receivers have good cosine response. One problem inherent in black thermopile receivers is their apparent optical instability to extended exposure, making periodic recalibration mandatory.

5.3.2.2 Photovoltaic Detectors The sensitive element of a photovoltaic detector used in solar radiation measurements is usually a silicon solar cell that is operated at short-circuit current. Silicon photovoltaic and photodiode detectors belong to the group known as junction semiconductor devices and are based on the so-called internal photoelectric effect. Figure 5.9 is a schematic of these photovoltaic devices. Incident radiation creates an electron-hole pair, which is separated at the junction of dissimilar materials. This separation of charge carriers results in a measured current in the unbiased mode and in a measured resistance in the reverse voltage biased, photoconductive mode. Silicon photovoltaic devices operated at short-circuit current are used to detect visible radiation, as are silicon photodiodes. The latter, however, are used in narrow-band radiometry and spectroradiometry because of their generally greater responsivity—especially, photodiodes with FET (field effect transistor) operational amplifiers.

Bulk semiconductor detectors are single material devices such as PbS and PbSe and are always biased to operate in the photoconductive mode only. They are used for detecting infrared radiation, and, except for their

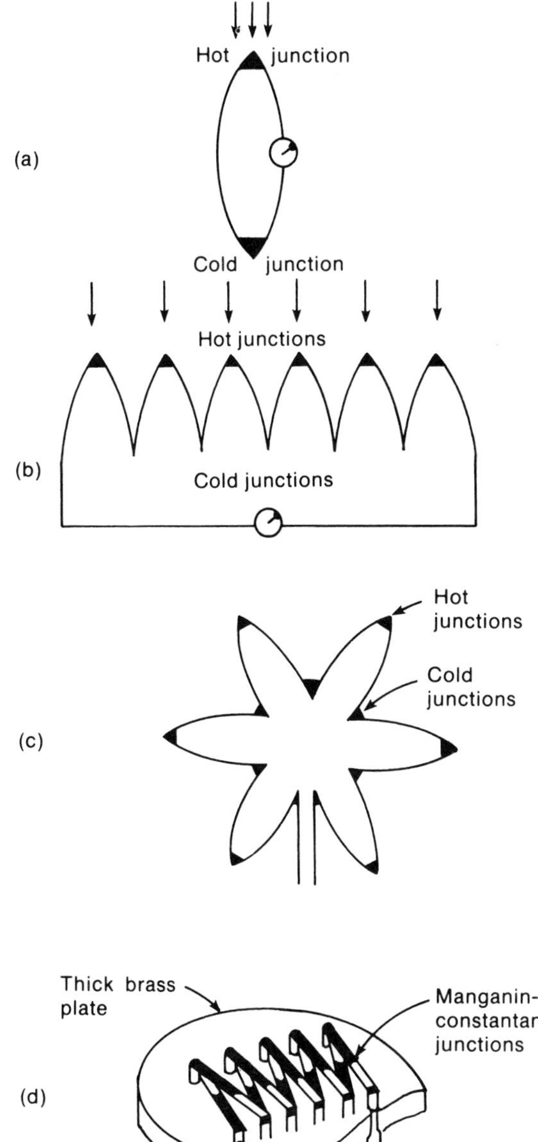

Figure 5.8
Various thermopile configurations. Source: Budde (1983).

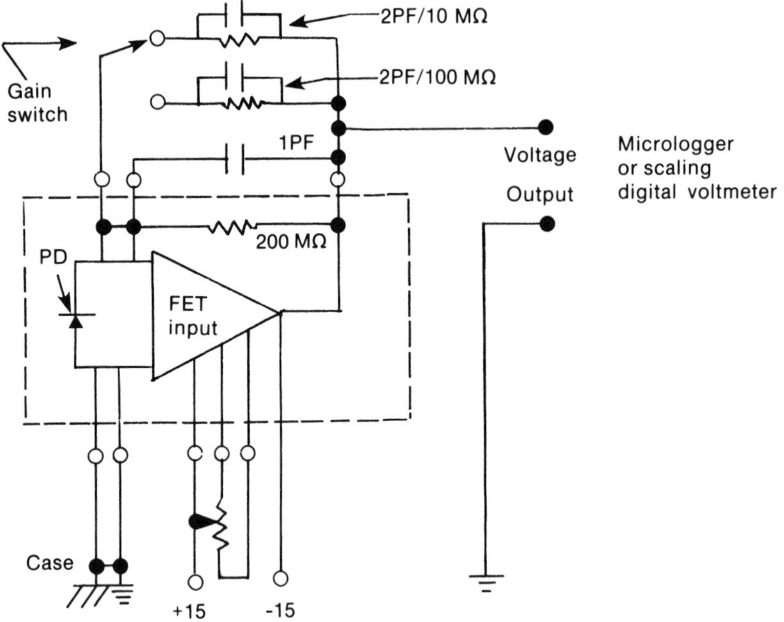

Figure 5.9
Equivalent circuit for an OP/AMP photodiode detector.

use in spectroradiometry (mentioned in section 5.4.4), further discussion of them is beyond the scope of this chapter. Budde (1983) gives a complete treatise on semiconductor detectors of optical radiation.

The advantages of photovoltaic detectors are (1) the linearity of their short-circuit current response over large ranges of radiant flux intensity, (2) their insensitivity to tilt, and (3) their generally high response factors (amp/W). The principal problems of silicon solar cell detectors, discussed previously, are their spectral selectivity and their inherent sensitivity to off-angle illumination in instruments lacking cosine correcting diffusers.

5.3.3 Pyrheliometers

5.3.3.1 General Considerations The geometrical considerations that apply to pyrheliometer design are described by figure 5.10, where

$$Z_o(\text{the opening angle}) = \tan^{-1} R/L, \tag{5.3}$$

$$Z_p(\text{the slope angle}) = \tan^{-1}[(R - r)/L], \tag{5.4}$$

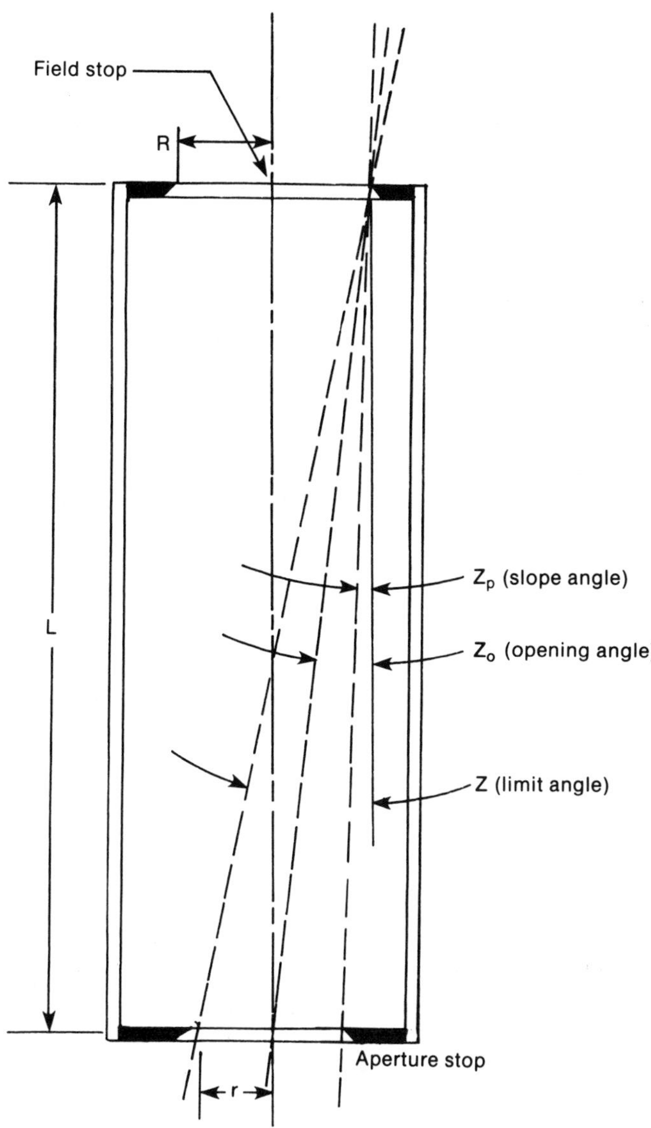

Figure 5.10
Sky-occluding geometry of a pyrheliometer.

$$Z(\text{the limit angle}) = \tan^{-1}[(R + r)/L], \tag{5.5}$$

and the field of view is $2Z_o$, or twice the opening angle. The half-angle values historically recommended by WMO are $Z_o \leq 4°$ and $1° \leq Z_p \leq 2°$ (1971). However, in the new guide (1983), WMO simply recommends that Z_o, the opening angle, be 2.5° (5×10^{-3} sr) and that Z_p, the slope angle, be 1°.

Obstacles to the successful use of pyrheliometers in continuously measuring direct solar radiation are imprecision in tracking the sun and the varying contribution of circumsolar radiation—especially between pyrheliometers with slightly different viewing optics. A number of equatorial mount trackers are available from pyrheliometer manufacturers; the author has successfully employed altazimuthal tracking mounts with bidirectional detectors in comparison testing of pyrheliometers.

In terms of the atmospheric and meteorological sciences, the measurement of beam radiation seldom needs to be confined to the 0.5° solar disk. However, the previously mentioned work of Grether et al. (1980, 1981) addressed this question, although the impetus for the studies was the need for assessing the beam component in support of the concept of developing central solar receiver facilities. Most pyrheliometers have fields of view of 5–6° and include significant amounts of circumsolar radiation in the direct beam assessment. This charactristic is fortuitous for many solar applications because the optical system of many parabolic solar concentrators have fields of view of 4–7°.

5.3.3.2 Eppley-Ångström Compensation Pyrheliometer Until after 1975, when absolute cavity pyrheliometers began to replace them, Ångström electrical compensation pyrheliometers were the primary reference instruments used for calibrating secondary pyrheliometers. The Eppley model of the Ångström instrument, shown schematically in figure 5.11, is the most widely used in the United States, Canada, and several other countries. The Ångström compensation pyrheliometer was designed by Knut Ångström and is still used throughout the world.

The Ångström pyrheliometer has a mean effective acceptance angle of 5° that is a result of a selectively shaded rectangular aperture and the use of two blackened, thin Manganin strips (2×20 mm^2) as sensors (Eppley Laboratory, n.d.1). Direct beam irradiance is computed from equation (5.6), where i is the electrical compensation current in amperes and k is the so-called instrument constant determined from first principles:

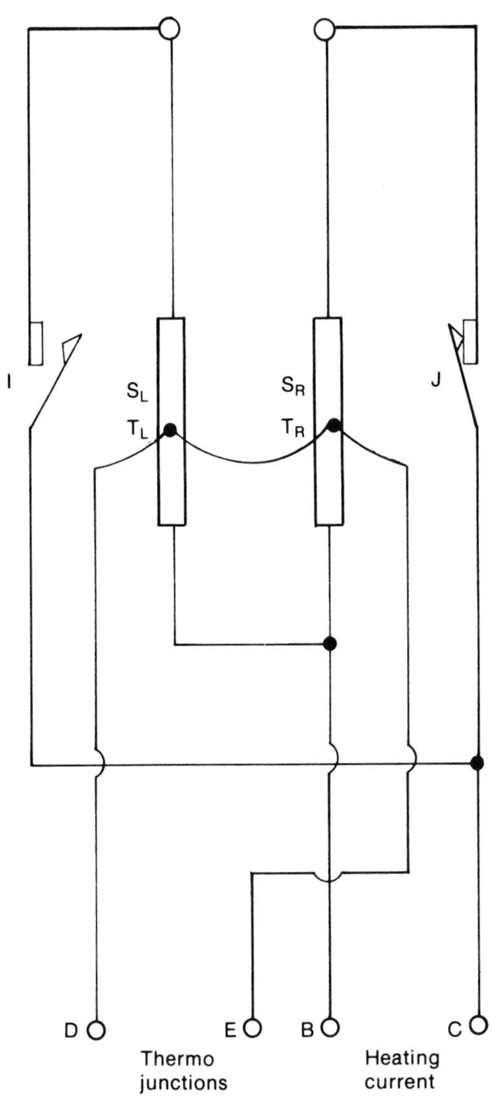

Figure 5.11
Electrical schematic of the Eppley-Ångström compensation pyrheliometer (when S_R is shaded). Source: Eppley Laboratory (n.d.1).

$$I = ki^2 . \tag{5.6}$$

A heating current is applied to the shaded strip to produce a temperature identical to that measured by the illuminated (unshaded) strip. Hence, the solar heating of one strip may be equated to electrical heating of the other strip as

$$\alpha_s A I = \frac{rli^2}{4.187}, \tag{5.7}$$

where α_s is the solar absorptance of the black coating (Parson's Black lacquer in the Eppley model), A is the area of the strip, r is the resistance of the strip in ohms per centimeters, l is the length of the strip in centimeters, and 4.187 is the electrical equivalent (in watts) of heat (in calories per second). Hence, the instrument constant is

$$k = \frac{rl}{4.187\alpha_s A}. \tag{5.8}$$

When characterized and used this way (the Ångström pyrheliometer is considered to be an absolute instrument), the value of k has most often been determined by comparison to a fully characterized Ångström, after which it is used to calibrate field pyrheliometers.

5.3.3.3 Kipp and Zonen/Linke-Fuessner Pyrheliometer The Kipp and Zonen version of the Linke-Fuessner pyrheliometer (commonly called the CM-1 Actinometer) is less than 6 in. (150 mm) long and uses a Moll-type thermopile as the detector (see figure 5.8). The CM-1 is more widely used in Europe and Africa than in the United States. The constantan-Manganin strips are covered with a thin sheet of paper and blackened with 3M Velvet Black™ 101-C10 (Kipp and Zonen, n.d.1).

Figure 5.12 is a cross-sectional diagram of the Kipp and Zonen CM-1 Actinometer, showing the copper block from which the internal apertures are constructed. This heat sink minimizes the possible radiative and convective thermal excursions that might otherwise affect the thermopile stability. The CM-1 is provided with a built-in thermometer used to correct the thermopile reading in accordance with the expression $[1 + 0.002 (t - 20°)]$. A typical temperature dependence plot is given in figure 5.13.

CM-1 Actinometers are both characterized and calibrated using a horizontal optical bench (that is, they are calibrated with the axis horizontal). They are calibrated at 500 W/m^2 (159 Btu/ft^2 h) using a 3,000-W xenon

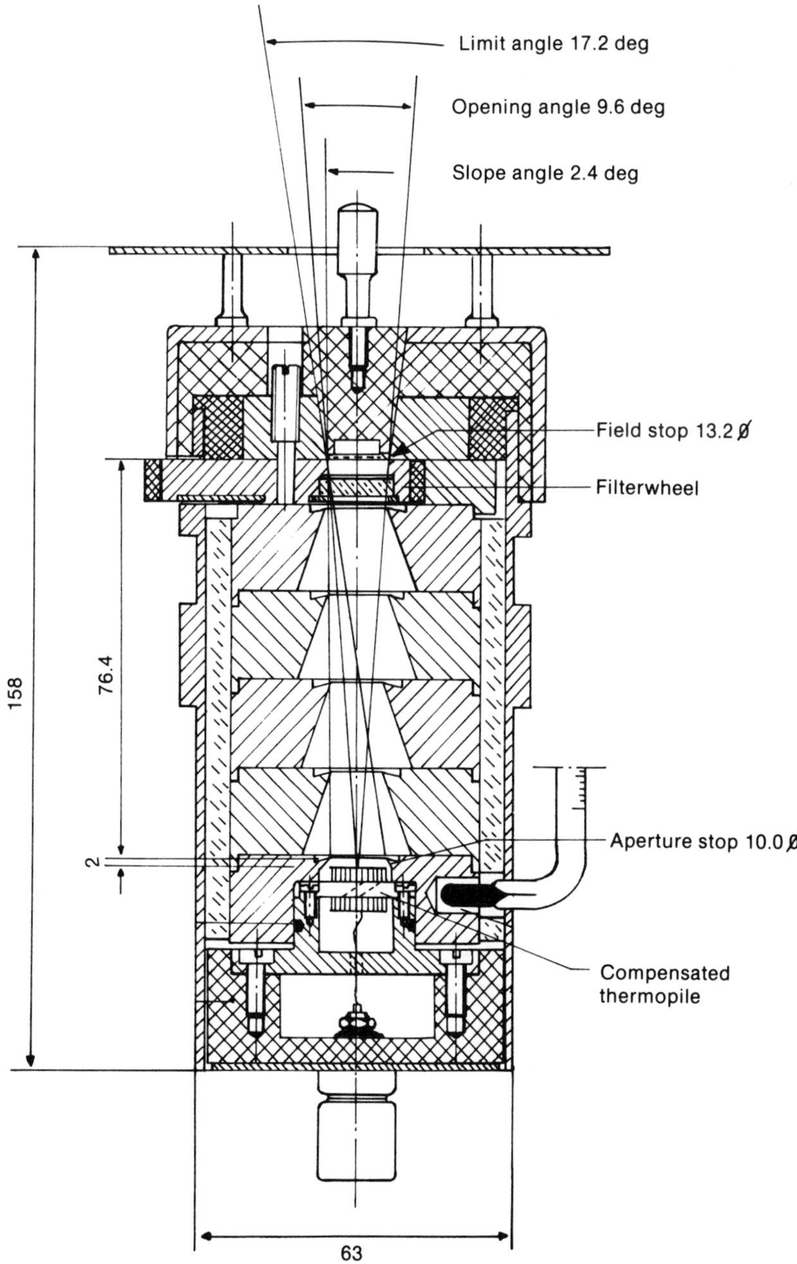

Figure 5.12
Mechanical schematic of the Kipp and Zonen Model CM-1 actinometer. Source: Kipp and Zonen (n.d.1).

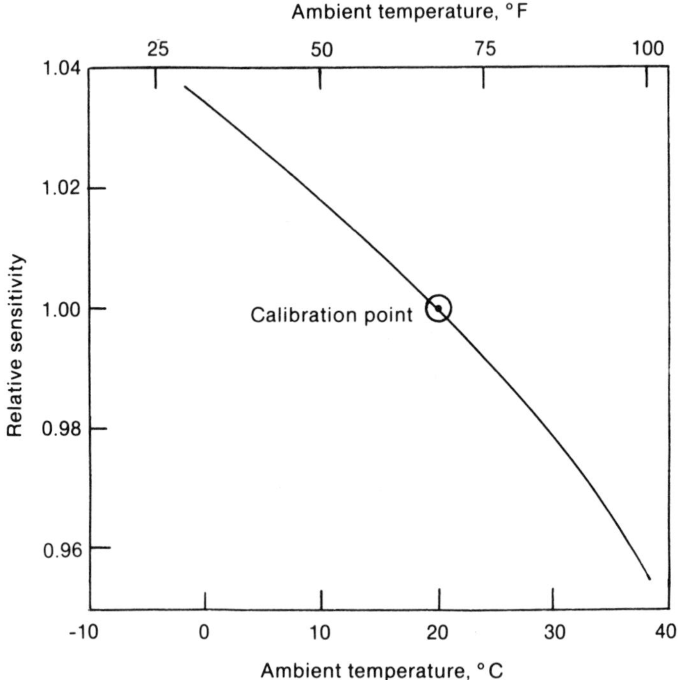

Figure 5.13
Typical temperature dependence plot for the Kipp and Zonen Model CM-1 actinometer. Source: Kipp and Zonen (n.d.1).

as a light source; typically they exhibit tilt sensitivities at 1,000 W/m^2 of +1% from tilts of 50° to 0°, and linearity sensitivities −0.8% from 500 to 1,000 W/m^2.

Instrument constants of CM-1 pyrheliometers are transferred on the optical bench by comparison with Kipp and Zonen's working standard CM-1, No. 820418, which was itself compared with Ångström pyrheliometer No. 559 at the Royal Dutch Weather Service in 1982 (Kipp and Zonen, n.d.2).

5.3.3.4 Eppley Normal Incident Pyrheliometer The Eppley Model NIP is not only the principal field pyrheliometer used throughout the world but also a carefully maintained instrument used as a secondary and working standard reference instrument by many organizations. The instrument has an excellent reputation for reliability, stability of its instrument constant to aging, ease of operation, and response time.

The instrument, shown in figure 5.14, is 280 mm (11 in.) long and uses a wire-wound thermopile electroplated with copper constantan. Instruments may be ordered with a temperature compensating circuit consisting of a thermistor embedded in the thermopile heat sink. The thermopile receiver is currently reported to be coated with 3M Velvet Black™, which is no longer manufactured (Eppley Laboratory, n.d.2).

The ratio of the diameter of the NIP's field stop to the distance between the field and aperture stop (as defined by figure 5.10) is 1 to 10, giving the NIP a 5° 43′30″ field of view (twice the opening angle). The diaphragmed NIP tube is sealed with a 1-mm-thick crystal quartz window and filled with dry air (Eppley Laboratory, n.d.2).

All Model NIP pyrheliometers have been calibrated to the Absolute Radiation Scale since late 1977. Calibrations are performed outdoors against the Eppley group of standard reference pyrheliometers, their primary reference being an Eppley Model H-F Absolute Cavity SN 14195. All Model NIP calibration transfers are performed outdoors.

5.3.3.5 Absolute Cavity Radiometry

History. Absolute cavity pyrheliometers derive from early satellite cavity radiometers having fields of view of approximately 80° (Plamondon, 1969). Subsequent spacecraft cavity radiometers were used to measure the solar constant. They were manufactured by the Jet Propulsion Laboratory (JPL) and The Eppley Laboratory, Inc. Early terrestrial versions of these spacecraft radiometers were the Kendall PACRAD (Primary Absolute Cavity Radiometers), manufactured by the Eppley Laboratory and JPL (Kendall, 1969) and the Willson ACR (Active Cavity Radiometer) manufactured by JPL and California Measurements, Inc. (Willson, 1973).

State-of-the-art absolute cavity radiometers currently manufactured in the United States are the Model H-F Absolute Cavity Pyrheliometer manufactured by Eppley Laboratory, Inc. (1977) and the Mark VI Kendall Radiometer manufactured by Technical Measurements, Inc. (TMI) (Kendall and Berdahl, 1970). Also available is the PMO 6 Abolute Radiometer (Compagnie Industrielle Radioelectrique, 1980), manufactured by Compagnie Industrielle Radioelectrique (Switzerland). Two other absolute cavity radiometers that were registered participants at the International Pyrheliometer Comparisons (IPC IV and V) held in Davos, Switzerland, in October 1975 and 1980 are the Model PVS absolute cavity radiometer, designed by Professor Yu. A. Sklarov of Saratov University in the Soviet

Figure 5.14
Eppley Model NIP normal incidence pyrheliometer. Source: Eppley Laboratory (n.d.2).

Union, and the CROM-series radiometer, designed by Professor D. Crommelynck of the Royal Meteorological Institute (Belgium).

Most absolute cavity radiometers are optically and thermodynamically similar, and their terrestrial versions are now used throughout the world as primary reference instruments. They are characterized by generally identical pyrheliometer-type viewing optics and are self-calibrating and absolute in the sense of using electrical substitution for the solar-radiation-induced detector signal. However, they all generally require continuous attention while in operation.

Physical Principles and Construction. Figure 5.15 shows a schematic of a typical absolute cavity pyrheliometer. Two opposing inverted cones contain the absorber, or receiver, surfaces of the radiometer. The cones reradiate to the sensitive element of the detector, either a thermopile or a platinum resistance thermometer.

As figure 5.15 shows, the cold junctions of both the receiver and the compensating thermopiles are heat sunk (A), and the compensating cavity (rear cavity) is viewed by a heat-sunk blackbody. The receiver, or forward, cone and its cylinder are wound with an electrical heater. In the Eppley Model H-F cavity radiometer, the rear, or compensating, cavity and cone are also wound with an electrical heater. Both the PMO-series cavities employed at the World Radiation Center (WRC) in Davos and the Willson ACR use an electrically calibrated differential heat flux transducer with platinum resistance thermometers as the detector.

The entire receiver and compensation cavity pair is mounted in a collimator tube, which limits the field of view as a function of the field and aperture stops (J and K in figure 5.15). The instruments registered in the Davos and New River comparisons have the collimator angles shown in table 5.1.

Operation. Absolute cavity radiometers are often called self-calibrating radiometers because of their electrical substitution mode of operations; they may be operated in either the passive or active mode. In the passive mode, the front heater is adjusted to produce the same signal from the detector as when illuminated with the field stop open and the heater completely off. Irradiance is computed simply as the power per unit area (of receiver). In the active mode, servo-type controls are used to adjust the heater power to a level necessary to maintain a specific temperature, or heat flux level, during exposure to mechanically chopped radiation. The power is continuously monitored, and the power required to reach the higher

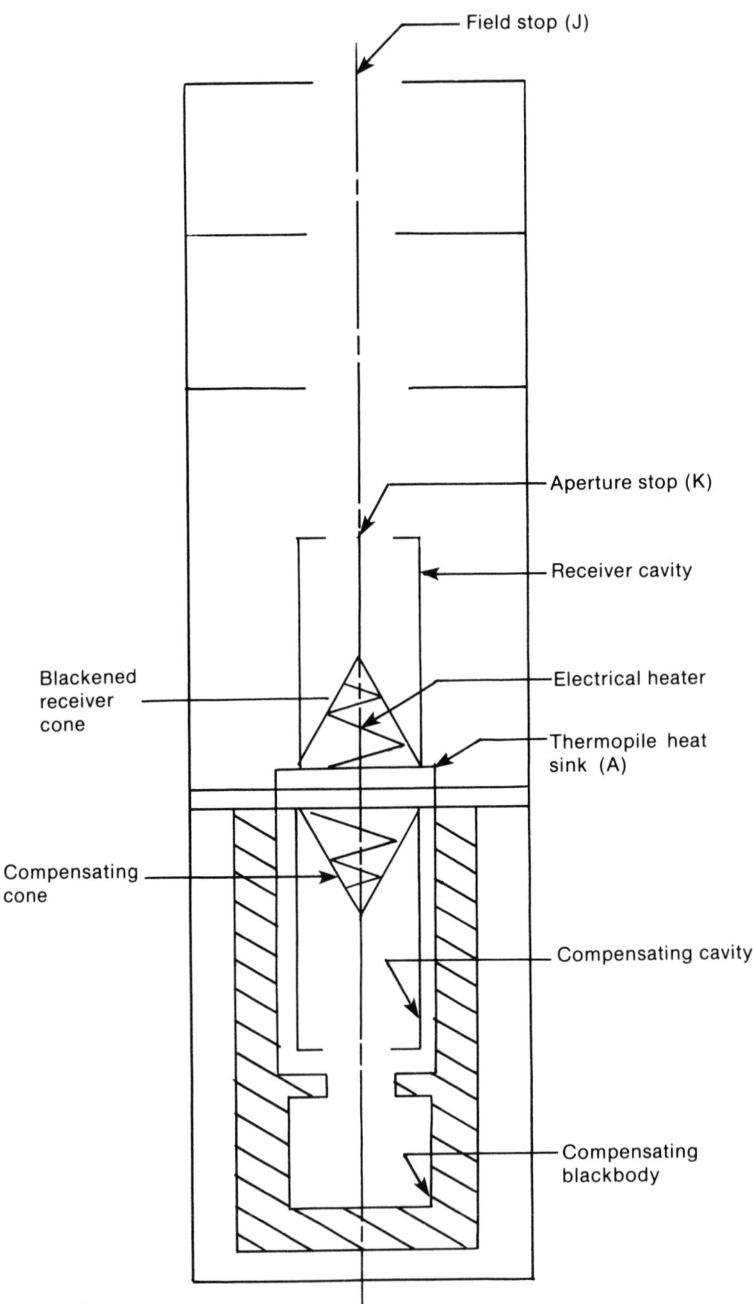

Figure 5.15
Schematic of a typical absolute cavity pyrheliometer.

Table 5.1
Optical geometry of absolute cavities used in the New River and Davos comparisons

Instrument	Model	Slope angle (deg)	Opening angle (deg)
Eppley	H-F	0.804	2.50
TMI	Mark VI	0.761	2.46
ACR	V; VI	0.939	2.49
Eppley	EPAC	0.816	2.50
PMO	PMO-2	0.741	2.43

preset level provides the power equivalent of the solar irradiance. The Willson ACR and the WRC PMO-series radiometers are operated in the active mode (Willson, 1973; CIR, 1980).

Most absolute cavity radiometers are operated in the so-called PACRAD mode. That is, they use the front heater to substitute electrically the heat flux incident on the detector. However, the Eppley H-F cavity pyrheliometer can use the rear heater to duplicate the power equivalent of the solar irradiance on the front cavity. This is done with the field stop continuously open (unshuttered) and the power to the rear heater adjusted to nullify exactly the signal developed when the front cavity is heated up. This is the so-called Ångström mode. This technique is amenable to active operation insofar as the rear heater is easily servo-controlled to nullify exactly the signal resulting from solar irradiation of the front cavity (Eppley Laboratory, 1977).

Note that all absolute cavity radiometers require careful characterization because of the small but still significant nonequivalence between electrical and radiant heating, the absorptivity of the cavity that is less than unity, uncertainties in the aperture area, a stray light factor for the pyrheliometer tube, the voltage drop in the cavity lead wires. Although these factors are all small, they all may be measured based on first principles, wherein the equation for the passive PACRAD mode is

$$H_s = kI(V - IR_c)\,, \tag{5.9}$$

where H_s is the solar irradiance, I is the measured heater current, V is the measured heater voltage, R_c is the measured lead correction resistance (IR_c is the voltage loss), and k is the various instrument constants (including the area of the receiver). For active PACRAD operation, the relationship is

$$H_s = k(I_0 V_0 - I_i V_i)\,, \tag{5.10}$$

where $I_0 V_0$ is the power applied to the heater with the cavity shaded and $I_i V_i$ is the power with the cavity irradiated.

Two mechanisms determine the constant k of an absolute cavity radiometer: characterization by the laborious measurement of the various instrument constants embodied in k, and characterization by comparison to a fully characterized instrument or group of instruments. A considerable body of opinion claims that only full characterization is appropriate and that comparisons to determine k do not serve to maintain the absolute radiation scale (WRR). Indeed, the new WMO (1983) guide classifies absolute cavity radiometers by the extent of characterization. In the Davos intercomparisons IPC IV, V, and VI and in the New River Intercomparisons of Absolute Cavity Pyrheliometers (NRIP) I–V, no attempts were made to adjust the k of participating instruments. It is nevertheless obvious that instruments exhibiting significant deviations from, for example, the mean of a large group of instruments with which they have been historically compared may very well be overdue for a complete recharacterization—that is, redefining k.

5.3.4 Pyranometers

5.3.4.1 The Eppley Model 8-48 One of the most widely used pyranometers is the Eppley Model 8-48 "black and white" pyranometer. This instrument is a direct descendant of the Eppley 50-junction bulb pyranometer, also known as the 180° pyrheliometer. It offers improvements in bulb (dome) design, absorptive and reflective paints employed, and detector design. Unlike its predecessors, the Model 8-48 has a built-in temperature compensation circuit.

The Model 8-48's dome is a precision-ground Schott WG295 hemisphere and possesses excellent transmission from 285-nm to 2,800-nm wavelengths. The detector is a radial differential thermopile with the hot junction coated with 3M Velvet Black™ and the cold junction coated with a white barium sulfate paint. The cast-aluminum body contains a sealed receptacle for desiccant; a spirit level is mounted in the instrument's base. The Model 8-48 is known as a Class II instrument in accordance with the old WMO guide (WMO, 1971) or as a Second Class instrument as defined by the new guide (WMO, 1983).

According to the manufacturer, in the spring of 1977, beginning with

Figure 5.16A
Eppley Model PSP precision spectral pyranometer. Source: Eppley Laboratory (n.d.3).

Serial Number SN 15007, Model 8-48s were modified largely to eliminate a serious tilt error in the model (Hickey, 1984). This error, which I first observed in 1975 and which many others subsequently have noted, is as great as −18% at a tilt of 45° from horizontal. The error in measuring irradiance on tilted surfaces has been ascribed to thermal convection effects in the thermopile housing that differentially affected the hot and cold junctions (Hickey, 1984). In an unpublished investigation, I determined that Model 8-48 instruments with serial numbers greater than SN 15007 do indeed exhibit little tilt effect, being at most 0.5% for the few instruments examined. However, Flowers (1985b) observed that Model 8-48 of higher serial number (SN 15896) still exhibited a tilt error of between 5% and 10%.

5.3.4.2 The Eppley Model PSP Pyranometer The Eppley Precision Spectral Pyranometer (PSP) is known as a WMO First Class pyranometer (figures 5.16A and 5.16B). Compared to the Model 8-48, its thermopile is an improved circular-wound type with blackened hot junctions and heat-sunk cold junctions; and its double concentric domes provide improved cosine acceptance, improved convective heat transfer characteristics, and

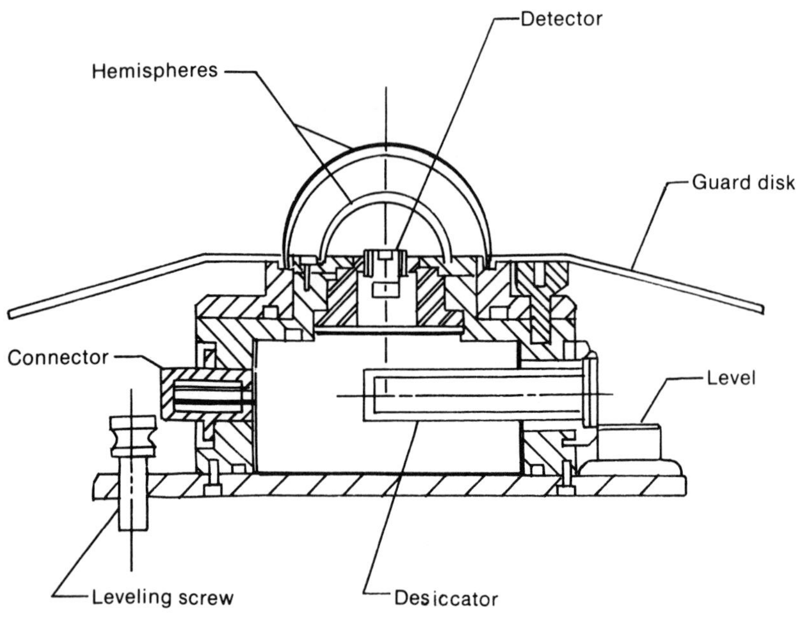

Figure 5.16B
Mechanical schematic of the Eppley Model PSP precision spectral pyranometer. Source: Eppley Laboratory (n.d.3).

the capability of spectrally isolating wide bands of solar radiation using outer domes of hemispherical filters.

The published specifications indicate that generic instruments meet the new WMO requirements for a Secondary Standard Pyranometer. However, as noted previously, Secondary Standard Pyranometers must be selected on the basis of characterization of individual instruments. As is the case with the Model 8-48, the domes are removable, but they are ground from WG295 glass and have high transmission of light for wavelengths from 285 to 2,800 nm. Parson's Black is used as the solar receiver in the PSP. Also, the cast-bronze case is provided with a desiccant chamber and a spirit level (Eppley Laboratory, n.d.3). The Model PSP has a white sunshade to preclude illumination of the case (heat sink) when the pyranometer is mounted in the horizontal position.

Calibrations of Model PSPs are performed by comparing their signal to the Eppley Dome Standard, SN 13055, in the Eppley 4-ft-diameter integrating hemisphere using a diffuse, artificial solar irradiance of about

700W/m^2 (222 Btu/ft^{2}h) (Hickey, 1981). The temperature sensitivity profile is determined for each instrument and is furnished with the calibration certificate. The working standard is in turn maintained in calibration by the shading disk method using one of a group of Model H-F absolute cavity pyrheliometers, chief among which is SN 14915.

5.3.4.3 Kipp and Zonen Models CM5 and CM6 Pyranometers The Models CM5 and CM6 are identical except that the latter possesses a white shield to shade the body of the instrument from differential illumination and a mounting base containing a spirit level. This Second Class pyranometer uses a Moll thermopile consisting of 14 constantan-Manganin junctions and a pair of Schott K5 glass domes. The transmittance spectra of K5 exhibit a cut-on[3] limit at 310 nm in the ultraviolet and a cut-off limit at 2,800 nm. A desiccant cartridge is mounted in the core of the radiometer. The Models CM5 and CM6 pyranometers are calibrated using the same protocols as those used for the new Models CM10 and CM11 instruments.

5.3.4.4 Kipp and Zonen Models CM11 Pyranometer The Model CM11 Solarimeter is a comparatively new First Class instrument and is a complete redesign of the earlier Model CM6. Like the Eppley Model PSP, the Model CM11 essentially meets the new WMO requirements for a Secondary Standard Pyranometer according to the specification sheet (Kipp and Zonen, n.d.). The Model CM11 (figure 5.17) uses a new type of constantan-Manganin thermopile, which consists of 100 junctions imprinted on an alumina ceramic substrate in a circular pattern to minimize azimuthal sensitivity. 3M Velvet Black™ is used on the receiver surfaces. Like the Model CM6, the Model CM11 has two concentric hemispherical domes ground from Schott K5 glass. Figure 5.18 is a schematic of the Model CM11. The compensation element is a second alumina substrate that along with the compensation circuit prevents temperature-induced instrument zeros.

Each instrument is characterized for cosine errors at four zenith angles in addition to calibration. Instruments with errors greater than $\pm 1\%$ at 40 and 60° zenith angles, $\pm 2\%$ at 70°, and $\pm 3\%$ at 80° are rejected (Kipp and Zonen, n.d.2).

Model CM11 pyranometers are calibrated and characterized indoors with a diaphragmed 1,000-W tungsten-halogen lamp (color temperature 3,300K). The instruments are calibrated by transferring sensitivity from a Model CM11 working standard, SN 800065, that is simultaneously illumi-

Figure 5.17
Kipp and Zonen Model CM11 pyranometer. Source: Kipp and Zonen (n.d.2).

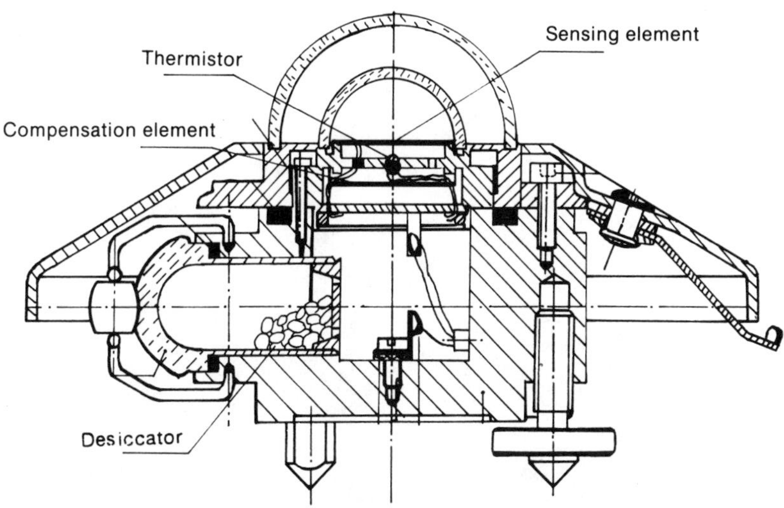

Figure 5.18
Mechanical schematic of the Kipp and Zonen Model CM11 pyranometer. Source: Kipp and Zonen (n.d.2).

nated at about 500 W/m^2 (158.6 Btu/ft^2 h) irradiance. Working Standard SN 800065 was calibrated at Davos in June 1981 against the WRC reference pyranometer [which is in turn calibrated by the shade method against the PMO-2 Absolute Cavity Radiometer (CIR, 1980)]. Two other Model CM11 reference instruments are also used at Kipp and Zonen—SN 800074 (calibrated indoors at Hamburg in April 1981 against Model CM10/SN 790057) and SN 820078 (intercompared at DSET Laboratories, Inc., in the May 1982 New River Comparisons of Pyranometers [NRCP II]).

A xenon light source is used with an optical bench to determine the cosine and azimuthal sensitivities of Model CM11 instruments (mounted vertically on the optical bench).

5.3.4.5 Phillip Schenk Model 8101 Starpyranometer The Black and White pyranometer, also known as the Dirmhirn-Sauberer pyranometer, uses a radial 72-junction nickel-constantan thermopile and a 12-segment receiver arranged as alternating black and white segments, each subtending 30° (Dirmhirn, 1958). The signal obtained is in direct relationship to the temperature difference between the black and white segments. The instrument uses a single ground-crystal glass dome manufactured from apparatus glass (Schenk, n.d.).

Figure 5.19
Swissteco Model SS-25 pyranometer. Source: Swissteco (1983).

The Starpyranometers are calibrated to the WRR by Zentralanstalt für Meteorologie in Vienna by transfer from working standard pyranometers. These reference pyranometers are themselves maintained in calibration by the shade method with one or more of the following: a Kendall (TMI) Mark VI absolute cavity pyrheliometer, an Ångström pyrheliometer, or a Linke-Feussner pyrheliometer. These are periodically compared in the pyrheliometric comparisons held in Davos. All primary calibrations are performed outdoors by the shade method.

5.3.4.6 Swissteco Model SS-25 This pyranometer is a new series from Swissteco with the thermopile sensor protected against indirect lighting. According to the manufacturer, it conforms to the old WMO specification for a Class I instrument (Swissteco, 1983). The double glass domes are ground and polished Schott WG-295. The 100-junction thermopiles are constructed from copper-constantan wire. The instrument is desiccated with dual silica-gel chambers. Figure 5.19 shows the Model SS-25.

Model SS-25 pyranometers are calibrated by reference to three working standard Model SS-25 pyranometers that are themselves calibrated by the

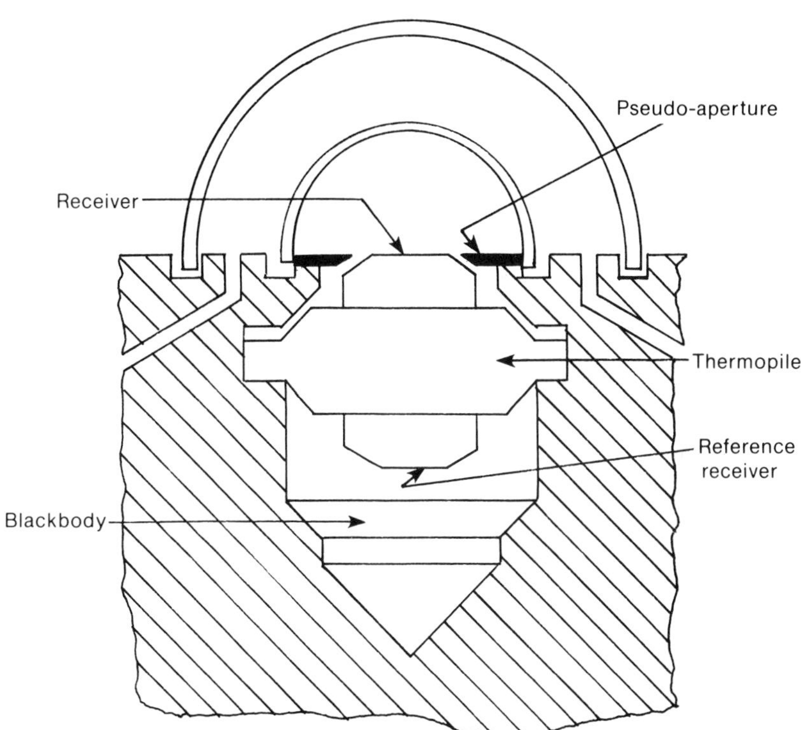

Figure 5.20
Schematic of Eppley Model SCP self-calibrating pyranometer. Source: Karoli, Hickey, and Frieden (1984).

shade method against a Linke-Feussner pyrheliometer, the traditional intermittent shade/unshade technique, and by the ratio method developed by Forgan (1983) where the comparison is against two reference pyranometers, one of which is continuously shaded. The Linke-Feussner pyrheliometer is twice yearly calibrated at the WMO Regional Center in Aspendale, Victoria, Australia.

5.3.4.7 The Eppley Model SCP Pyranometer The new Model SCP self-calibrating pyranometer is thoroughly discussed both in principle and in detail in a recent paper by Karoli, Hickey, and Frieden (1984). Also, I have worked with an early serial number of this new series; my results are noted in section 5.7.2 (figure 5.48). The Model SCP schematic is shown in figure 5.20 and consists of double glass domes enclosing an inverted conical receiver that thermally irradiates a plated copper-constantan thermopile. Like the Model H-F cavity pyrheliometer discussed in section 3.3, the Model SCP pyranometer has a reference receiver. Both receivers have electrical heater wires wound on them to provide the self-calibrating aspect of the radiometer by the electrical substitution method. A precision Invar[4] aperture is placed over the receiving cavity. The major problem with the concept of a cavity pyranometer is the conduction and convection losses between the cavity and aperture that prevent the instrument from having a complete 180° field of view. In this respect, the instrument described by figure 5.20 is not a true, absolute cavity pyranometer and therefore is designated as a "self-calibrating pyranometer (SCP) with shaped receivers to provide good cosine response"; that is, it has an essentially full 180° field of view.

5.3.4.8 Other Thermal Pyranometers Although I have obtained information on the Trickett-Proctor pyranometer (Proctor and Trickett, 1982), it is not currently manufactured and no additional technical information is available. The instrument consists of a single glass dome enclosing a new sensor that is a thermoelectric generator in which the solar flux incident on a black chrome receiver induces a potential difference across a p-n semiconductor junction proportional to the solar flux.

Another instrument is the Model NP-42 Eko Black and White Star pyranometer manufactured in Japan (EKO Instruments Trading Co., n.d.). This single-dome thermopile instrument possesses a 39-junction copper-constantan thermopile and six black and white sectors. Like other black and white pyranometers, the hot junction receivers are black (Parson's) and the cold junction receivers are white (barium sulfate paint). A ventilated version (with fan) is designated as Model NP-43F.

Still another thermopile pyranometer is the comparatively new Hollis Model MR-5, which uses a selective black receiver coating that presumably reflects unwanted infrared radiation. The Model MR-5 is temperature-compensated from −20° to +40°C and, according to the Hollis technical bulletin [MR-3(A)], has a cosine response of ±2% from 0° to 70° (Hollis Geosystems Corp., 1978).

5.3.4.9 Solid-State Pyranometers The LiCor Model LI-200SB silicon photodiode pyranometer and the Hollis Model TR-201 silicon photovoltaic insolameter are the two most widely used solid-state pyranometers.

The Model LI-200SB pyranometer weighs only 28 g, has an aluminum housing and an acrylic eye, and is fully cosine corrected with an external translucent acrylic diffusing window (LiCor, 1982a). LiCor pyranometers are calibrated against an Eppley Model PSP pyranometer.

The Hollis Model MR-5 photovoltaic pyranometer is based on a temperature-compensated silicon solar cell and is cosine-corrected with an external diffusing window. The case is provided with a built-in spirit level and microcoulombmeter to continuously integrate the measured irradiance over time. The manufacturers claim a 3% resolution (Hollis Geosystems Corp., 1978).

5.3.4.10 Summary of Pyranometer Characteristics The most often specified and quoted characteristics of pyranometer are compiled in table 5.2. The values given are a combination of those quoted by manufacturers in product literature, those quoted in correspondence with the author, and those based on laboratory and field experience as reported by various workers. Table 5.2 presents the most reliable and up-to-date data available.

5.4 Instrumentation for Measuring Solar Spectral Irradiance

5.4.1 Wide-Band Spectral Measurements

Many instruments measuring total solar radiation have optional cut-off filters to isolate wide spectral bands. In fact, wide-band spectral measurements can be made more easily than scanning spectroradiometric measurements and have the advantage of being readily adapted to continuous monitoring.

Perhaps the most widely used sensor for filter measurements is the Eppley Model NIP pyrheliometer, which, when equipped with the standard four-aperture filter wheel, can resolve direct beam radiation into three bands using Schott OG1, RG2, and RG8 filters (leaving one aperture for the total measurement). The new designations for these filters are, respectively, Schott OG530 (orange), RG630 (red), and RG695 (dark red), having cut-offs at 530, 630, and 695 nm, respectively. Other cut-off filters may be used to give different band selectivity. Spectral transmittance curves of typically used filters are shown in figure 5.21.

Similarly, glass domes ground from these Schott filters, as well as other types, may be used to isolate various bands of the global, or hemispherical, solar spectrum. These filters have been used as the outer dome on such pyranometers as the Eppley Model PSP, Kipp and Zonen Models CM6

Table 5.2
Summary of pyranometer characteristics

Manufacturer	Model	Cosine response (% deviation to 70°)	Temperature sensitivity, −20°/+40°C (%)	Tilt error (%)	Response time 1/e (s)	Spectral range (nm)	Linearity with intensity to 1,000 W m̄	Stability change (% per year)	Sensitivity
Eppley Labs	8-48	±2[a]	±1.5	<0.5[b]	5	285–2,800	0.99		11–12
Eppley Labs	PSP	1[c]	±1	<0.5	1	285–2,800	0.995	[d]	8–10
Eppley Labs	SCP	0	[e]	0	~1	<285–>2,800	>0.995	[f]	0.1–0.15
Kipp and Zonen	CM6	[f]	−5	[f]	3	305–2,800	0.985	[f]	11–12
Kipp and Zonen	CM11	<3	±1	<0.5	5	305–2,800	0.995	[f]	5–6
Phillip Schenk	T-8101	±2	~2	1	20	300–3,000	0.99	~3	15
Swissteco	SS-85	[f]	~1	[f]	5	280–2,900	[f]	[f]	16
Eko	MS-42	~5	~3.5	1.5	[f]	[f]	[f]	[f]	7–8
LiCor	200SB	~3	[f]	0	~0.1	400–1,100	>0.99	~1	8–10^{-3} μA
Hollis Obs.	TR-201	1.5	~1.5	0	~0.1	400–1,100	0.99	<2	[g]

a. Greater than 2% measured.
b. Value quoted for all instruments of or greater than SN 15007 for prior SNs.
c. Can be as great as 8% on aged receivers.
d. As great as 2%/year for higher SNs (in southern climatic regions; continuously exposed).
e. Not applicable.
f. Unknown.
g. Displays integrated irradiance.

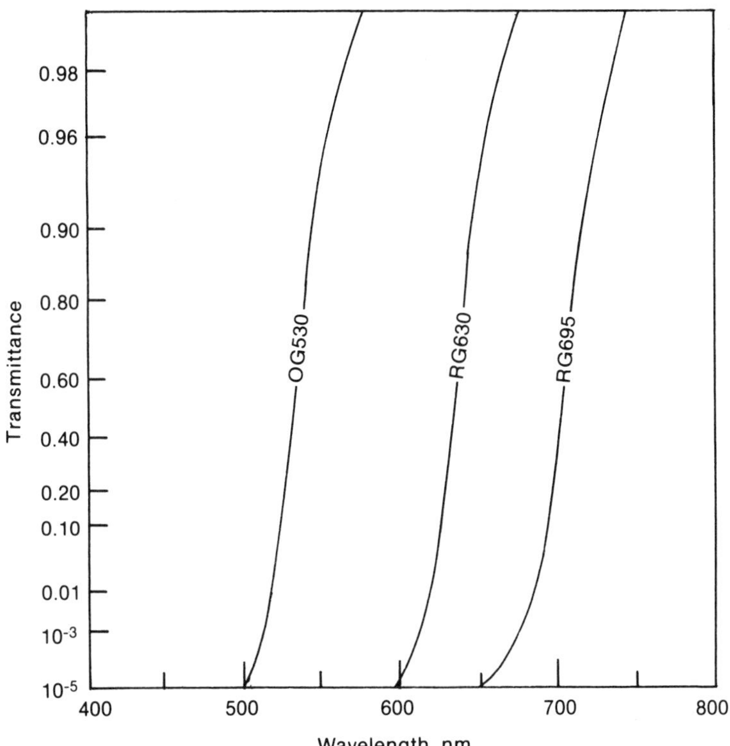

Figure 5.21
Spectral transmittance of Schott OG530, RG630, and RG695 cut-off filters typically used in spectral solar radiometry using pyranometers and pyrheliometers.

and CM11, and Swissteco Model SS-25. The WMO Guides (WMO, 1971, 1983) and the IGY (International Geophysical Year) Instruction Manual should be consulted when using broad-band filters for the spectral isolation of solar radiation. Crommelynck and Latimer (1980) published an excellent review of broad-band filter radiometry.

5.4.2 Infrared Measurements

Terrestrial and sky infrared measurements are important in at least three aspects of solar energy: use in passive building energy control, assessing active solar systems, and designing large solar simulators for testing solar collectors indoors. The last requirement relates to the ability to ensure that solar simulator results correlate well with outdoor tests. The thermal signature of both terrestrial surfaces and the atmosphere relate to the

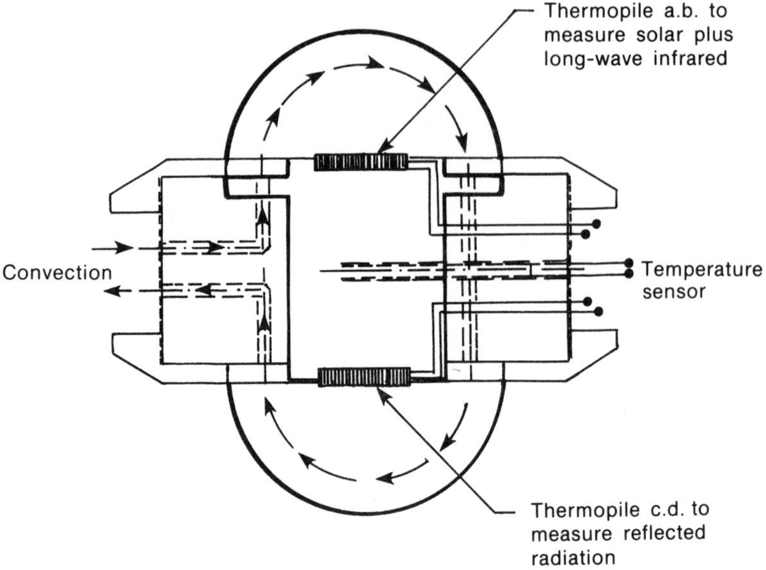

Figure 5.22
Schematic of Philipp Schenk Model 8111 pyrradiometer. Source: Schenk (n.d.).

absolute temperature of the surfaces (and the bulk for semitransparent surfaces) and to the spectral emittance characteristics of the constituents.

These radiometers are of two principal types—pyrradiometers that measure radiation from the sun and sky (or visible, near infrared, and far infrared) and pyrgeometers that measure only long-wavelength infrared radiation. A net pyrradiometer possesses two essentially identical sensors mounted back to back and measures in 4π sr as a net exchange between two hemispherical radiation quantities. Net pyrradiometers are of two types—unshielded plates with two exposed receiving and radiating surfaces and shielded, usually hemispherically, windowed sensors.

The best known unshielded net pyrradiometer is of the Gier Dunkel[5] design (manufactured by Teledyne Geotech). Net pyrradiometers of the shielded type are typified by the Schenk Model 8111 and the Swissteco Type S-1 instruments. The Schenk Model 8111 pyrradiometer schematic is shown in figure 5.22, including the polyethylene[6] domes encasing a 66-junction thermopile. A photograph of the Swissteco Type S-1 net pyrradiometer is shown in figure 5.23. This instrument is provided with nitrogen gas ventilation and inside inflation. These are but 2 of some 15 available commercial instruments. Martin and Anson (1980) give an excellent discus-

Figure 5.23
Swissteco Model S-1 net pyrradiometer. Source: Swissteco (1983).

sion, commissioned by the International Energy Agency (IEA), of net pyrradiometry and a complete summary of their characteristics.

The only pyrgeometer currently manufactured is the Eppley Precision IR Radiometer, a development derived from the well-known Model PSP (Eppley Laboratory, n.d.4). This instrument is used unidirectionally only and possesses the same type of temperature-compensated, wire-wound thermopile used in the Model PSP (see figure 5.24). Because of the need to eliminate solar shortwave radiation to measure daytime sky infrared radiation, a tellurium and zinc selendide-coated silicon hemisphere is used. The coated window cuts on at about 4 μm and is transparent to 50 μm, the long-wave cut-off of the silicon hemisphere. I use an Eppley pyrgeometer at my solar collector test facility in New River, AZ, to monitor the long-wave contribution to total absorbed radiation—espcially in the solar simulator facility.

5.4.3 Narrow-Band Filter Radiometers

A number of filter and spectral radiometers based on Schott- and interference-type filters have been reported. Stair (1953) discussed the general

Figure 5.24
Eppley Model PIR precision infrared radiometer (a pyrgeometer). Source: Eppley Laboratory (n.d.4.).

application of filter radiometry, and numerous papers from the Eppley Laboratory (Ångström and Drummond, 1961; Drummond and Roche, 1965; Drummond and Hickey, 1968) offer an excellent resource for this technique. Drummond and Hickey (1968) describe the Eppley Multichannel Spectroradiometer, which was designed for high-altitude solar spectral measurements. This radiometer consisted of eight interference filters peaking at 300, 360, 405, 430, 457, 485, 540, and 604 nm, four cut-off filters (GG338, OG562, RG633, and RG690), and a total transmittance filter. A special 13-channel variation of this instrument has been used in terrestrial measurements. More recently, Shaw, Reagan, and Herman (1973) presented the results of determining atmospheric extinction using a multiple filter radiometer.

Actually, filter radiometers are usually constructed for very specific purposes and are generally not commercial off-the-shelf items. An exception is the Eppley Model TUVR ultraviolet radiometer that has been used in several networks and in the recently deactiviated U.S. Department of

Energy's Solar Energy Meteorological Research and Training Sites Program. Also, I use the Eppley Model TUVR total ultraviolet radiometer to measure continuously total ultraviolet radiation in a mini-network aimed at developing ultraviolet sky models for various climatic regions of the United States. The Model TUVR, shown in figure 5.25, consists of a selenium barrier-layer photoelectric cell and a pair of cut-on and cut-off filters that isolates the ultraviolet portion of the sun's wavelength spectrum from 295 nm to 385 nm. A beveled opaque quartz[7] diffusing window is used both to reduce the flux to acceptable levels and to provide good cosine response. Unfortunately, as will be discussed later, the calibration of ultraviolet radiometers is very difficult, and the Model TUVR is no exception. I have observed differences in measured ultraviolet irradiance as great as 8% in new instruments checked side by side. Furthermore, four out of five instruments have exhibited exposure-induced changes of between 10% and 15% per year, prompting triannual recalibration in our laboratories. This is presumably caused by either a loss of inert gas from the selenium barrier-layer cell or solarization of the quartz diffuser. Replacement diffusing windows are available from the Eppley Laboratory (Hickey 1984).

The only other commercially available ultraviolet radiometer I know of is the Model MS-140 UV Pyranometer manufactured by Eko (Japan). Light passing through the quartz glass hemisphere is first diffused and filtered to eliminate all radiation except that in the band from 300 μm to 400 μm. The MS-140 uses a silicon photodiode detector (Eko Instruments Trading Co., n.d.).

I recently constructed 10 narrow-band ultraviolet radiometers for isolation and continuous measurement of specific regions of ultraviolet radiation. The radiometers will be used in further refining the ultraviolet sky models mentioned earlier in this section. (The sky models are an integral element in developing prediction methods for exposure-induced material deterioration.) The radiometer is based on an operational amplifier/photodiode, suitable solar blind filter pairs and a Teflon®-coated cosine-correcting diffusing window. Figure 5.26 shows one of the radiometers. The DSET Model NBUV narrow-band ultraviolet radiometer is currently being exposure-tested for durability of the instrument constant—that is, for stability of the optical components.

5.4.4 Scanning Solar Spectroradiometer

5.4.4.1 General Considerations Precise scanning solar spectroradiometry is always difficult and costly. Because of the special design of precision

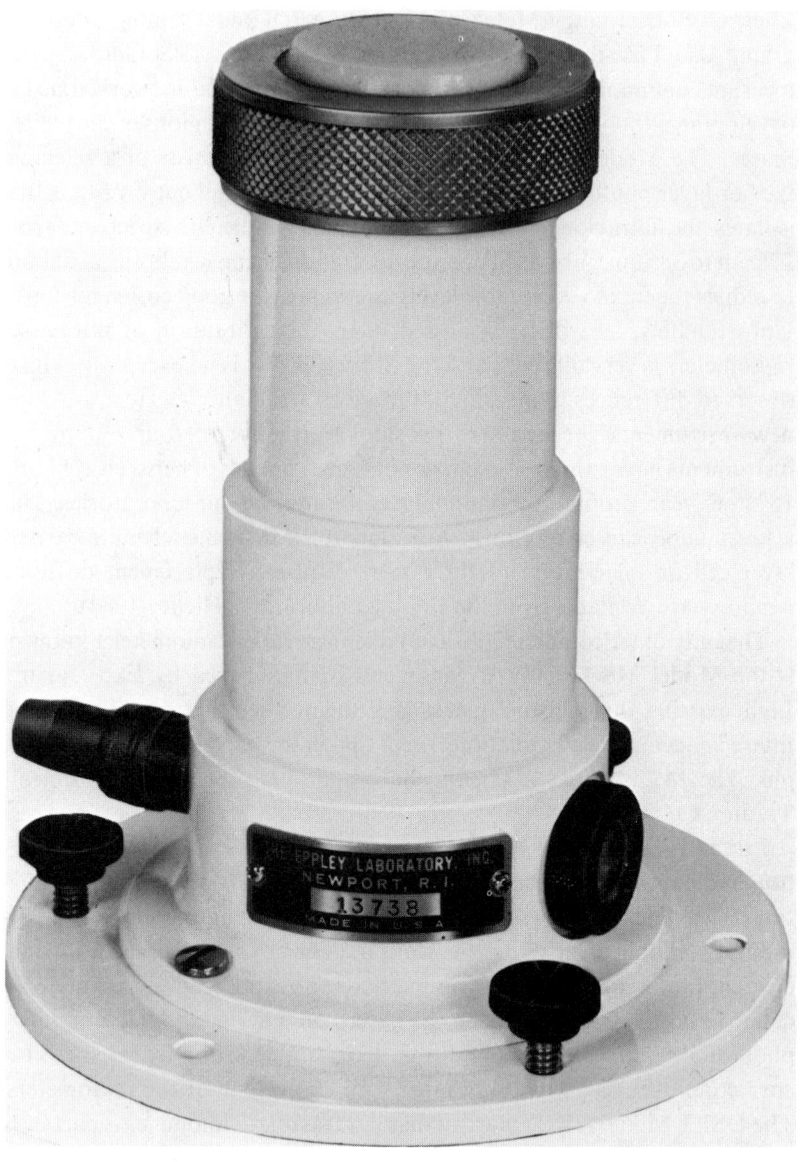

Figure 5.25
Eppley Model TUVR total ultraviolet radiometer. Source: Hickey (1984).

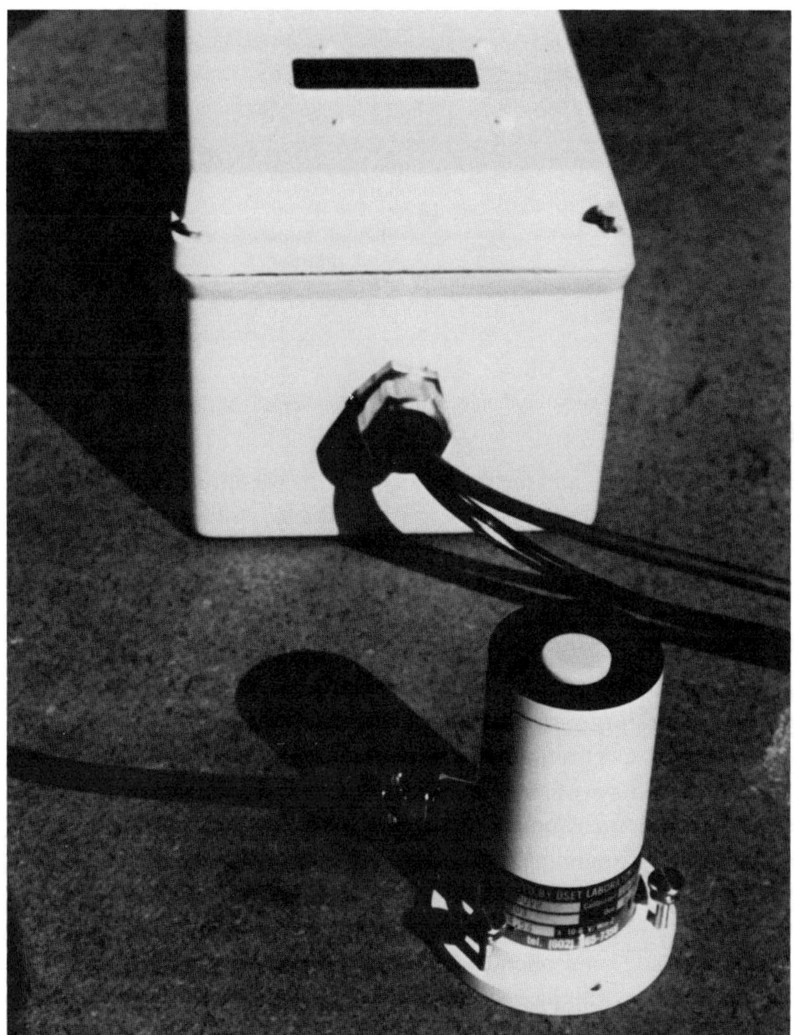

Figure 5.26
DSET Model NBUV narrow-band ultraviolet radiometer.

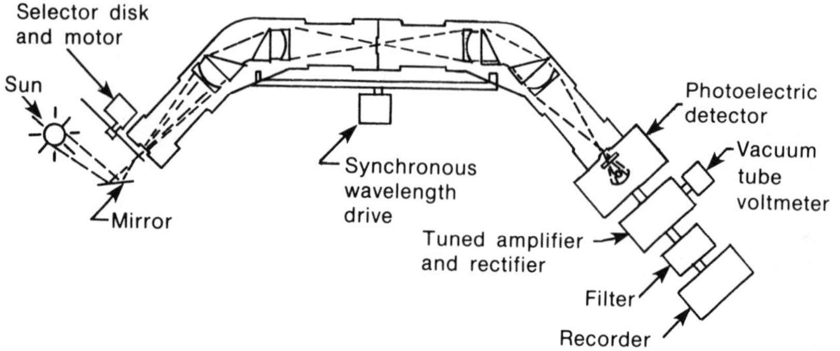

Figure 5.27
Schematic of the Stair spectroradiometer. Source: Stair (1951).

spectroradiometers and their need for constant attention during operation, it is understandable that commercially packaged, full-spectrum scanning solar spectroradiometers are not available as off-the-shelf instruments. One of the principal problems in ultraviolet spectral radiometry is the lack of good transfer standards of irradiance with adequate intensity to compensate for low signal noise ratio of available detectors below 330 nm. Another is the maintenance of wavelength calibration outside of the calibration laboratory—especially at ambient temperatures above 35°C (95°F). However, for spectral radiometry work relating to many fields of study, these disadvantages are overshadowed by the advantages of scanning spectroradiometry. Compared to most multichannel filter radiometers, scanning spectroradiometry can characterize both Fraunhofer and atmospheric absorption bands much more accurately; can determine precisely the ozone cut-off region; can differentiate easily the relationship between direct beam and hemispherical (global) spectral radiation; and, finally, can interrogate, for example, stored 1-nm band-pass data to provide a wide range of spectral analyses even months and years later.

Modern solar spectral radiometry owes much to the pioneering terrestrial measurements of Stair (1951, 1952), who in his early work used a Farrand double quartz prism monochromator and a heliostat to measure direct solar ultraviolet from 299 to 535 nm. Figure 5.27 is a schematic of his instrument. The 1952 data were obtained at an altitude of 3.4 km (11,000 ft) at AM1.40 to provide Langley regression plots to determine atmospheric ozone. His work laid some of the groundwork for dealing with the problems peculiar to field measurement of solar spectral radiation. In 1953,

Stair and his coworkers (1954, 1956) constructed a spectroradiometer based on the well-known (but no longer manufactured) Leiss double quartz prism monochromator. In these studies they used a partially automated monochromator for the first time. Although they made total solar spectral measurements to the 2,500-nm wavelength, the principal objective was to determine atmospheric transmittance and ozone (computed to be 0.21 cm NTP in mid-July from Sacramento Peak in New Mexico).

In 1959, Dunkelman and Scolnik used a Leiss monochromator converted to automatic scanning and equipped with a siderostat to focus direct sunlight onto the entrance slits. Their measurements were made from Mount Lemmon near Tucson, AZ., in October and agreed closely with the ultraviolet and visible measurements of Stair and Johnston (1956).

The currently accepted absolute values of the Fraunhofer lines and the terrestrially derived solar constant are based on the extensive work of Labs and Neckel (1962a, b, 1967, 1968, 1970) at 3.6 km (12,000 ft) on the Jungfrau near Grindelwald, Switzerland. This work was performed using a pair of Dzerny-Turner grating monochromators employing Bausch and Lomb gratings ruled at 600 l/mm, blazed at 5,000 Å, and used in the first order. Generally, the Labs and Neckel data disagree with that of Stair and Johnston and of Dunkelman and Scolnik, which Labs and Neckel (1968) attribute largely to calibration errors in both. These errors relate to use of incorrect comparison lamp intensities in both, and additionally to calibration techniques in the Stair and Johnston spectra.

In 1977, I constructed a scanning spectroradiometer, also based on the Leiss quartz prism monochromator, to obtain precise field measurements of the direct beam and hemispherical (global) solar spectrum at or near sea level (Zerlaut, 1981b; Zerlaut and Maybee, 1983). More recently, the Baird Corporation designed and constructed a two-monochromator (grating) spectroradiometer for the Resource Assessment and Instrumentation Branch of the Solar Energy Research Institute (Bird and Hulstrom, 1981; Kliman and Eldering, 1981). Both of these instruments are discussed later.

Scanning solar spectroradiometry is needed in several areas of solar energy development: ultraviolet for determining accurate durability characteristics of materials exposed to long-term weathering, visible and near infrared for determining the operational efficiency of flat-plate and concentrating photovoltaic systems, and overall solar spectral irradiance for assessing efficiency characteristics of concentrating solar devices under urban and near-urban environments.

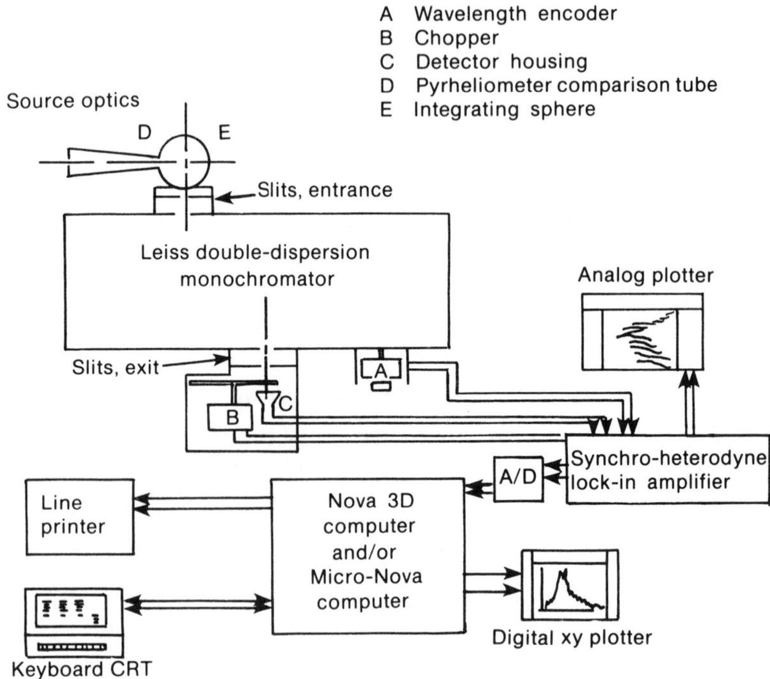

Figure 5.28
System schematic of DSET's Model SSR solar scanning spectroradiometer.

5.4.4.2 The DSET Scanning Solar Spectroradiometer (SSR) Constructed in 1977–1979, the DSET SSR is based on a Leiss Model 9100 double-dispersion, quartz prism monochromator using a silicon photodiode and lead sulfide detectors. Figure 5.28 shows a system schematic. The criteria used in designing the SSR were

- full spectrum capability from 280 to 2,500 nm,
- primary area of interest UV/VIS to 1,100 nm with greatest spectral resolution in the ultraviolet,
- grating interchange deemed not desirable forcing a quartz prism single dispersing element,
- capability of field utilization and transportability,
- one-person operability, and
- required computer-analyzed spectra.

Partial financial support was received from the Arizona Solar Energy Commission in support of measurements at Sky Harbor International Airport (Zerlaut and Maybee, 1983) and from JPL of the California Institute of Technology in support of software design (Zerlaut et al., 1980).

Source Optics: Solar radiation is collected by an integrating-sphere and a pyrheliometer-comparison-tube combination that furnishes nonpolarized light to the monochromator entrance slit. The entire optical instrument, along with the lock-in amplifiers and other electronics, is mounted on a follow-the-sun azimuth tracker. The pyrheliometer tube is attached to the entrance aperture of a 6-in.-diameter integrating sphere. The sphere, with or without the pyrheliometer comparison tube, is mounted to rotate precisely about the exit aperture to provide solar altitude tracking. The sphere is painted with five coats of Eastman 6080 barium sulfate paint.

The pyrheliometer comparison, or collimator, tube is 15 in. long and has six apertures, two of which are the precision field aperture and the limiting aperture (which is the knife-edge aperture to the sphere). The other apertures are stray light baffles. The geometry is 10:1 to give optical characteristics identical to the Eppley Model NIP (with a field of view of 5.72°).

With the collimator tube removed, the instrument measures the spectral distribution of global solar radiation at any angle from the horizontal (including 0° horizontal) and at any azimuth orientation.

Monochromator/Detector-Chopper Assembly: The Leiss Model 9100 double Infrasil® prism monochromator was chosen because of its high resolution, spectral purity, low stray light, and high transmittance (especially in the ultraviolet). The wavelength is computer operated using a stepping motor and a 100-segment sector wheel. Ultraviolet and visible irradiances are detected with an ultraviolet-enhanced EG&G Model UV-444B silicon photodiode with a current sensitive preamplifier. The infrared sensor is a biased, thermoelectrically cooled Optoelectronics lead sulfide detector (Model OTC-225-85-TC). The exit signal is chopped at 260 Hz with an EG&G Model 194A chopper connected to a Princeton Applied Research (PAR) Model 186A Syncro-Het® lock-in amplifier.

Signal Conditioning/Computer Package: Analog-to-digital signals are converted by a dual slope integration technique using a $3\frac{1}{2}$ digit DataTech Model 5313 panel meter. The digital panel meter is interconnected between the lock-in amplifier and the TTL (transistor-transistor logic) input/output board of a field-packaged Data General MicroNova minicomputer. As

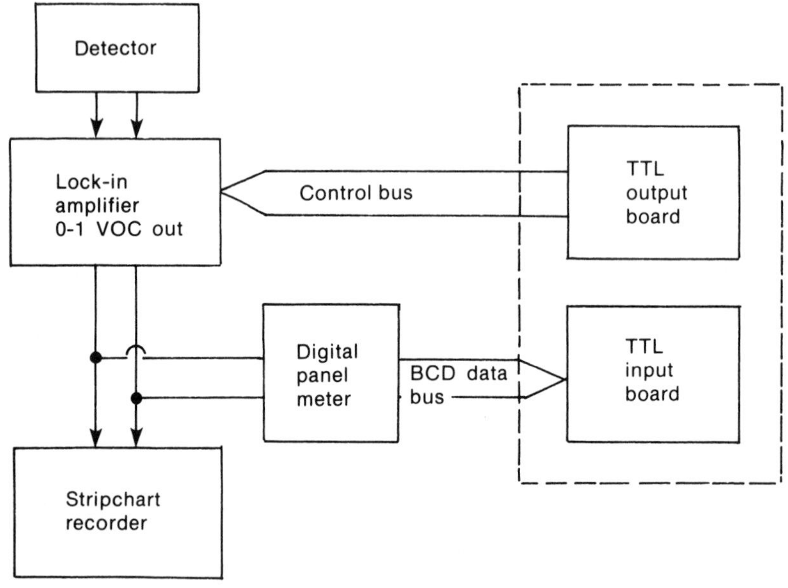

Figure 5.29
System block diagram for DSET Model SSR spectroradiometer.

shown in the system interconnect block diagram of figure 5.29, the lock-in amplifier and the event counting sector wheel are under control of the computer through the TTL output board.

Calibration and Computation: Wavelength is calibrated using a combination of standard lamps (Oriel Corporation Model 6053), didymium and holmium oxide filters, trichlorobenzene, internal Fraunhofer absorption assignments in the solar ultraviolet, and a 0.01-W neon/helium 632.8-nm laser for field checking. From the wavelength assignments in each electronic "event," a computer file plot is maintained, and wavelengths are continuously recorded by interpolation. Intensity calibration is performed at each event traceable to the WRR (IPC IV) from 280 nm to 1,600 nm (and IPS 1956 from 1,600 nm to 2,500 nm). This is done by measuring the output of an NBS 1,000-W quartz iodine lamp standard of spectral irradiance. All intensity calibrations are performed with the collimator tube removed from the integrating sphere.

All spectral irradiance data are stored on dual flexible disks. Computer-drawn spectral plots include full spectrum and regional fine structure plots

Figure 5.30
Use of DSET Model SSR spectroradiometer in the field.

and are done with a Houston COMPLOT digital plotter. Reports of 1 and 10 nm are also printed.

Spectra have been obtained with the SSR at New River, AZ; Compton, CA; Huntsville, AL; Homestead, FL; and at 9,500 ft at Humphreys Peak near Flagstaff, AZ. Hundreds of spectra have been taken at New River over a 3-year period, the analysis of which has permitted the development of a direct-beam ultraviolet histogram for that climatic region (Zerlaut and Robbins, 1984). The SSR, shown in field operation in figure 5.30, was used to obtain the spectra presented in Figures 5.4, 5.5, and 5.6.

5.4.4.3 The SERI Solar Spectroradiometer The highly sophisticated Bird/SERI spectroradiometer is similar in some respects to the DSET instrument. The SERI instrument uses integrating sphere source optics, a lead sulfide detector for the infrared, and a computer-controlled wavelength drive. It uses two Czerny-Turner type grating monochromators, each of which views the integrating sphere wall, one for visible and the other for infrared; a cooled gallium arsenide photomultiplier tube for the visible wavelengths; a 1° field of view collimator for direct-beam spectral measurements; and continuous dynamic spectral and wavelength field calibration. The order-separating filters located at the exit slit of the monochromators are controlled by a Digital Equipment Company Model PDP 11V03-L computer. Another feature of the SERI spectroradiometer is its ability to measure spectral diffuse irradiance directly; a shade disk is used for this measurement. Altazimuth tracking to an accuracy of 0.02° is necessary because of the 1° field of view measurement of the spectral beam component.

Unique to the spectroradiometer is a spherical chopper that sequentially activates the entrance aperture to the sphere, the internal irradiance reference, and the internal wavelength reference. A dark measurement is also made to faciliate signal processing (Kliman and Eldering, 1981). Schematics of the spectroradiometer are shown in figures 5.31 and 5.32.

5.4.4.4 LiCor Model LI-1800 Portable Sepctroradiometer The Model LI-1800 is one of the few prepackaged spectroradiometers available and is unique in its capabilities—precision high-resolution measurements, portability, graphical output capabilities, and advanced microcomputer control. The Model LI-1800 is available in two wavelength options—300–850 nm with a 4-nm band pass and 300–1,100 nm with a 6-nm band pass (LiCor, 1982b). The manufacturer quotes a wavelength accuracy of ± 2 nm, a scan rate of 30 nm/s, and an intensity calibration accuracy of $\pm 10\%$ in the ultraviolet and 3–5% in the remaining spectral regions. The Model LI-1800 is based on a holographic grating monochromator with a silicon photodiode detector. The Model LI-1800 uses a seven-filter wheel order separator. RS-232 interfaces permit terminal control and compatible printing and plotting.

5.5 Maintaining the World Radiation Reference Scale

5.5.1 Radiation Measurement Scales Prior to 1977

The original radiation scale was adopted in 1905 as the Ångström scale (known as ÅS 1905). This scale was based on the Ångström electrical compensation pyrheliometer described in section 5.3.3.2. Measurements made with this scale experience a 2% negative error because the surface

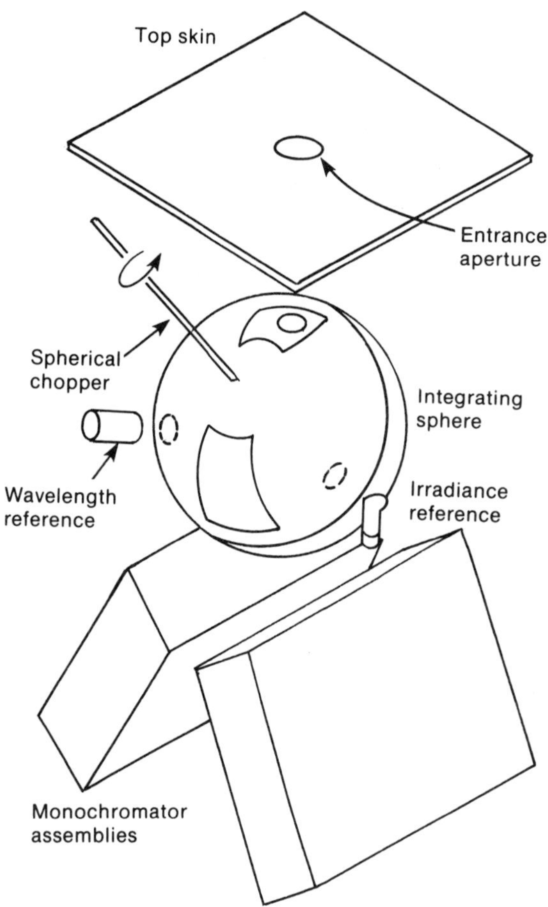

Figure 5.31
Schematic of SERI SSR spectroradiometer showing the integrating sphere in relation to the two monochromators. Source: Kliman and Eldering (1981).

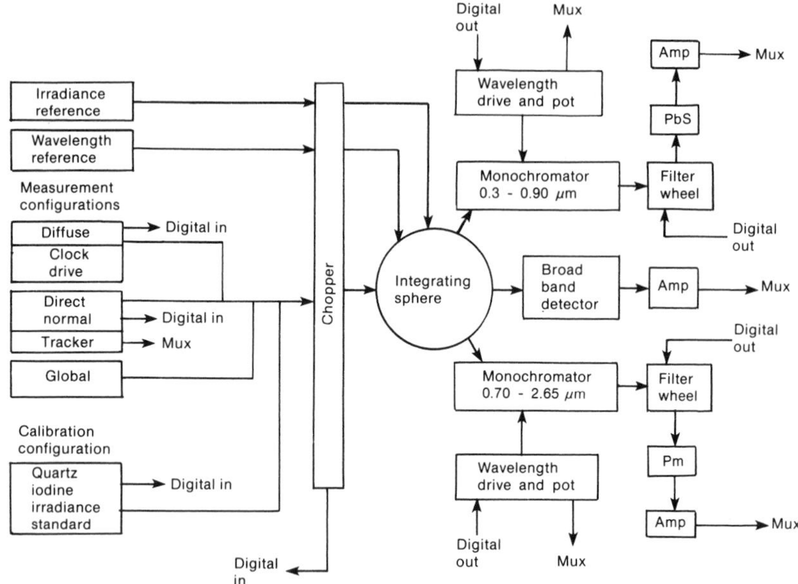

Figure 5.32
The sensor block diagram of the SERI SSR spectroradiometer. Source: Kliman and Eldering (1981).

area of the manganin strip is not uniformly irradiated and the electrically heated shaded strip loses heat uniformly (Iqbal, 1983).

In 1913, the Smithsonian scale (SS 1913) was developed based largely on intercomparisons including the Abbott single-chamber water flow pyrheliometer. In 1932, use of the New Abbott double-chamber pyrheliometer showed that the SS 1913 scale was 2.5–3% too high (Abbott and Aldrich, 1932). Thus the ÅS 1905 and SS 1913 scales differed by nearly 5%.

In 1956 at Davos, Switzerland, the International Radiation Commission adopted the International Pyrheliometric Scale (IPS 1956) as a compromise between the Ångström and Smithsonian scales of 1905 and 1913. The resulting IPS 1956 scale was then about 0.4% greater than AS 1905 and about 4.4% below SS 1913 (WMO, 1983).

The accuracy of the IPS 1956 scale was subsequently called into serious question after researchers determined that the Ångström pyrheliometers used to maintain the scale differed among themselves by as much as 3% (Thekaehara, 1966; Drummond, 1966). The IPS 1956 scale has now been replaced by the WRR as a result of the 1975 and 1980[8] comparison of absolute cavity pyrheliometers.

5.5.2 World Meteorological Organization and Physikalisch-Meteorologisches Observatorium (Davos)

The creation of the United Nations provided a new framework for international collaboration in various areas, including technical and scientific fields. One such collaboration was the adoption of the World Meteorological Convention in Washington, D.C., in 1947, establishing a new organization founded on a formal agreement among governments. This convention was ratified by many countries, and in 1951, the new WMO replaced the former organization. In December 1951, the General Assembly of the United Nations approved an agreement between the United Nations and WMO recognizing the latter as a specialized agency of the United Nations (Johnson, 1984). WMO's purpose is to establish uniform procedures for measuring meteorological parameters, to provide for the calibration of instrumentation within the worldwide network of solar radiation measurements, to provide for and host international intercomparisons, to develop special instrumentation (such as absolute cavity radiometers and sun photometers), and to measure the spectral aspects of the atomsphere and its relation to worldwide weather patterns.

WMO has two World Radiation Centers (WRCs), located at Davos, Switzerland, and Leningrad, USSR, and numerous regional centers. Region IV, the region that includes the United States, has its regional centers in Toronto, Ontario, Canada, and Boulder, CO. (The world and regional radiation centers are listed in the appendix.)

The two WRCs are equipped with at least three of the most accurate primary standard pyrheliometers, as well as auxiliary radiometers and measuring equipment for the maintenance of the WRR scale of 1975. In addition, they serve as centers of interregional and international comparisons of radiation, closely follow all developments in measuring solar radiation, have scientists with wide experience in solar radiation, and train specialists in solar radiation (WMO, 1971, 1983).

WRC in Davos is operated for WMO by the Physikalisch-Meteorologisches Observatorium (PMOD). As noted previously, one responsibility of PMOD is to maintain the internationally accepted radiation scale, currently defined by WRR using a group of absolute cavity radiometers known as the World Standard Group (WSG). PMOD is also responsible for making these radiometers and the scientific facilities of WRC available for comparison with the primary standard radiometers of the regional centers, of the official meteorological institutes of various member countries, and of various other worldwide scientific laboratories working in solar radiometry. Such comparisons are accomplished primarily by convening the International Pyrheliometer Comparisons at Davos about once every 5 years.

5.5.3 International Comparison of Regional Pyrheliometers and Absolute Cavity Radiometers

5.5.3.1 The Fourth IPC (1975) IPC IV was held in October 1975 at WRC at PMOD in Davos. Over a 5-day period, 34 Ångström and 1 Abbott (silver disk) pyrheliometers, 1 Linke-Feussner and 1 Eppley Model NIP pyrheliometer, and 11 absolute cavity radiometers were intercompared. The 11 absolute cavity radiometers first compared at IPC IV (Anon., 1977), listed in table 5.3, are essential to the maintenance of the absolute WRR scale. (The terrestrial version of the Eppley Model H-F dual cavity radiometer was not available for IPC IV.)

Because PACRAD III, manufactured by J. Kendall of the Jet Propulsion Laboratory (JPL) of the California Institute of Technology, was observed to be very stable and was the only absolute cavity radiometer compared in IPC III (1970), it was chosen as the reference instrument for IPC IV. Table 5.4 presents the results of the nearly 2,000 instantaneous observations. The results are given as the mean ratio of the irradiance measured by the stated instrument to that measured by PACRAD III for the number of observations shown.

The weighted average of all instruments for all observations was 1.0017 with a standard error of only 0.18% among the instruments. The excellent agreement among all the cavity radiometers thus justified the establishment of the absolute WRR scale as a result of IPC IV.

The absolute WRR scale is based on the fact that the cavity radiometers employed in IPC IV (and subsequent comparisons) use Standard International units of measure for length, temperature, and electrical measurements of the various official national standards institutes, associations, and organizations of the country of manufacture. These intercomparisons establish the relationships among all such instruments, which, when in agreement, establish the scale defined by the intercomparisons. Traceability to the absolute WRR scale is achieved and maintained either by periodic intercomparison or by substantial comparison to an absolute cavity radi-

Table 5.3
Absolute cavity radiometers compared at IPC IV, (Davos, November 1975)

Type	Serial number	Designer	Owner
PACRAD (original)	III	J. Kendall, Sr.	JPL, Pasadena, CA
ACR (active cavity)	311	R. C. Willson	R. C. Willson, Altadena, CA
ACR (active cavity)	701	R. C. Willson	CSIRO, Australia
Eppley (PACRAD type)	11402	J. Kendall/Eppley Lab.	Eppley Lab., Newport, RI
Eppley (PACRAD type)	12843	J. Kendall/Eppley Lab.	NOAA/GMCC, Boulder, CO
Eppley (PACRAD type)	13617	J. Kendall/Eppley Lab.	U. Bergen, Norway
PMO (active type)	2	R. W. Brusa	PMO/WRC, Davos, Switzerland
PMO (active type)	3	R. W. Brusa	PMO/WRC, Davos, Switzerland
PMO (active type)	5	R. W. Brusa	PMO/WRC, Davos, Switzerland
PVS	5	Yu. A. Skliarov	U. Saratov, USSR
TMI (PACRAD type)	67502	J. Kendall/C. M. Berdahl	NOAA/ERL, Boulder, CO

Table 5.4
Summary of comparison results of absolute radiometers in IPC IV

Instrument	Owner	Mean ratio	SDS	Number
PACRAD III	JPL, U.S.	1.0000	(Reference instrument)	1,993
ACR 311	Willson, U.S.	1.0027	0.0042	1,993
ACR 701	CSIRO, Australia	1.0010	0.0079	1,992
EPAC 11402	Eppley, U.S.	1.0020	0.0022	1,642
EPAC 12843	NOAA, U.S.	1.0039	0.0016	980
EPAC 13617	U. Bergen, Norway	0.9987	0.0019	1,610
PMO-2	PMOD, Switzerland	1.0029	0.0028	1,218
PMO-3	PMOD, Switzerland	1.0052	0.0032	1,813
PMO-5	PMOD, Switzerland	1.0017	0.0039	1,974
PVS-5	U. Saratov, USSR	0.9971	0.0045	27
TMI 67502	NOAA, U.S.	1.0004	0.0021	1,678

Source: Anon. (1977).

ometer that is itself periodically intercompared with one or more international or national absolute radiometers having similar histories.

5.5.3.2 The Fifth IPC (1980) In early October 1980, at WRC/PMO in Davos, 24 absolute cavity radiometers and 36 pyrheliometers (Ångström, Model NIP, etc.) were intercompared on two different days at IPC V. Actually, some of the radiometers were intercompared on only one day, the second clear day having occurred after the formal close of IPC V. Because PMOD desired to establish as the reference instrument for IPC V an absolute cavity radiometer developed at the WRC/Davos, they chose the Model PMO-2 absolute radiometer that had participated in IPC IV. Table 5.5 presents the instruments, their owners, and summary of results (Anon., 1981). Of the 24 absolute radiometers, the 6 absolute cavity radiometers noted with asterisks were intercompared in IPC IV. Of these, Models PMO-2 and EPAC 13617 exhibited essentially no change over the 5 intervening years when normalized to PACRAD III.

Again, these tests demonstrated excellent agreement between absolute instrument with an average ratio (to Model PMO-2) of 0.9995 and a standard deviation between instruments of only $\pm 0.26\%$.

5.5.4 The New WMO Guide

5.5.4.1 The World Radiometric Reference WRR was adopted by the World Meteorological Congress in 1979 and is the result of numerous

Table 5.5
Summary of comparison results of absolute radiometers in IPC V

Instrument	Serial	Owner	Mean ratio	SDS	Number
PMO	PMO-2[a]	PMOD,[b] WRC	1.00000	(Reference instrument)	162
PMO	PMO-5[a]	PMOD, WRC	0.99740	0.09	78
PMO	PMO-6D	Obs. Hamburg	0.99315	0.07	81
PMO	PMO-6G	NBS, U.S.	1.00400	0.06	81
PMO	PMO-6C	PMOD/WRC	1.00071	0.07	81
Eplab-HF	14915	Eppley, U.S.	1.00111	0.05	162
Eplab-HF	15744	NTI, Boras, Sweden	0.99922	0.07	96
Eplab-HF	17142	DSET, U.S.	1.00106	0.06	150
Eplab-HF	18747	AES, Canada	0.99790	0.20	88
Eplab-HF	19744	CNR-IEA, Italy	0.99874	0.04	96
Eplab-HF	19746	Met. Inst., Hungary	1.00034	0.06	150
EPAC-Kendall	13219	Met. Inst., Italy	0.99532	0.12	18
EPAC-Kendall	13617[a]	U. Bergen, Norway	0.99617	0.93	77
PACRAD	II	JPL, U.S.	0.99952	0.13	108
PACRAD	III[a]	JPL/PMOD	0.99779	0.10	150
TMI-MK VI	67401	TMI, U.S.	1.00101	0.07	96
TMI-MK VI	67502[a]	NOAA/SRF, U.S.	1.00057	0.06	96
TMI-MK VI	67604	Met. Off., UK	0.99749	0.04	96
TMI-MK VI	67702	JPL, U.S.	1.00105	0.07	96
TMI-MK VI	67814	SERI, U.S.	0.99800	0.04	96
TMI-MK VI	68016	Met. Serv., France	0.99837	0.07	96

Table 5.5 (continued)

Instrument	Serial	Owner	Mean ratio	SDS	Number
ACR IV	401	JPL, U.S.	1.00426	0.07	75
ACR IV	403	JPL, U.S. (Willson)	1.00553	0.09	75
ACR	701[a]	CSIRO, Australia	0.99232	0.14	45
CROM	2L	Crommelynck, Belgium	0.99599	0.08	45
CROM	3R	Crommelynck, Belgium	0.99876	0.44	41

Source: Anon. (1981).

a. Participated in IPC IV (1975).

b. PMOD—Physicalish Meteorologicalisch Observatorium-Davos; WRC—World Radiation Center; Obs.—Observatorium; NBS—National Bureau of Standards; NTI—National Testing Institute; DSET—DSET Laboratories, Inc., AES—Atmospheric Environmental Services; Met. Inst.—Meteorological Institute; JPL—Jet Propulsion Laboratory; TMI—Technical Measurements, Inc.; NOAA—National Oceanic & Atmospheric Administration; Met. Off.—Meteorological Office; SERI—Solar Energy Research Institute; Met. Serv.—Meteorological Service.

comparisons of 15 individual absolute cavity pyrheliometers representing 10 different designs. WRR is maintained by a group of not less than 4 absolute cavity radiometers, known as the World Standard Group (WSG). WSG currently comprises 5 absolute cavity pyrheliometers that are designated as primary standard pyrheliometers: Models PMO2, CROM2, PACRAD III, ACR310, and ACR311. To be included in WSG, a radiometer must have a long-term stability of better than $\pm 0.2\%$, its accuracy and precision must be within the limits of uncertainty of the WRR ($\pm 0.3\%$), and its design must differ from that of other instruments. During international intercomparisons, the WRR value is calculated from the unweighted mean of at least three WSG pyrheliometers. The relationship of WRR to old scales is

WRR = 1.026 (ÅS 1905),

WRR = 0.977 (SS 1913 REV),

WRR = 1.022 (IPS 1956).

We note that while WRR is 2.2% higher than IPS 1956, it is 2.3% lower than the SS 1913.

5.5.4.2 Classification of Pyrheliometers Primary standard pyrheliometers are absolute cavity radiometers that meet certain rigid criteria: (1) one of the design series must have been fully characterized to show a root mean square uncertainty of less than $\pm 0.25\%$ at 1 kW/m^2, (2) each instrument of the series must be compared with a fully characterized instrument and may not deviate more than the root mean square uncertainty of $\pm 0.25\%$, and (3) each instrument must be traceable to WRR, which must lie within the root mean square uncertainty as previously determined.

Current rules require that, if the above specifications are not met, the pyrheliometer may be used as a secondary standard if it is calibrated by comparison to the WSG.

The new WMO guide (1983) lists four absolute cavity pyrheliometers that meet the requirements to be designated primary standard instruments: Models PACRAD, ACR, CROM, and PMO (see tables 5.3, 5.4, and 5.5). Although the Eppley Model H-F (Hickey-Frieden) and the TMI Model MK VI pyrheliometers, both manufactured in the United States, are not mentioned in the new WMO guide, both are now represented in the WSG. The Eppley Model H-F with SN 18748, which participated in NRIP III, IV, and VII, was loaned by The Eppley Laboratory, Inc., to NOAA for inclusion in the WSG. Similarly, the TMI Mk VI with SN 67814, which participated in all seven NRIPs and IPC V and VI, was loaned by the Solar Energy Research Institute to NOAA for inclusion in the WSG. These instruments have since been contributed by NOAA to the World Radiation Center-Davos and will reside in perpetuity at PMOD for that purpose.

The new guide specifies the Ångström compensation and the silver disk pyrheliometers as Secondary Standard pyrheliometers but does not specifically require that they be calibrated against the WSG to be considered such. Thermopile pyrheliometers such as the Eppley Model NIP are considered as First Class instruments. Second Class instruments are distinguished from First Class as having time constants of 30 s maximum compared with 10 s for First Class.

5.6 Maintaining the Absolute WRR Scale in the United States (The New River Intercomparisons)

5.6.1 Genesis of the New River Intercomparisons

Largely as a result of their early assessment of the disparate results of the first NBS-sponsored Round Robin thermal performance tests of solar collectors (Streed et al., 1978) and the observed differences in pyranometer-to-pyranometer irradiance measurements (even for WMO Class I instruments), DSET Laboratories purchased an Eppley Model H-F absolute cavity pyrheliometer in January 1977. This instrument was the first terrestrial version of the Hickey-Frieden absolute cavity radiometer that was developed for the Nimbus satellite series. In early 1979, DSET obtained a second Model H-F absolute cavity radiometer.

By the summer of 1977, DSET began periodic shading disk calibrations of their family of Eppley Model PSP pyranometers at normal incidence—that is, in the exact mode at which they were mounted to test solar collectors for thermal performance. Transfers were made directly from the Eppley Model H-F cavity radiometer.

Because of the need for a U.S. reference base that functions independently of and is convened more frequently than the WRC/PMOD-hosted comparisons at Davos every 5 years, DSET proposed hosting the first of seven national, regional, and international intercomparisons of absolute cavity radiometers, commencing in November 1978. A reference base was needed to provide material continuity and legal traceability for the then-emerging

Table 5.6
NRIP instrument breakdown

Intercomparison	Date	Registered instruments		Solar observations
		New	Total	
NRIP I	Nov. 1–5, 1978	14	14	315
NRIP II	May 2–5, 1979	4	12	441
NRIP III	Nov. 5–9, 1979	4	17	550
NRIP IV	Nov. 17–19, 1980	4	15	465
NRIP V	May 3–5, 1982	2	7	798
NRIP VI	Nov. 14–18, 1983	3	16	525
NRIP VII[a]	Nov. 18–21, 1985	3	19	693

a. NRIP VII data have only recently been published, and are available from NOAA (Nelson et al., 1987).

national solar device testing and product certification and labeling program. These 3-5-day intercomparisons, known as the New River Intercomparisons of Absolute Cavity Pyrheliometers (NRIPs), have been cosponsored by DSET, NOAA/SRF, the Solar Energy Research Institute (SERI), and the Department of Energy (DOE). In May 1982, with NRIP V, the New River intercomparisons became partially self-supporting on a participating fee basis. The success of the NRIP has been largely a result of the helpful support and enthusiasm of E. Flowers of NOAA/SRF, J. Hickey of The Eppley Laboratory, J. Kendall of the Jet Propulsion Laboratory, and C. Wells of SERI. Their success has also been a result of the superior environmental conditions that prevail at DSET's New River, AZ, site near Phoenix.

5.6.2 Operation of the NRIP

More than 50,000 individual instantaneous irradiance readings have been accumulated by 34 different absolute cavity radiometers in NRIP I–NRIP VII, as shown in table 5.6. Five different cavity designs from the United States, Canada, Switzerland, Italy, and the People's Republic of China have contributed to this data base.

Dates for the intercomparisons are selected to ensure the greatest probability of cloud- and haze-free conditions. The site at New River, AZ, is distinguished by a combination of moderate temperatures, low average relative humidity, clear-sky conditions, and high average daily sunshine for the times of year chosen.

Figure 5.33
Some of the participants and instruments assemble for NRIP V.

Outdoor laboratory facilities were constructed for the NRIP experiments. They consist of three permanent double-tiered instrument benches with the upper table serving as the cavity mounting deck, which in turn shades a lower electronic shelf that also doubles as working desk space. Each table accommodates eight experimenters, making it possible to intercompare 24 absolute cavity radiometers at a time. Figure 5.33 shows a typical NRIP array of absolute cavity radiometers and participants.

Instantaneous readings were organized into 10-min sequences of 21 readings at 30-s intervals (ACR-type instruments occasionally obtained only 11 readings at 1-min intervals). Readings for all instruments are sensed within less than 1 s of each other to minimize scatter between instruments caused by rapidly changing sky conditions. Calibrations were usually performed before and after each experimental sequence with the apertures shuttered. A description of the NRIP comparisons has been published by Estey and Seaman (1981) and Zerlaut (1984).

5.6.3 Summary of Results of NRIP I–NRIP VI

Data for each intercomparison are obtained by making ratios of the irradiance of each instrument to that of the agreed-upon reference instrument NOAA/SRF's Model SN 67502. I have further analyzed the results of Estey and Seaman (1981), and have reproduced them in table 5.7 (Zerlaut, 1984). The standard deviation shown is the statistical average of the standard deviation computed for each intercomparison in which the instrument was involved. The standard deviation for all readings (weighted average of the individual standard deviations) was ± 0.0011. Results are also obtained by making a ratio of the irradiance of each individual instrument to the mean irradiance of all participating pyrheliometers.

The excellent agreement obtained among all absolute cavity instruments (agreement within 0.25% for more than 40,000 instantaneous measurements) is attributed principally to (1) the excellent quality of currently available absolute cavity instruments, (2) the quality of the environmental conditions obtained at New River, AZ, for the times of the year chosen for the intercomparisons, and (3) the improved techniques with which the experimenters themselves operate their respective instruments as a result of the experience gained.

5.6.4 Comparison of NRIP Instruments to Model SN 67502

Table 5.8 gives the set values for these instruments that were intercompared in most of the NRIP. Also the ratios to Model SN 67502 are given for the results of IPC V for those instruments that participated in both. This was done by normalizing IPC V data to Model SN 67502 results for these instruments.

These data are plotted in figures 5.34 and 5.35 to show instrument trends. All instruments show a general trend to greater sensitivity compared to Model SN 67502. The marked similarities in data for NRIP I, II, and III lead to speculation that the reference instrument Model SN 67502 exhibited a decrease in sensitivity of about 0.15% in the period since May 1979. Also, one of the most important attributes of the frequency and procedures used in the NRIP is manifest in the behavior of Model SN 17142 (my instrument). As a result of the nearly 0.5% apparent decrease in sensitivity of this instrument in NRIP V, the cavity was examined and several large dust particles were observed. After we removed them using an optics bulb, the apparent sensitivity increased and the instrument again assumed its historical position in its peer group.

Table 5.7
Results of NRIP I—NRIP VI with ratio of each instrument to group mean

Model	SN	Participant	NRIPs	N[a]	Weighted mean ratio	S.D.
EPAC	11399	AES,[b] Canada	1, 3	40	0.9997	0.0021[c]
EPAC[d]	11402	The Eppley Lab.	1	15	1.0008	0.0018
EPAC[d]	12843	NOAA (GMCC)	3, 4	50	0.9968	0.0042
H-F[c]	14915	The Eppley Lab.	1 thru 6	140	1.0010	0.0009
H-F	14917	NASA-Lewis	2	22	0.9995	0.0004
H-F[d]	15745	NOAA (SRF)	1, 2, 3, 4, 6	122	0.9973	0.0007
H-F[c]	17142	DSET Labs	2 thru 6	137	1.0006	0.0013
H-F[c]	18747	AES, Canada	3	25	0.9990	0.0007
H-F	18748	The Eppley Lab.	3, 4	44	1.0026	0.0009
H-F	19743	Peop. Rep. China Met.	6	22	0.9973	0.0008
H-F	20294	Peop. Rep. China Met.	6	25	0.9992	0.0010
ACR	501	Boeing Aerospace	1, 5, 6	59	0.9966	0.0027
ACR	601	NOAA (GMCC)	3	24	1.0067	0.0028
ACR	904	Battelle Mem. Inst.	5	23	1.0027	0.0018
ACR	1104	Lawrence Berkeley Lab.	4	31	1.0079	0.0028
PMO[c,d]	PMO2	PMOD/Davos, Switzerland	3	29	1.0008	0.0007
MK VI	001	JPL	2, 3, 4	74	1.0027	0.0017
MK VI[c]	67401	Tech. Measurements	1, 2, 3, 4, 6	117	0.9997	0.0010
MK VI[c,d]	67502	NOAA (SRF)	1 thru 6	160	1.0005	0.0010
MK VI	67603	Sandia Natl. Lab.	1, 2, 3, 6	84	0.9999	0.0011
MK VI[c]	67702	JPL	1, 2, 3, 4, 6	116	1.0011	0.0009
MK VI	67706	So. Cal. Edison	1, 2, 3, 4, 6	118	1.0004	0.0010
MK VI	67707	Utah State Univ.	1	15	1.0000	0.0011
MK VI	67811	Sandia Natl. Lab.	2	11	0.9997	0.0006
MK VI	67812	Ga. Institute of Tech.	1, 3	40	0.9980	0.0007
MK VI[c]	67814	SERI	1–6	162	0.9990	0.0006
MK VI	68017	SERI	4	27	0.9989	0.0006
MK VI	68018	SERI	4, 5, 6	94	1.0017	0.0009
MK VI	68020	Martin Marietta	4	31	0.9990	0.0006
MK VI	68021	Stazione Astron. Italy	6	24	0.9978	0.0009
MK VI	68022	Sandia Natl. Lab.	6	13	1.0012	0.0005

a. N = number of 21-reading sets (runs) in which each instrument was intercompared.
b. AES—Atmospheric Environmental Services; DSET—DSET Laboratories, Inc.; NOAA—National Oceanic & Atmospheric Administration; NASA—National Aeronautics and Space Administration; PMOD—Physicalish Meteorologicalisch Observatiorium-Davos; SERI—Solar Energy Research Institute.
c. Also compared in IPC V, Davos, Nov. 1980.
d. Also compared in IPC IV, Davos, Nov. 1975.

Table 5.8
Ratio of each instrument to the NOAA/SRL reference cavity, SN 67502

Instrument SN	NRIP 1	NRIP 2	NRIP 3	Davos IPC V	NRIP 4	NRIP 5	NRIP 6
14915	0.9984	0.9984	1.0013	1.0005	1.0011	1.0016	1.0006
15745	0.9963	0.9955	0.9971	—	0.9971	—	0.9975
17142	—	1.0008	1.0012	1.0005	1.0019	0.9971	1.0009
67401	0.9968	0.9956	1.0019	1.0004	0.9996	—	1.0006
67502	1.0000	1.0000	1.0000	1.0000	1.0000	1.0000	1.0000
67603	0.9966	0.9980	0.9999	—	—	—	0.9998
67702	0.9940	0.9996	1.0023	1.0005	1.0019	—	1.0020
67706	0.9992	0.9983	1.0003	—	1.0002	—	1.0010
67814	9.9978	0.9968	0.9989	0.9974	0.9988	0.9989	0.9986
68018	—	—	—	—	1.0021	1.0013	1.0014

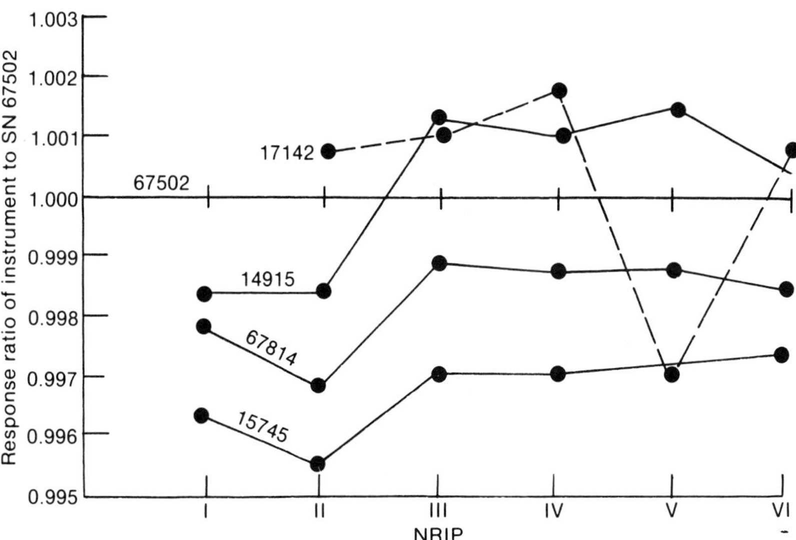

Figure 5.34
Comparison of several NRIP cavities to the reference cavity, NOAA's Model SN 67502 (showing deviation in Model SN 17142 for NRIP V).

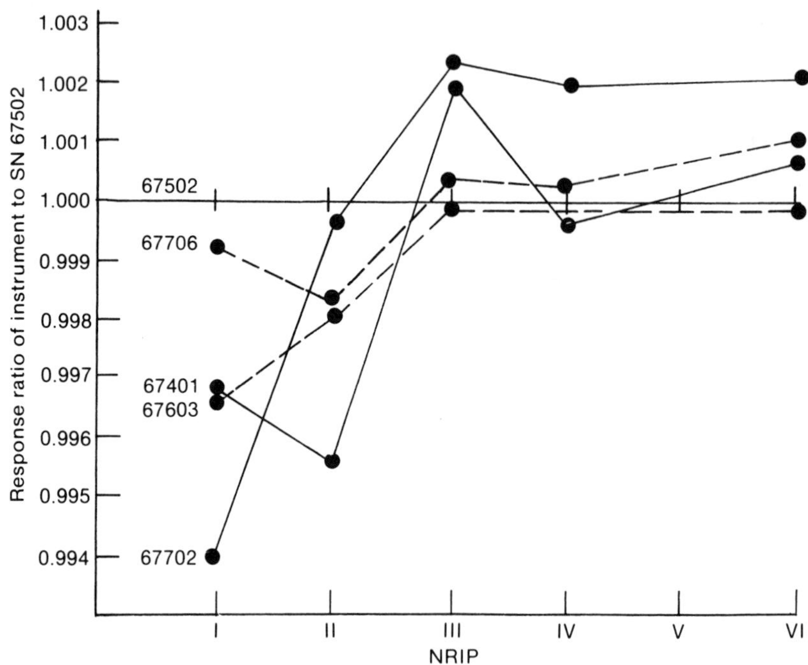

Figure 5.35
Comparison of several NRIP cavities to the reference cavity, NOAA's Model SN 67502 (showing an apparent change in Model SN 67502 between NRIP II and III).

5.6.5 Comparison of the NRIP with IPC IV and V

Unlike the WMO-sanctioned intercomparisons held every 5 years at Davos, Switzerland, where only the cavity itself is compared,[9] the New River Intercomparisons compare the cavity and its associated electronic data logging and read-out equipment. The advantage of the NRIP comparisons is that the actual equipment compared is the same as that employed by the participants when using the absolute cavity radiometers as primary reference instruments in their own laboratories. Hence, the performance of individual instruments compared in NRIP has greater reference and legal significance, if not greater credibility, than the same instruments compared at Davos. From a legal point of view, the NRIPs do not have a broken chain of evidence. As such, they are "legally referenceable." At Davos, prior to IPC 1985, traceability to the IPCs would not be legally referenceable due to the fact that only the cavity radiometers were intercompared (their respective read-out instruments were not).

Wells (1981) has discussed these differences in detail, giving the advantages and disadvantages of each approach. He concluded that the principal advantages of the New River Intercomparisons were the following:

1. Because instrumentation, techniques, and operators are essentially unchanged between those in end-use applications and those of the intercomparisons, the significance and traceability of the intercomparison may have more significance when the instrument is used outside the framework of peer grouping.

2. There are no difficulties in compatibility between the electronics of a given radiometer and the data accumulation techniques of the NRIP.

3. Operational problems either are weather-related or affect only a specific instrument at a time.

4. Protocol changes are readily made to accommodate new or different instruments.

Wells cites some disadvantages of the New River Intercomparisons, the most significant of which are

1. A large investment in each operator's time is required to record initially the solar irradiance of each instrument.

2. The formatting of these data for subsequent computer analysis is also very labor intensive.

3. There may be slight timing errors caused by the operators not reading their instruments simultaneously.

Advantages of the WRC/IPC method and protocols are

1. The data are collected with significantly fewer recording errors and with little effort.

2. First (non-quality-controlled) results can be distributed within hours of the comparisons.

3. Operators are free to watch for subtle instrument problems without the tedium of continuous data readings.

The principal disadvantages of the WRC/IPC methods are

1. The data acquisition and sequence of operational steps used in the IPC are not necessarily the methods normally used with the absolute radiometers.

2. As a result, the question is whether or not the data from the IPC (that is, the irradiance values) are identical to those that would be obtained in normal use.

3. Changes in procedure and protocols are more difficult, and changes in zero or system gain cannot be determined at the end of each run, so linear corrections with time cannot be determined and applied.

In summary, the advantages of the NRIP relate to the de facto maintenance of the SI Radiometric Scale (WRR) in the United States by this group of instruments, to the reference quality of the intercomparisons for citation and traceability purposes, and to the availability of representative instruments of the grouping for the calibration and quality control purposes of any given organization.

5.7 National and International Activities in Calibrating of Pyranometers

5.7.1 International Energy Agency Activities

Recognizing the need for a coordinated effort to resolve certain energy problems facing the petroleum importing countries, the member countries of the Organization for Economic Cooperation and Development (OECD) established in 1974 the International Energy Agency (IEA). An international treaty organization, IEA's funding is obtained from the member countries.

In 1975, IEA developed a solar heating and cooling program as one of 16 technology areas of concern to OECD. Fifteen countries and the Commission of European Communities support this program. Table 5.9 lists nine tasks, their current status, and the country responsible for each.

5.7.1.1 Initial Activities in Pyranometry As a consequence of results of an international Round Robin program on testing thermal solar collectors sponsored by Task III (Talarek, 1979), an intercomparison of 21 pyranometers was held at WRC/PMOD in Davos, Switzerland, in March 1980 (Talarek and Froelich, 1980). Nine Eppley Model PSPs, 10 Kipp and Zonen Model CM5s, 1 Kipp and Zonen Model CM 10, and 1 Schenk instrument were compared with WRC's Model 6703A (PMOD) reference pyranometer at 0° horizontal.

The results of this intercomparison were quite disturbing to the participants (Talarek and Froelich, 1980). The average deviations between the

Table 5.9
Task responsibilities within the IEA Solar Heating and Cooling Program

Task	Subject	Current status	Operating agent
I	Investigation of the performance of solar heating and cooling systems	Completed	Denmark
II	Coordination of R&D on solar heating and cooling components	Active	Japan (AIST)
III	Performance testing of solar collectors	Active	Federal Republic of Germany (KFA)
IV	Development of an insolation handbook and instrument package	Completed	U.S.
V	Use of existing meteorological information for solar energy application	Completed	Sweden
VI	Performance of solar heating, cooling, and hot water systems using evacuated collectors	Active	U.S. (DOE)
VII	Central solar heating with seasonal storage	Active	Sweden (Council of Building Research).
VIII	Passive and hybrid solar low-energy buildings	—	U.S. (DOE)
IX	Solar radiation and pyranometry studies	—	Canada-Atmospheric Environmental Services

Kipp and Zonen Model CM 10 and Eppley Model PSP instruments were 3% and 2%, respectively, using instrument constants supplied with the instruments. The maximum deviations for these two groups were an alarming 10.2% and 6.3%, respectively. These results prompted a series of Round Robin investigations to determine the cause of the disparities.

5.7.1.2 Round Robin I Pyranometer Comparison These disparities could not be explained by the characteristic differences of the primary absolute cavity reference instruments to which they were traceable. The absolute cavity radiometers all agreed to within a standard deviation of about ± 0.002 solar constants (as shown in tables 5.7 and 5.8 of section 5.6.3). The owners of three of the instruments involved subsequently agreed that they would have a separate Round Robin intercomparison performed in an attempt to explain the reasons for these discrepancies. Table 5.10 shows the instruments, the owners, and the ratios to PMOD. The ratios shown are based on instrument constants furnished to PMOD by the owners. The calibrations were performed in order by DSET Laboratories, NOAA/SRF (Boulder), and The Eppley Laboratory.

Table 5.10
Ratio of three pyranometers to the PMOD reference pyranometer

Manufacturer	Serial number	Ratio	Owner
Eppley	14806	0.9378	National Bureau of Standards, U.S.
Eppley	19129	0.9468	DSET Laboratories, Inc., U.S.
Kipp and Zonen	774120	0.9159	Kernforschungsanlage, Jülich, FRG

Horizontal Shading Disk: The three laboratories calibrated the instruments using the shading disk method and transfer from a reference pyranometer that had itself been calibrated using the shading disk technique. Selected results are shown in table 5.11 (Zerlaut, 1981c).

Excellent agreement between laboratories was obtained for the Eppley Model PSP instruments when calibrated against absolute cavity pyrheliometers using the shading disk method, even though DSET used a 30 s/30 s and NOAA a 5 min/6 min sequence for the shaded/unshaded measurements. The DSET shading calibrations were performed at an average solar elevation of 64° and 60° for the NOAA measurements. The Eppley calibrations were obtained during the winter months when the average solar elevation was only 25°.

Horizontal Transfer of Calibration: Good agreement was obtained for the horizontal measurements referenced against the Eppley Model PSP pyranometers. The agreement between DSET and NOAA was exceptionally good, with values differing by only 0.17%, 0.21%, and 0.41% for Models SN 19129, SN 14806, and SN 774120, respectively. The Eppley integrating hemisphere calibrations averaged about 1.3% higher than the outdoor calibrations. The reference pyranometers at Eppley and NOAA were their respective primary working standard pyranometers. However, at DSET the NBS instrument (SN 14806) was checked against the horizontal shading disk calibration of the DSET instrument (SN 19129), the DSET instrument (SN 19129) was checked against the horizontal shading disk calibration of the NBS instrument (SN 14806), and the value for the K & Z instrument (SN 774120) was the average obtained when checked against SN 19129 and SN 14806.

Calibrations at Tilt: Excellent agreement was obtained between Eppley and DSET in normal incidence calibrations of SN 19129 using the shading disk method. The DSET data were obtained at a tilt of 30° (summer months)

Table 5.11
Summary of DSET/NOAA/Eppley Round Robin results

			Instrument constants (in μV/W m^{-2})		
			Eppley PSPs		K & Z
Test mode	Lab	Reference	SN 19129	SN 14806	SN 774120
Horizontal	DSET	ACP[a] 64° sun	10.427	9.834	—
Shade disk	NOAA	ACP 60° sun	10.500	9.840	—
	EPLAB	ACP 25° sun	10.290	9.290	—
Horizontal	DSET	PSP outdoor	10.570	9.910	12.820
	NOAA	PSP outdoor	10.588	9.889	12.873
	EPLAB	PSP hemisphere	10.640	10.070	13.090
Tilt, normal	DSET	ACP 30°	10.410	9.832	—
	DSET	ACP 60°	10.330	—	—
	EPLAB	ACP 60°	10.340	—	—
Tilt, off	DSET	ACP 30°	10.470	9.837	—
Tilt, normal	NOAA	PSP 40°	10.496	9.884	—

a. ACP = absolute cavity pyrheliometer.

and the Eppley data were obtained at a tilt of 60° (early winter). On return to DSET, SN19129 was recalibrated by the shade method at normal incidence, and a value of 10.33 μV/W m^{-2} was obtained. Note that the tilt required to achieve normal incidence in the second DSET shading disk calibration (then wintertime) of SN 19129 was 60° from the horizontal—identical to that obtained at 60° tilt during the wintertime normal incidence calibration at The Eppley Laboratory.

Pyranometer SN 19129 was subsequently recalibrated at a tilt of about 30° using the shading disk technique checked against the Model H-F absolute cavity pyrheliometer. The new value for SN 19129 was 10.37 μV/W m^{-2}, indicating an approximate 0.5% tilt effect now believed to be caused largely by the effects of incident angle and horizon interactions.

5.7.1.3 Rehabilitation of Davos Results The ratios between the radiation measured by each of the three Round Robin instruments to that measured by the Davos reference pyranometer PMOD/SN 6703A (Talarek and Froelich, 1980) are given in table 5.12, along with the average instrument constant determined by DSET and NOAA and cosine and temperature corrected values. As table 5.12 shows, the average deviation from the reference instrument was 6.6% (column 3), and after recalibration in this

Table 5.12
Cosine and temperature corrections to the Davos comparison ratios

Instrument nameplate	Davos I.C.[a]	Davos Ratio	RR recalibration I.C.[a]	RR recalibration Ratio	Cosine correction	Cost/temperature correction
KZ/SN 774120	13.70	0.9159	12.84	0.9772	0.9834	0.9834[b]
EP/SN 14806	10.02	0.9378	9.84	0.9550	0.9871	0.9871
EP/SN 19129	10.76	0.9468	10.46	0.9740	0.9768	0.9827

a. Instrument constant is in $\mu V/W\ m^{-2}$.
b. Temperature correction not applied.

study, it was still 3.1% (column 5). The next step was to use the cosine and temperature compensation corrections for the Davos data as defined by the declination angle δ of $-6.37°$ and the solar noon sun elevation of 37.1° for Davos on March 5, 1980, and a temperature of 0°C. These corrections are taken from DSET data and the report by E. Flowers of NOAA presented by Riches, Stoffel, and Wells (1982); they are presented in the last two columns of table 5.12. The temperature correction for SN 14806 is unity based on the difference between 26°C (the nameplate temperature) and 0°C as determined by its compensation curve. No correction was made for SN 774120 since we have no knowledge of the temperature at which the "original" instrument constant was determined.

These manipulations showed that agreement between the three instruments compared with PMOD/SN 6703A could be significantly improved using carefully determined instrument constants and could be further improved by using cosine and temperature corrections. As seen in the last column of table 5.12, the three corrected instruments agreed closely with each other, although they differ from PMOD/SN 6703A by about 1.6%. PMOD's reference instrument was subsequently determined to err by 2.6% and has been adjusted accordingly.

5.7.1.4 The Second IEA Round Robin A second Round Robin pyranometer study was initiated by Task V on meteorology in early 1981. Twenty-two instruments were first compared by the National Atmospheric Radiation Center (NARC) of Canada, NOAA/SRF, and SERI. Table 5.13 presents results of the horizontal calibrations performed by NARC and NOAA/SRF for several selected instruments (Riches, Stoffel, and Wells, 1982). The NOAA/SRF results are compared with the NARC values. Also, the original PMOD values (Talarek and Froelich, 1980) were multiplied by

Table 5.13
Selected results of IEA Round Robin II pyranometer calibrations and comparisons

Model	Calibration organization	NARC[a] NOAA/ SRF	Ratio of stated calibrations to that of NARC	
			Revised	PMOD revised[b]
14806	NBS, U.S.	9.66	1.007	0.998
15834	SNTI, Sweden	8.74	1.026	0.991
16692	TIL, Denmark	9.55	0.992	0.992
17750	NARC, Canada	9.24	1.002	1.007
17823	KFA, Jülich, FRG	8.67	1.010	1.002
K & Z CM6				
75-2438	ITE, Stuttgart, FRG	10.45	1.014	1.049
77-4120	KFA, Jülich, FRG	12.56	1.025	1.061
78-4750	Facultie Blytech., Belgium	10.81	0.998	1.039

a. Instrument constant is in $\mu V/W\ m^{-2}$.
b. Original PMOD values (Talarek and Froelich, 1980) multiplied by 1.026.

the factor 1.026, and ratios were made using the NARC calibration values. This scalar corresponds to the 2.6% increase in the calibration factor assigned by PMOD to the reference pyranometer that was used in the original Davos comparisons.

Although the difference between the revised PMOD results and the Canadian NARC values is less than between the NOAA/SRF and NARC values, the agreement between PMOD and NARC in these results is mostly fortuitous (Riches, Stoffel, and Wells 1982). PMOD and NARC do not actually correspond because of the wide difference between the techniques used in the two cases—the NARC calibrations were performed by direct comparison with a reference pyranometer in an integrating sphere, and the PMOD values were obtained outdoors at low sun angles with an uneven envelope of snow in the north field of view of the pyranometers (including the PMOD reference).

The SERI studies were largely confined to comparing the irradiance measured by test instruments with those of SN 17860 as a function of the angle of incidence of the direct beam component at tilts from the horizontal of 0°, 20°, 40°, 60°, and 90°. Selected results are shown in figures 5.36 and 5.37.

The disparate results of Round Robin II represented by the differences between calibration results and among pyranometers themselves simply

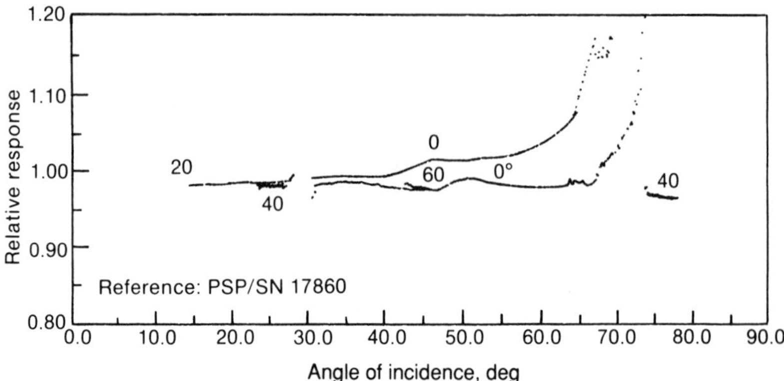

Figure 5.36
Ratio of irradiance of Eppley Model PSP SN 17823 to reference PSP as a function of angle of incidence at several tilt angles. Source: Riches, Stoffel, and Wells (1982).

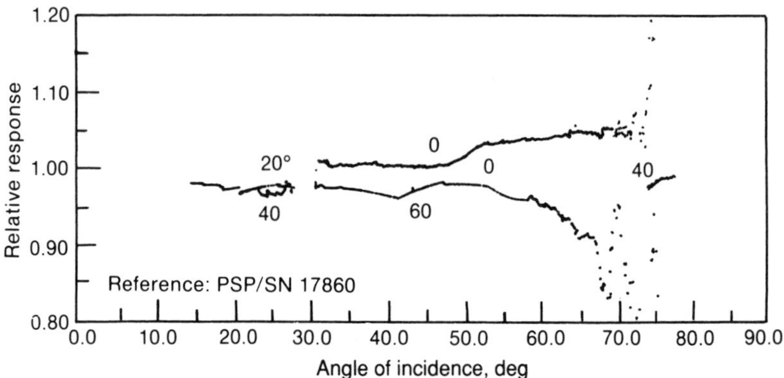

Figure 5.37
Ratio of irradiance of Kipp and Zonen Model CM5 SN 78-4750 to reference PSP as a function of angle of incidence at several tilt angles. Source: Riches, Stoffel, and Wells (1982).

amplify the need to calibrate pyranometers carefully that will be used in testing solar collectors and in determining instantaneous solar irradiance.

5.7.1.5 IEA Outdoor/Indoor Comparisons A comprehensive comparison of the outdoor and indoor performance characteristics of a broad spectrum of pyranometers was conducted in 1981 under IEA's auspices (Ambrosetti et al., 1984). Sponsored by PMOD and the Swedish National Testing Institute (SNTI) in Borås, the objective of the comparison was to evaluate a significant sample of representative instruments and to compare indoor characterization with outdoor performance.

The outdoor comparisons were performed at PMOD. The following 31 pyranometers were investigated as follows:

6	Eppley Model PSP	6 Eko Star
5	Kipp and Zonen Model CM10	3 Swissteco
5	Kipp and Zonen Model CM5	1 PMOD (WRC)
5	Schenk Star	

The principal reference instruments were two PMOD absolute cavity pyrheliometers (PMO1 and PMO2) for the beam component and the shaded PMO pyranometer (6703A) for the diffuse component. The mean of four reference pyranometers (two PSPs and two CM10s) was used on days where uncertainties in the beam component precluded use of the more desirable absolute cavity reference.

Tilt effects in groups of instruments were examined by checking daylong output at an inclination of 30° from the horizontal against the daylong output relative to the experimental reference instruments discussed in the previous paragraph. From these experiments, Ambrosetti et al. (1984) concluded that the mean daily irradiance values of the Eppley Model PSP and Kipp and Zonen Model CM10 pyranometers were unaffected by tilt. The Kipp and Zonen Model CM5 instruments showed a small decrease in sensitivity at tilt, but the Schenk and the Eko Star black and white pyranometers displayed significant tilt effects of more than 3%. A strong incident angle dependence was observed for the Swissteco instruments.

The authors cited (Ambrosetti et al., 1984) have drawn the following conclusions from their studies: (1) that only the Eppley Model PSP and

Kipp and Zonen Model CM10 be employed in collector testing, stating that an accuracy of 97% can be achieved for these two instruments, (2) that pyranometers exhibiting a tilt error should not be used in solar energy applications, (3) that purchasers of instruments should obtain temperature-dependence plots (or temperature coefficients) and tilt and linearity statements from the manufacturers, (4) that uncertainties in determining directional response of pyranometers requires further research, and day-long variability during calibration indicates the ultimate usefulness of a pyranometer for solar energy applications, and (5) that correction for environmental and geometrical conditions of measurement is not a viable option for improving the accuracy of pyranometric measurements.

The authors cited (Ambrosetti et al., 1984) imply that pyranometers destined for solar collector testing should be preselected, which I have also suggested (Zerlaut and Maybee, 1984). However, my experience indicates that uncertainties in the irradiance of a well calibrated Model PSP or Model CM10 can be $\pm 3\%$ to 4% when temperature and seasonal corrections (which include geometric considerations) are not applied. Also, the application of incident angle and more precise temperature corrections can reduce uncertainties to between 1% and 2%. My bias for such precise measurements derives from a founding association with U.S. national solar collector product testing, certification, and labeling programs in which solarization projects were won or lost on the basis of a 2–3% difference in the rated efficiency of a solar collector.

5.7.2 The New River Pyranometer Comparisons (NRCP)

5.7.2.1 Purpose The New River pyranometer comparisons were designed to determine whether a set of instruments, largely manufactured and calibrated in the United States, showed the same deviations from each other as observed in the March 1980 Davos comparisons of pyranometers having various manufacturers and different calibration histories (Zerlaut, 1981c; Talarek and Froelich, 1980). The comparisons have evolved into a task to determine the instrument constant under three instrument and sky geometries—horizontal, south-facing 45° tilt from horizontal, and exactly normal incidence (altazimuth, follow-the-sun)—and to determine the extent to which the instruments checked deviate from the cosine law at 0° horizontal and 45° tilt for the solar elevation at the time of year of the experiment.

Table 5.14
Pertinent experimental test parameters for the New River pyranometer comparisons

	Average temperature (°C)			Average humidity (% RH)		
Orientation	Nov. 1981 I	May 1982 II	Dec. 1983 III	Nov. 1981 I	May 1982 II	Dec. 1983 III
0° horizontal	27	33	18	19	7	53
45° south	27	34	20	19	7	49
Normal incidence	23	30	15	16	8	55

5.7.2.2 Experimental

Site and Test Parameters: The facilities are located at DSET Laboratories in New River, AZ, approximately 20 miles north of Phoenix. The geography is desert plateau with an elevation of 610 m (2,000 ft) at 33°51′N latitude and 112°10′W longitude. Table 5.14 gives the dates, average temperature, and average humidity for NRCP I, II, and III (NRCP IV was held in early December 1985; data are not available at the time of completion of this chapter).

Facilities and Mounting: The pyranometers were affixed to a common platform and spaced approximately 15 in. apart center to center. They were each leveled with the platform mounted in an exactly horizontal plane on a multiposition, multitilt exposure rack. All 0° horizontal comparisons were made in this position; the 45°S comparisons were performed by tilting the mount table to exactly 45° from the horizontal with the normal to the plane of the table facing exactly south (180°). The platform containing the mounted and exactly coaligned pyranometers was removed from the multitilt exposure rack and placed on a large, precision altazimuth test mount for the normal incident comparisons (that is, on a sun-following rack).

Organization of the Comparisons: The comparisons were organized into 10-min data acquisition periods during which 21 instantaneous millivolt output signals were recorded from all pyrheliometers and pyranometers every 30 s. The comparisons were performed in compliance with ASTM Standard E-824 (n.d.2). All channels were measured by the data logger in a 70 ms interval; the buffered data were recorded on magnetic tape.

Data Reduction: The instantaneous irradiance measured by each instru-

Table 5.15
Instrument constants determined by shade disk method in one summer and two winter comparisons[a]

	0° horizontal			45° S			Normal incidence		
SN	W_1	S	W_2	W_1	S	W_2	W_1	S	W_2
19129	10.29	10.33	10.14	10.29	10.30	10.32	10.34	10.34	10.33
19917	10.19	9.92	9.72	9.96	9.75	9.88	9.92	9.90	9.90
21017	9.59	9.75	9.48	9.77	9.63	9.80	9.82	9.80	9.79

a. W_1 = November 1981 results; W_2 = December 1983 results; S = May 1982 results.

ment was computed based on the instrument constant furnished by the owner. The irradiance values were averaged for each set of 21 observations. Ratios were made between the average irradiance of each instrument for that set and the average irradiance for the set obtained for all instruments and then to the average irradiance of the three reference pyranometers chosen. The average ratios and standard deviations were then obtained for all sets within a calibration angle (for example, at tilts of 0° horizontal, 45°S, and normal incidence). New instrument constants were computed with the reference irradiance being the average determined by the three reference instruments.

Computer plot routines were used to reduce the comparison results of the ratio of irradiance of each instrument to the mean irradiance of the three reference instruments as a function of angle of incidence.

Selection of Reference Pyranometers: The reference instruments in all three comparisons have been Eppley Model PSPs: DSET's SN 19129, NOAA/SRF's SN 19917, and The Eppley Laboratory's SN 21017. These instruments were chosen because of their traceability and previous comparison history and for their comparative lack of deviation from cosine response. Their instrument constants were determined using careful shade disk techniques during the comparison and using an Eppley Model H-F absolute cavity pyrheliometer (SN 17142) in accordance with ASTM Standards E913 and E941 (ASTM, n.d.3; ASTM, n.d.4).

5.7.2.3 Results

Shade Calibration of Reference Pyranometers. If we examine the temperature-corrected instrument constants derived from the December 1983 (III) experiments with those from the November 1981 (I) and May 1982 (II) comparisons (table 5.15), we observe two interesting trends: (1) essentially each of the three pyranometers have identical values at normal incidence in spite of different seasonal tilts and slightly different procedures, and (2) the close agreement between the normal incident values and the values for the seasonal tilt most closely resemble the mean tilt at normal incidence.

These relationships show that 0° horizontal shade calibrations in the winter and 45°S calibrations in the summer result in the greatest uncertainties.

Note that the normal incident calibration data encompassed summer tilts of 15–20° versus winter tilts of 60–65° from the horizontal. The agreement in the normal incident values for each of the three reference pyranometers proves that tilt effects in all three pyranometers are essentially nonexistent through 45°. Hence, we conclude that the seasonal differences in the fixed horizontal and 45° tilt results are primarily caused by incident angle and horizon effects (since temperature differences were presumably normalized out). This conclusion is not at all surprising, considering that seasonally different geometrical relationships apply to pyranometers oriented at 0° horizontal and at tilts of 45° from the horizontal, as shown in the somewhat exaggerated figures 5.38 and 5.39. These figures show that the beam track on the dome is near to normal incidence (at solar noon) in the summer at horizontal orientation and in the winter at a tilt of 45°. Conversely, the greatest angular disparities for the beam component can occur at horizontal orientation in winter months (trace W in figure 5.38); the summer beam trace for pyranometers oriented at 45° tilt occurs entirely in the pyranometer dome's northern hemisphere (trace S in figure 5.39).

Comparison of Pyranometers: In general, these comparisons, checked against the mean of the three reference pyranometers, resulted in data similar to those observed in the shading disk calibrations: (1) the calibration factors for normal incidence were in closer agreement with those of 45°S in the two winter comparisons and with 0° horizontal results for the May comparisons, (2) the two winter calibrations at all three tilt modes were themselves in closer agreement than either was with the May results, and (3) the most disparate data were from the winter data for 0° horizontal and May data for 45°S comparisons.

A plotter routine was used to compute the angle of incidence from exact solar time (logged as input data) and to plot the ratio of instrument response to the mean response of the three reference pyranometers for the mean angle of incidence of each set. Examination of these incident angle

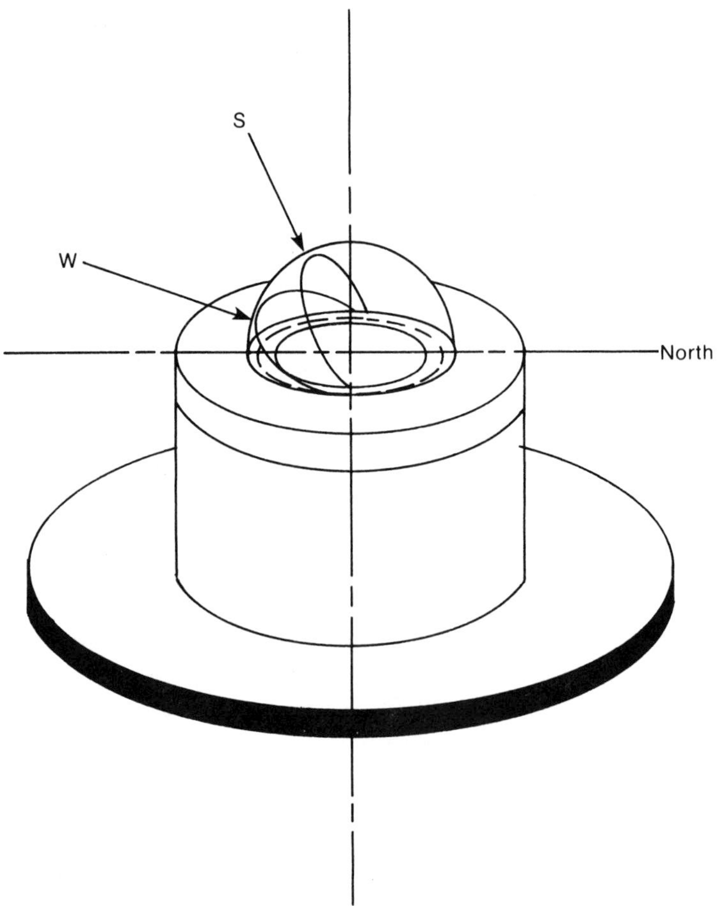

Figure 5.38
Beam tracks on the dome of a pyranometer with horizontal orientation.

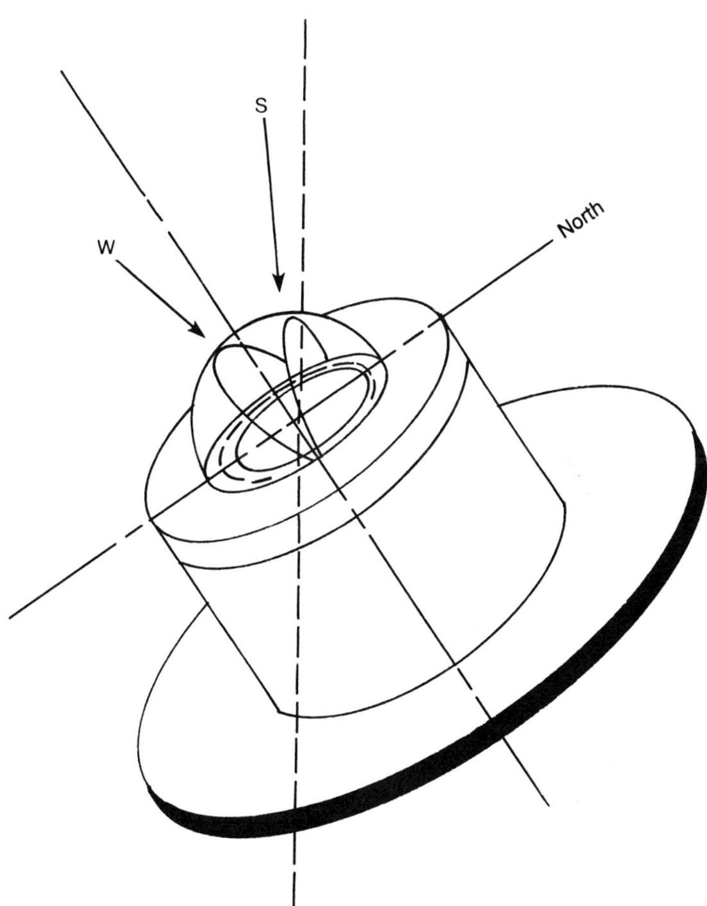

Figure 5.39
Beam tracks on the dome of a pyranometer with 45° tilt from the horizontal.

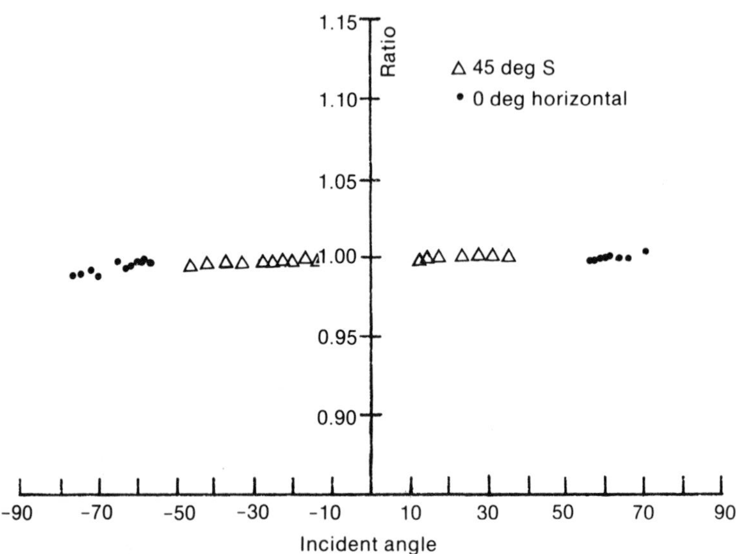

Figure 5.40
Ratio of response of Eppley Model PSP SN 19129 to three reference PSPs as a function of angle of incidence (NRCP III).

plots further amplifies the usefulness and need for developing seasonal pyranometer instrument constants for applications requiring highly accurate irradiance measurements.

First, figures 5.40, 5.41, and 5.42 show that the three pyranometers selected for reference indeed have little cosine deviation relative to each other (these are reference instruments used at DSET, Eppley, and NOAA and, therefore, not subjected to continuous exposure). Although their agreement could indeed be fortuitous, all other field pyranometers showed considerable deviations relative to the three reference instruments—making the assumption that the reference instruments indeed have no significant incident angle effects an entirely logical one.

Several of the field pyranometers examined (figures 5.43, 5.44, and 5.45) exhibited significant sensitivity losses of up to 8% at incident angles of ±60% during wintertime exposure for a horizontally mounted pyranometer. Note that instruments SN 14230 and SN 15659 were evaluated in previous comparison, and the previous incident angle plots for SN 15659 are presented in figure 5.46. The difference between summer and winter values can be better understood by examining their behavior in relation to figures 5.38 and 5.39.

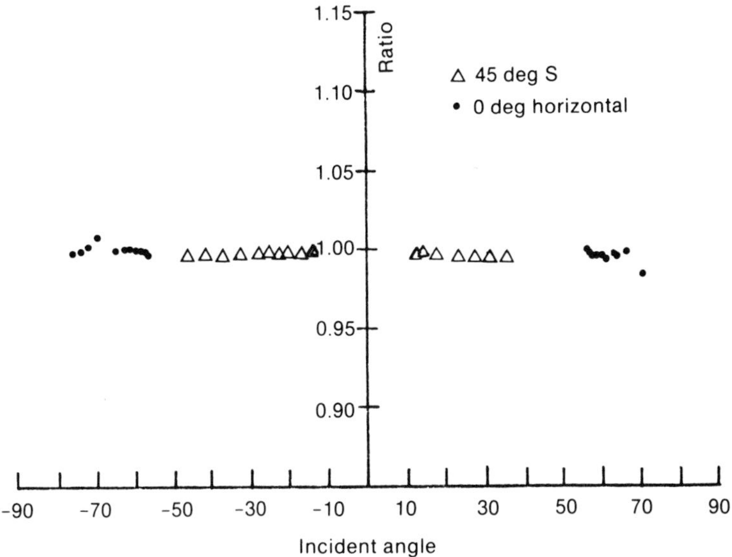

Figure 5.41
Ratio of response of Eppley Model PSP SN 21017 to three reference PSPs as a function of angle of incidence (NRCP III).

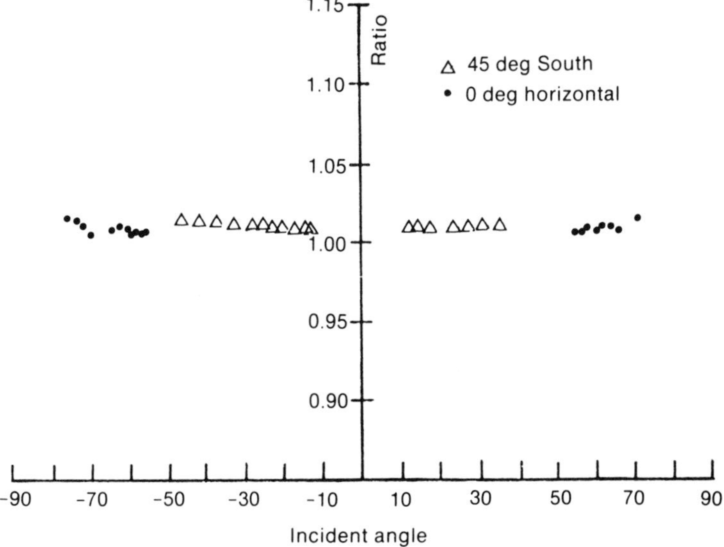

Figure 5.42
Ratio of response of Eppley Model PSP SN 19917 to three reference PSPs as a function of angle of incidence (NRCP III).

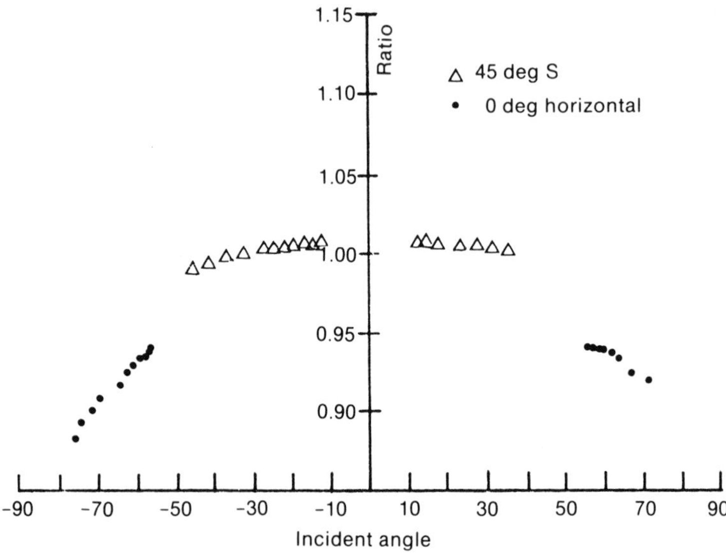

Figure 5.43
Ratio of response of Eppley Model PSP SN 15659 to three reference PSPs as a function of angle of incidence (NRCP III).

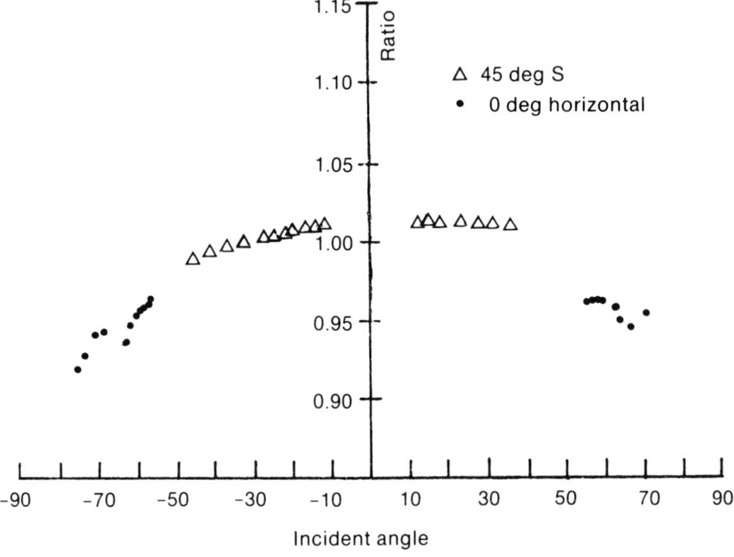

Figure 5.44
Ratio of response of Eppley Model PSP SN 14230 to three reference PSPs as a function of angle of incidence (NRCP III).

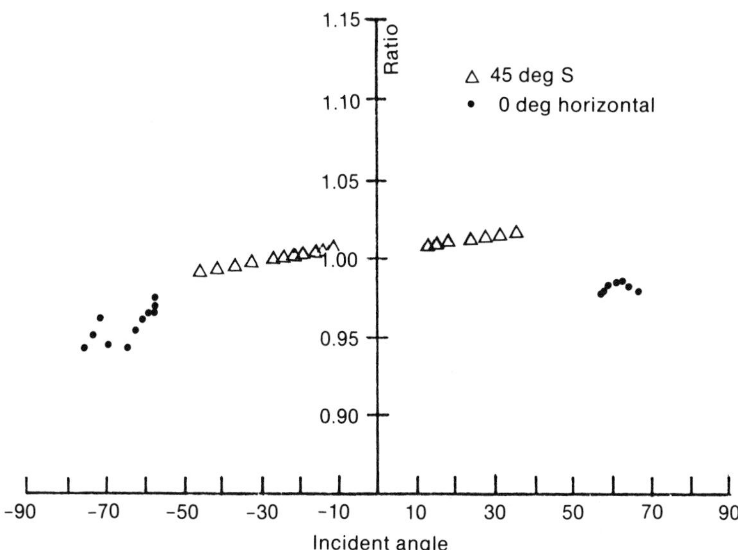

Figure 5.45
Ratio of response of Eppley Model PSP SN 22044 to three reference PSPs as a function of angle of incidence (NRCP III).

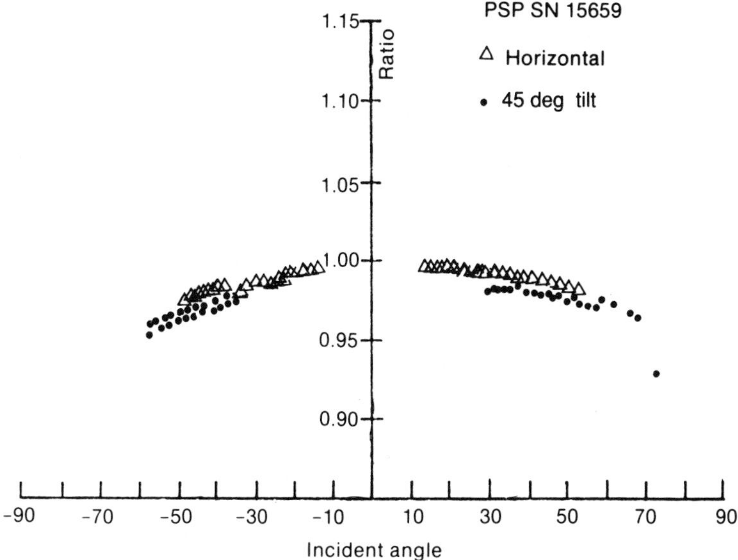

Figure 5.46
Ratio of response of Eppley Model PSP SN 15659 to three reference PSPs as a function of angle of incidence (NRCP II).

Three other observations are noteworthy: (1) the skewed plot shown in figure 5.45 for SN 22044 indicates that the plane of the black receiver may not be properly coaligned with the spirit level, (2) the incident angle plot for the Kipp and Zonen instrument SN 80-0065 (figure 5.47) is typical of the Model CM10 and CM11 pyranometers, and (3) the new Eppley Model SCP pyranometer (figure 5.48) exhibits a cosine response that is essentially identical to that of the mean of the three reference instruments.

5.8 National and International Standardization

5.8.1 ASTM Committee E-44 on Solar Energy Conversion

The American Society for Testing and Materials (ASTM), the largest voluntary consensus standards organization in the world, organized Committee E-44 on Solar Energy Conversion in June 1978 (Zerlaut, 1981a). The committee is organized into 12 subcommittees dealing with nomenclature, environmental parameters, safety, materials performance, heating and cooling subsystems and systems, process heating and thermal conversion power systems, photovoltaic electric power systems, wind driven power systems, ocean thermal power systems (inactive), biomass conversion systems, advanced energy systems (inactive), passive heating and cooling systems, and environmental and social impact of solar energy conversion systems. The Subcommittee on Environmental Parameters, E-44.02, is charged with identifying environmental parameters and establishing standard measuring and reporting procedures for data pertinent to solar energy conversion. As such, it is concerned with the promulgation of standards dealing with (1) the calibration and characterization of pyranometers and pyrheliometers, (2) the development of standard global and direct beam solar irradiance spectral distributions for AM1.5 conditions, and (3) the proper operation of solar radiometers.

Four methods dealing with the calibration of solar radiation measuring devices have become ASTM standards since May 1981. Two standards deal with the transfer of calibration from reference instruments to field pyrheliometers and pyranometers, respectively. The other two deal with the calibration of pyranometers by the shading disk technique with axes horizontal and axes tilted, respectively. However, they have recently sustained technical objections at the domestic (United States) and international level and are being reexamined in subcommittee 02.

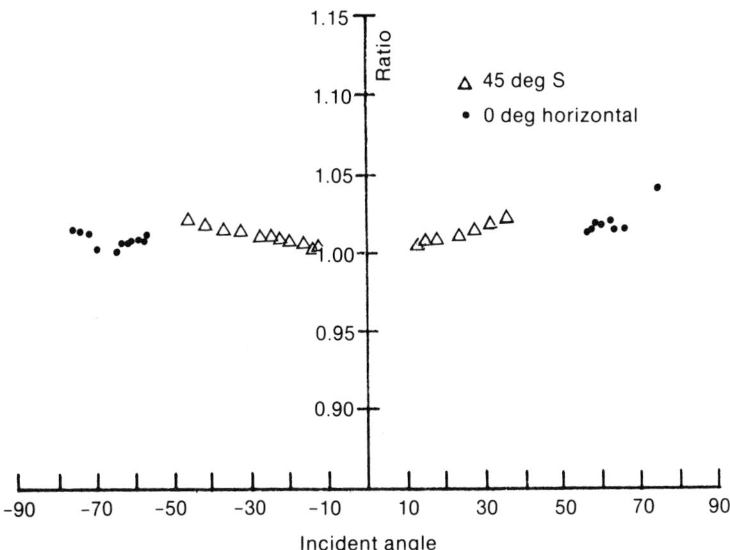

Figure 5.47
Ratio of response of Kipp and Zonen Model CM11 SN 80-0065 to three reference PSPs as a function of angle of incidence (NRCP III).

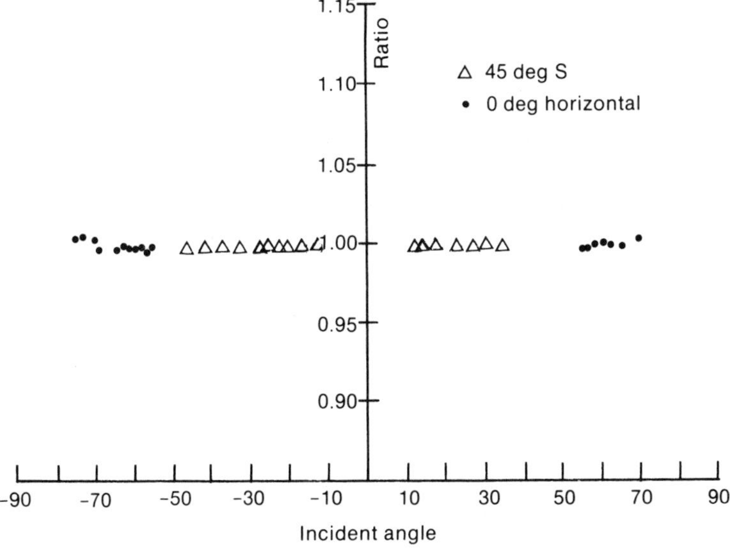

Figure 5.48
Ratio of response of Eppley Model SCP SN 21321 to three reference PSPs as a function of angle of incidence (NRCP III).

The task group is involved primarily with developing standard methods to characterize pyranometers. Their draft document, entitled *Standard Method for Calibration of Reference Pyranometers with Respect to Cosine, Azimuth and Tilt Errors*, represents the indoor laboratory approach to the problem. (Various aspects of this problem are covered in section 5.10.) The task group and its advisors from throughout the solar radiation community believe that a companion method is required for establishing incident angle and tilt effects by methods using natural sunshine outdoors.

5.8.2 International Standards Organization Activities

Headquartered in Geneva, the International Standards Organization (ISO) was founded in 1947 and issues voluntary international standards (IS) on nearly every technical field associated with world commerce. New ISO standards are derived, whenever possible, from existing national standards used by its member countries. However, methods are developed within the international technical community when necessary and usually involve interlaboratory testing prior to consensus. Documents developed by the various ISO committees are voted on by the member countries of any given technical committee, and finally by all ISO member countries. Negative votes are resolved to give as high a degree of consensus as possible.

International standards are one of the principal mechanisms for reducing certain international trade barriers. First, they result in and promote a common language for the establishment of trade, and second, they reduce the extent to which products must meet many often disparate requirements.

The Technical Committee on Solar Energy (Thermal Applications), ISO/TC 180, was organized at an international meeting held in Sydney, Australia, in May 1981 and is one of approximately 170 currently active technical committees of the ISO. Its secretariat is held by the Standards Association of Australia, which is the ISO member body in Australia. The organization meeting, attended by official representatives from Australia, Canada, Sweden, Germany, France, Israel, New Zealand, Japan, Kenya, the United Kingdom, and the United States, undertook the development of a program for one working group and four subcommittees. (In addition to the countries attending the organization meeting, other participating members are Austria, Belgium, Brazil, Denmark, Greece, Italy, Jamaica, Norway, Rumania, South Africa, Switzerland, and the USSR).

The current committee breakdown and the secretariat responsible for the individual programs of work are

Working Group	(WG) 1	Nomenclature—U.S.
Subcommittee	(SC) 1	Climate—Germany
Subcommittee	(SC) 3	Component—Thermal Performance/Reliability and Durability (Colors)—France
Subcommittee	(SC) 4	Systems—Thermal Performance/ Reliability and Durability—U.S.

SC 3 consists of two working groups—one on materials (France) and one on collector testing (Canada, SCC).

The program of work for SC 1 includes four high priorities:

1. specification and classification of instruments for measuring global and direct beam solar radiation,
2. calibration of a primary reference pyranometer using a primary reference pyrheliometer,
3. characterization of the primary reference pyranometer, and
4. transfer of calibration from primary reference pyranometers to field pyranometers.

Secondary priority has been given to four other standards development areas:

1. transfer of calibration from primary reference pyrheliometers to field pyrheliometers,
2. standard practice for use of field pyranometers,
3. standard reference solar spectrum for direct beam and global radiation, and
4. standard practice for measuring meteorological data in solar applications.

Under the auspices of the American National Standards Institute (ANSI), the ISO member body for the United States, a U.S. Technical Advisory Group (TAG) was organized in the spring of 1981 to represent the technical positions of the U.S. industry and the U.S. solar community of interest in ISO/TC 180 activities. The U.S. TAG is made up of technical experts selected from and appointed by the various standards organizations with pertinent interests in the fields represented by the TC 180 program of work. The U.S. TAG has members who represent the American Society of Heating, Refrigeration, and Air-Conditioning Engineers, ASTM (Committee E-44), the Solar Energy Industries Association, and the American Solar Energy Society. This representation ensures that the U.S. delegation to technical committee meetings and meetings of its subcommittees and working groups is properly briefed on U.S. positions as represented by constituent organizations.

The U.S. TAG is organized into Subcommittee Advisory Groups (SCAG), which possess the pertinent technical expertise of the TAG for each of the ISO/TC 180 subcommittees and working groups. Each SCAG must ascertain and develop a U.S. consensus position for each item of work on each of the subcommittees and working groups within TC 180; these SCAGs must develop positions essentially independent of the position of the member body of any other country. The delegation to ISO meetings, selected by TAG, must make whatever compromises are necessary and expedient as long as they are in the best interest of that country's industry position.

The relationship between domestic and international standards in solar radiometry dealing with solar energy use (thermal and photovoltaic applications) is presented in figure 5.49. In the United States the official member of ISO is ANSI, the coordinating organization for domestic standards in the United States. The counterparts to ANSI are, for example, the Deutsches Institut für Normung (DIN) in Germany, and British Standards Institution (BSI) in Great Britain.

Since the SCAG of the U.S. TAG for SC 1 (Climate) consists largely of members from ASTM's Committee E-44, a close liaison exists between the U.S.-developed documents within the relevant domestic consensus standards developing group, namely ASTM's Subcommittee E-44.02, and the U.S. TAG.

5.9 Calibration Standards for Pyrheliometry

5.9.1 Future Comparisons of Absolute Cavity Radiometers

Although the WMO-sanctioned International Pyrheliometer Comparisons can be expected to be held at Davos every 5 years, the participants of the New River intercomparisons decided that the NRIP should be continued

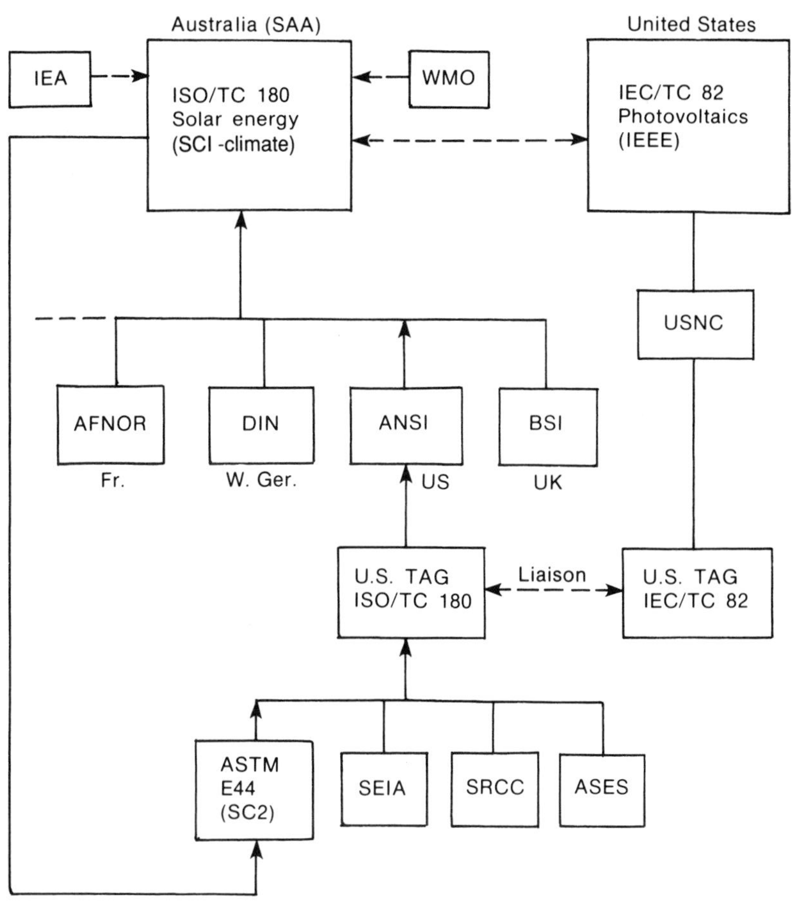

IEA - International Energy Agency
ISO - International Standards Organization
WMO - World Meteorological Organization
IEC - International Electrotechnical Commission
USNC - U.S. National Committee
AFNOR - Association Francais pour Normalization
DIN - Deutsches Institute fur Normung
ANSI - American National Standards Institute
BSI - British Standards Institution
ASTM - American Society for Testing and Materials
SEIA - Solar Energy Industries Association
SRCC - Solar Rating and Certification Corporation
ASES - American Solar Energy Society

Figure 5.49
Relationship between domestic and international standards in solar radiometry. Source: Zerlaut (1986).

every 2 years. (NRIP VIII, which was scheduled for October/November 1987, was not held because of funding limitations.)

The U.S.-based intercomparisons are important because they are increasingly cited for traceability purposes in the United States. The participants of NRIP I–NRIP VII have agreed that official recognition by WMO is desirable and that broader participation by South and Central American countries, as well as the United States and Canada, should be sought. Indeed, it is hoped that the designation of an NRIP as WMO Region IV Intercomparisons will engender greater participation from laboratories throughout the hemisphere. Even without official recognition, though, the protocols developed for the NRIP activities have in themselves become de facto standard procedures.

5.9.2 Transfer of Calibration from Reference to Field Pyrheliometers

The calibration of reference and field pyrheliometers is covered by ASTM Standard E 816 entitled *Calibration of Secondary Reference Pyrheliometers and Pyrheliometers for Field Use* (ASTM, n.d.1). The standard is limited to pyrheliometers with field angles of 5–6°; for calibration of secondary reference pyrheliometers, the primary reference instrument must be a self-calibrating absolute cavity radiometer directly traceable to the WRR through participation in an IPC (V or VI) or in one of the NRIP series intercomparisons. A secondary reference pyrheliometer so calibrated may be used to transfer calibrations to pyrheliometers for field use. Neither primary nor secondary reference instruments may be field pyrheliometers, and their exposure to sunlight must be limited to calibration or intercomparison activities.

The standard discusses such interferences and cautions as sky conditions, comparisons between instruments of different opening angles and spectral responses, and wind-induced errors.

The procedure calls for performing simultaneous instantaneous measurements of direct solar irradiance with identically tracked reference and field pyrheliometers. Measurements are taken every 30 s for 10 min to produce a 21-value measurement set. Five test sets are required on each of 3 days with data required 2 h before and after solar noon. The instrument constant of the test instrument is obtained simply by dividing the voltage output of the test instrument by the irradiance determined by the reference pyrheliometer.

An international standard dealing with the same general method has been proposed in ISO/TC 180 (solar energy), SC 1 (climate). The document has been prepared and is currently being circulated for ballot to experts from member countries as a Draft Standard (the first step in promulgation of an International Standard). It differs from the ASTM standard primarily in definitions and in the data-taking sequences specified.

5.10 Calibration Standards for Pyranometry

5.10.1 Problems in Pyranometry

A number of comprehensive reviews and symposia dealing with the calibration and characteriziation of solar radiation instruments has been reported in the past several years. Bahm and Nakos (1979), in a review sponsored by DOE, have discussed such aspects of calibration as linearity and stability of response, incident angle and spectral response characteristics, and temperature effects. In support of IEA's Task IV project on developing an insolation measurements package, DOE sponsored a compilation of papers dealing with the physics of solar radiation and atmospheric interaction, total and spectral irradiance methodology, and meteorological aspects related to solar energy (Riches, 1980). One of the most comprehensive seminars convened to investigate instrumental problems in pyranometry was the January 1984 Pyranometry Symposium, held under the auspices of IEA's Task IX Solar Radiation and Pyranometry Studies (Wardle and McKay, 1984). This symposium covered operational practices in national networks, laboratory characterization studies, the physics of pyranometers and field characteristics, and performance of pyranometers.

5.10.1.1 Incident Angle Effects Cosine and azimuth errors, or deviations, are simply two descriptions of the one effect of deviation from Lambert's cosine law at incident angles greater than 0°. Some researchers prefer to describe cosine response as the deviations from the cosine law in two orthogonal planes, each passing through the axis of the pyranometer, with azimuth response being the deviations from Lambert's cosine law at a given incident angle, or solar altitude, when the horizontally mounted pyranometer is turned through 360° (see figures 5.1 and 5.2). Others refer to cosine reponse as the deviation from the cosine law at the incident angle defined at solar noon for a given time of year or solar altitude. Azimuth response then is a set of response factors that track the pyranometer circumferentially for each incident angle of interest. Such azimuthal plots may then be used to characterize the pyranometer for all the solar altitudes it will encounter throughout the year, whether mounted horizontally or at various tilts from horizontal.

The laboratory assessment of cosine and azimuthal correlation may be performed on an optical bench with the pyranometer mounted in the vertical position on a goniometer stage. In this method, used by Statens Provningsanstalt (Borås, Sweden), the pyranometer is illuminated by a collimated light source. The pyranometer rotates on one of the vertical stages to give an azimuth response at the cosine angle selected (adjusted by the second, horizontal stage). A similar arrangement is proposed by ASTM Subcommittee E-44.02 on Environmental Parameters.

In the draft method, both the cosine and azimuthal corrections are determined on the apparatus depicted in figure 5.50. For cosine response measurements, the pyranometer is mounted with axis vertical on the turntable and is positioned so the source can be swung in a vertical plane through an arc centered on the pyranometer's receiver (or entrance aperture). With the source intensity held constant, the response is measured. The cosine error is computed as the ratio between the response at any angle divided by the product of the response on axis (at normal incidence) and the cosine of the angle of concern. Using the same apparatus, the pyranometer is positioned on the turntable, which is a precision rotating stage, so its vertical axis is precisely colinear with the axis of rotation of the stage. The source is swung in a vertical plane to a predetermined incident angle (representing the altitude of the sun for a horizontally mounted pyranometer). With the source intensity held constant, the pyranometer response is measured as a function of the azimuth position through the azimuthal angles -180° to $+180^\circ$, where 0° represents true south with the electrical connector positioned at $\pm 180^\circ$. The azimuth response is normalized to 0°.

Azimuthal response plots may be developed in both normal-normal and polar coordinates. Polar coordinates, in turn, may be done in several ways. The plot shown in figure 5.51 represents isopleths of azimuth response for a Kipp and Zonen Model CM5 pyranometer (McGreggor, 1982). A different polar representation is given in figure 5.52, where all data are normalized to the normal incident pyranometer response.

Actually, there is considerable disagreement within the solar radiation community regarding the appropriateness of the proposed indoor characterization as well as the actual usefulness of such data. Flowers (1984a) has stated that similar carefully done laboratory experiments by himself as well

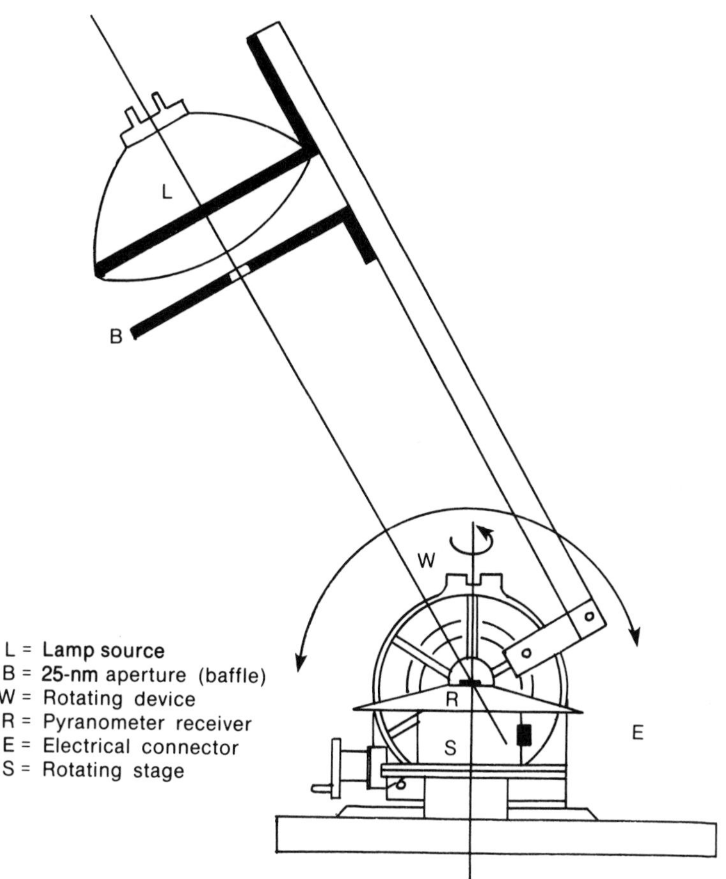

Figure 5.50
Apparatus for determining cosine and azimuth response correction factors to pyranometers.

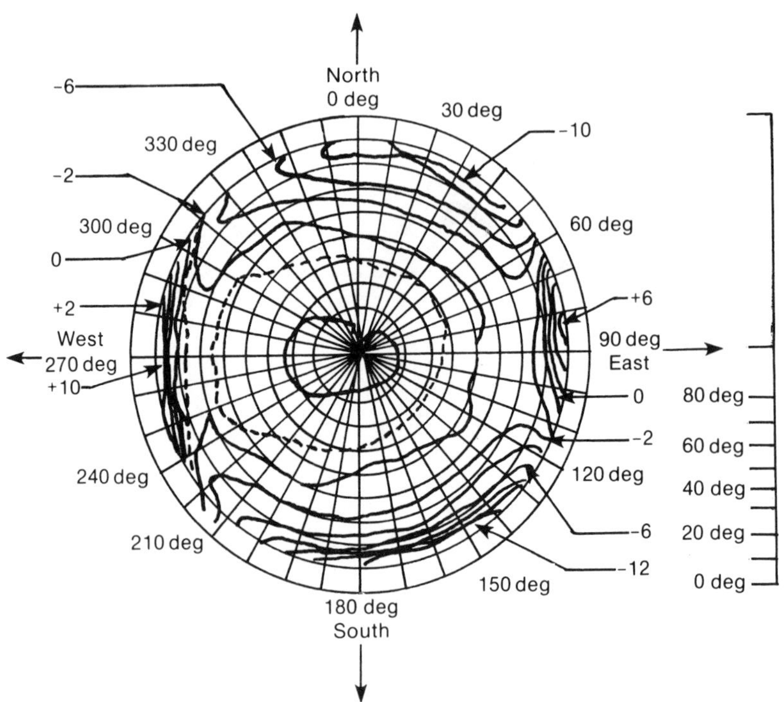

Figure 5.51
Combined azimuth and elevation response for Kipp and Zonen Model CM5 SN 76-3154—percentage response relative to zenith measurement. Source: McGreggor (1982).

as others have produced results that do not agree with results of outdoor experiments for certain instruments. He has pointed out that the source beam in figure 5.50 must be well-collimated and must cover the glass domes (since they contribute to cosine/azimuth response errors), that the precise alignment of the pyranometer axis and the source beam for all angles of incidence of interest is not a trivial problem, and that it is important that the color temperature of the source be as close to 5,700K as possible. Furthermore, this objection, coupled with the recognition that the standard ought not to be balloted until data are obtained using the apparatus of figure 5.50, has delayed further consideration of this method in the United States.

Andersson et al. (Ambrosetti et al., 1984; Andersson et al., 1981) have performed cosine and azimuth corrections to pyranometers and have reported reasonable agreement between laboratory and field behavior.

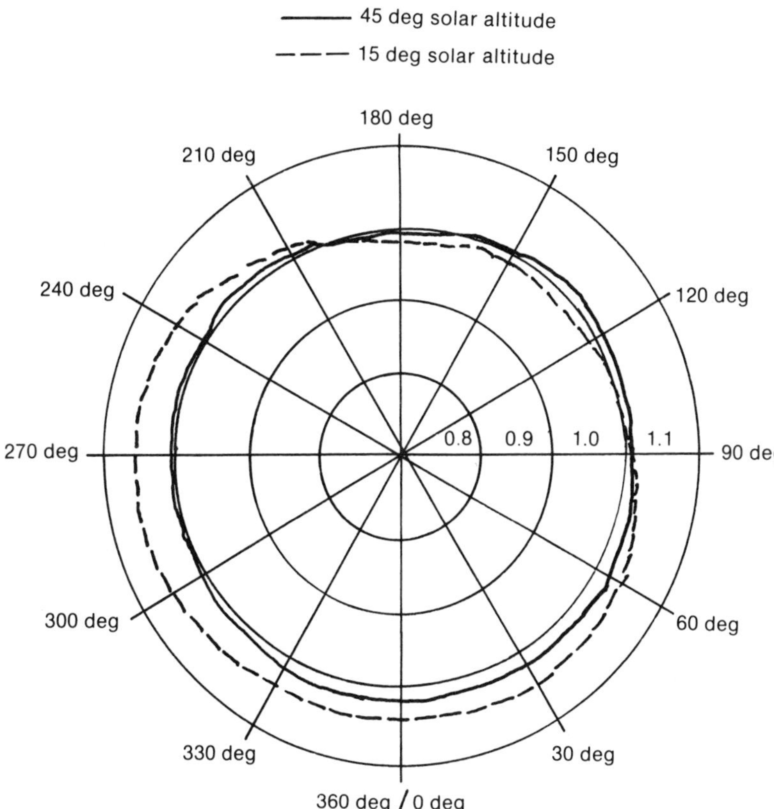

Figure 5.52
Response in polar coordinates of a typical Eppley Model PSP pyranometer in terms of normal incidence response.

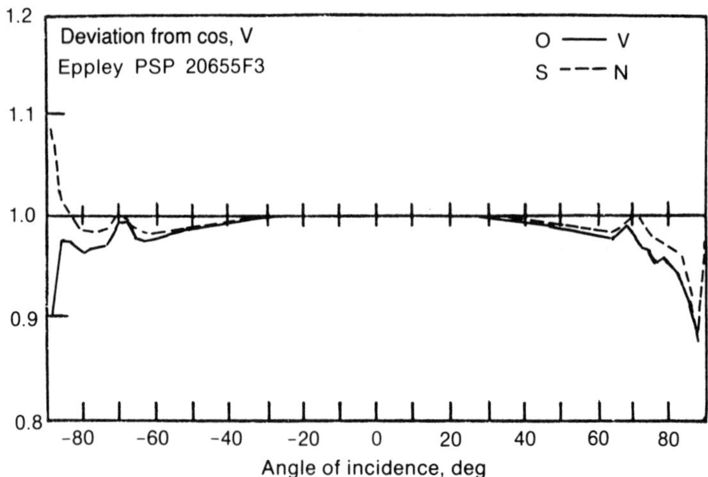

Figure 5.53
Deviation from cosine response for Eppley Model PSP SN 20655. Source: Andersson et al. (1981).

Cosine deviations for an Eppley Model PSP and a Kipp and Zonen Model CM5 are shown in figures 5.53 and 5.54, respectively; azimuth response at incident angles of 45° and 75°, corresponding to solar altitudes of 45° and 15°, is shown in figures 5.55 and 5.56, respectively, for these two pyranometers.

One of the problems elucidated at DSET is that using indoor techniques for the symmetrical characterization of pyranometers is inappropriate—no matter how accurate the family of azimuthal plots might be. Unless the complete set of characterization data is used as file information in the computer used to reduce solar collector efficiency test data, simple incident angle or even azimuthal plots of the response error as a function of season (at the range of solar altitudes throughout a test day for a given time of year) may be much more useful. For normal incident testing of solar collectors using an altazimuth follow-the-sun mount (as performed at DSET and several other solar collector test laboratories), the problem reduces to the accurate calibration of the pyranometer at normal incidence, and the characterization of the pyranometer as a function of angle of incidence in two orthogonal planes of the pyranometer at the tilts defined by the general time of year. This simplified cosine calibration is required for bidirectional incident angle modifiers [$k_{\tau,\alpha}$ in equation (5.1)] required by

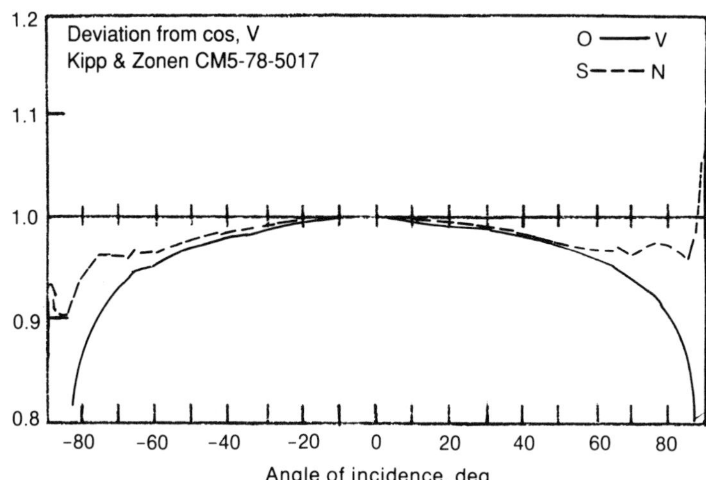

Figure 5.54
Deviation from cosine response for Kipp and Zonen Model CM5 SN 78-5017. Source: Andersson et al. (1981).

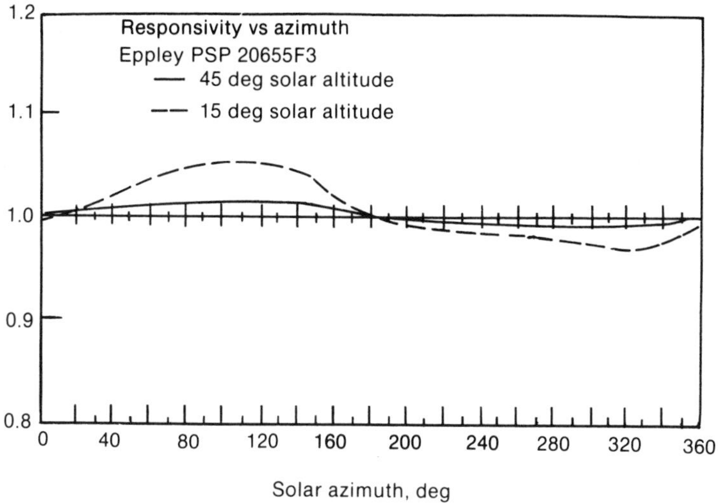

Figure 5.55
Responsivity of Eppley Model PSP SN 20655 as a function of azimuth angle relative to 180°S. Source: Andersson et al. (1981).

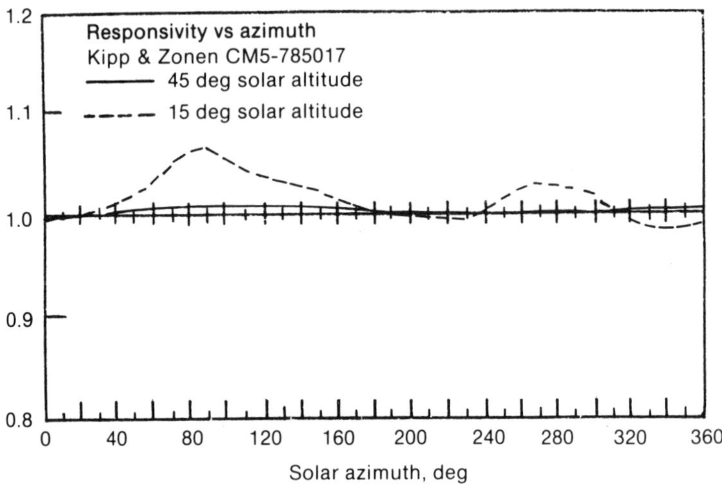

Figure 5.56
Responsivity of Kipp and Zonen Model CM5 SN 78-5017 as a function of azimuth angle relative to 180°S. Source: Andersson et al. (1981).

ASHRAE Standard 93-77 (1977). A more complete characterization of the pyranometer is required to determine the incident angle modifiers of flat-plate collectors outside the two orthogonal planes of the collector.

Such incident angle plots have been done at SERI in support of the IEA activities discussed in section 5.7.1.6 (Riches, Stoffel, and Wells, 1982), figures 5.36 and 5.37.

Data on relative incident angle response errors are now provided in DSET's calibration reports in the format presented in figures 5.40–5.48. The furnished graph plots the ratio of the test pyranometer's irradiance to that of the reference instrument as a function of incident angle at 0° horizontal and a tilt-from-horizontal of 45°. However, relative incident angle response curves can be misleading when the reference pyranometer exhibits significant sensitivity deviations at high incident angles. Fortunately, the component summation method discussed in section 5.10.2.4 permits one to determine absolute incident angle response curves with little difficulty.

5.10.1.2 Tilt Effects The effect associated with changes in pyranometer sensitivity as a function of tilt has become well known over the past several years. In the field, tilt can be separated from cosine response errors by accurate normal incidence shading disk calibrations at solar noon, when the solar elevation is greatest (during summer months in the Northern Hemisphere), compared with solar noon calibrations at other times of the year that define the tilt at normal incidence. Although one could perform normal incident shading disk calibrations at lower morning and afternoon solar elevations, errors may be introduced if the irradiance levels are less than about 70% of the near-horizontal values.

Tilt errors as high as 13–15% have been observed for certain WMO Class II pyranometers that have an alternating black-and-white, segmented receiver—the so-called star-types. These errors are largely believed to be caused by thermodynamic cooling of the thermopile hot junction as a result of an increase in convective heat transfer at tilts from the horizontal. Although Norris (1974) has reported tilt errors of 11% for Eppley Model PSPs at vertical (90° orientations), most researchers have observed errors of only about 1–2% at vertical orientations. In our own studies, we observe response errors of less than 0.5% for Model PSP pyranometers tilted at 45° from the horizontal. Loxom and Hogan (1981) have also identified another type of tilt effect as a result of solar heating of the pyranometer body because the sunshade at moderate solar elevations is ineffective. They claim to have solved this problem by placing a white insulating jacket around the pyranometer body.

The previously mentioned ASTM Committee E-44 draft standard (Document No. 142) contains a prescriptive procedure for measuring the effect of tilt from the horizontal on the response characteristics of pyranometers. This method provides for mounting the pyranometer and a strong source of solar radiation at opposite ends of a rigid tilt table, or bar, as shown in figure 5.57. With the tilt fixture assembled in a darkened room, the pyranometer and source are swung in a vertical plane. Pyranometer response readings are taken at appropriate angles of tilt and the tilt correction is computed as the ratio of the response at tilt to the response at exactly 0° horizontal—that is, with axis vertical. The ASTM subcommittee agrees on the need for performing the tilt test at three irradiance levels (500, 750, and 1,000 W/m^2) and also agrees on the need for matching the color temperature of the sun as closely as possible. However, there is general disagreement on the requirement for a baffle (noted in figure 5.57) to illuminate only the pyranometer dome. The most recent version of this draft standard cautions that selecting the source is critical since it is essential that the radiant characteristics of the source's burner, or filament, do not change as

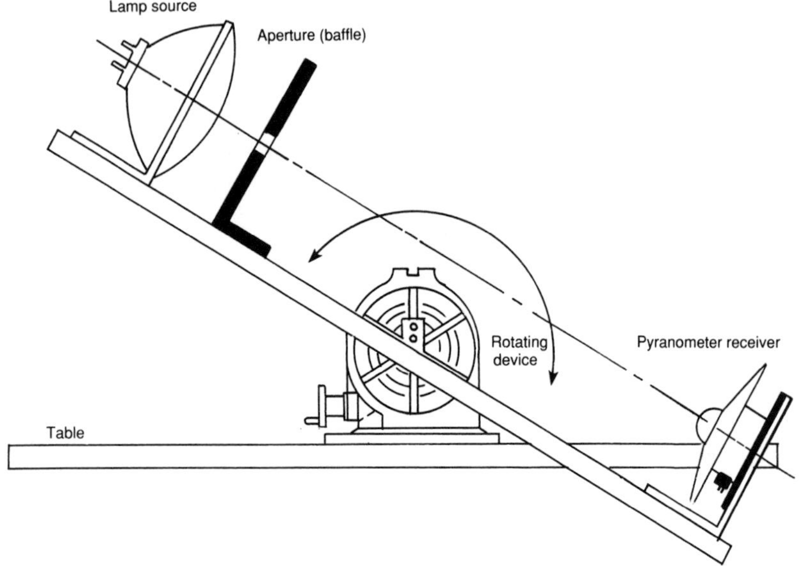

Figure 5.57
Apparatus for determining tilt correction factors of pyranometers.

a function of gravitational effects at tilt. The method requires that the source be monitored during the response measurements with a silicon solar-type cell detector not subject to tilt effects.

Andersson et al. (1981) have measured the tilt response of several pyranometers using a similar tilt table. Figures 5.58, 5.59, and 5.60 present selected data. They showed that Model PSP pyranometers have little error associated with tilt effects. At tilts greater than 45°, they observed nearly 2% tilt error in the two Kipp and Zonen Model CM5 instruments and nearly 4% error in the measurements from the two Schenk Star pyranometers. Using an enclosed tilt box similar to that depicted in figure 5.57, Flowers (1984b) measured the tilt response of a number of pyranometers at an irradiance level of about $450 W/m^2$. Figure 5.61 shows typical data for Eppley Model PSP, Kipp and Zonen Model CM5, and Spectrolab Model SR73 pyranometers. These data do not agree in sign (that is, direction) with that of Andersson for Eppley Model PSPs, but, again, the magnitudes are small and the difference might be accounted for by the significant difference in irradiance levels used by the two authors.

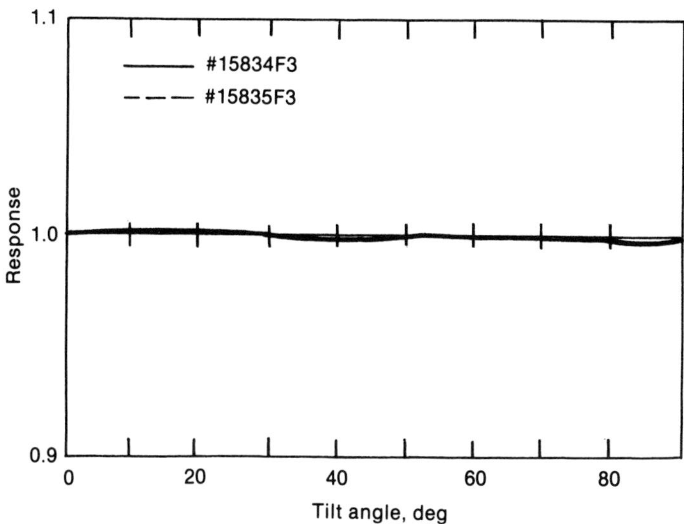

Figure 5.58
Variations in response of Eppley Model PSP pyranometers with tilt from the horizontal. Source: Andersson et al. (1981).

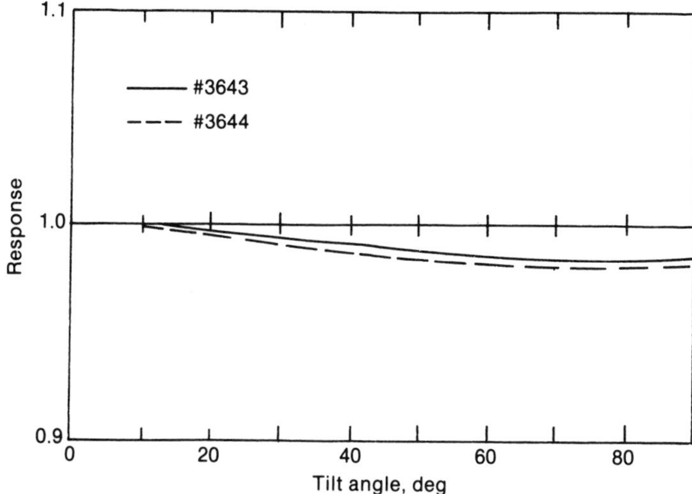

Figure 5.59
Variations in response of Kipp and Zonen Model CM5 pyranometers with tilt from the horizontal. Source: Andersson et al. (1981).

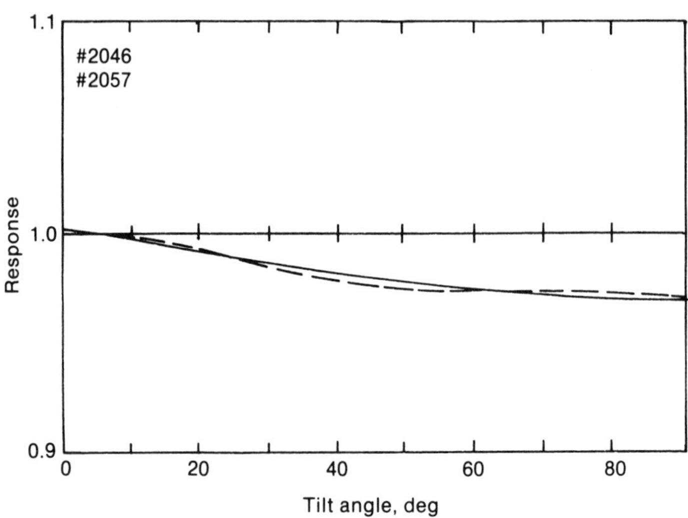

Figure 5.60
Variations in response of Schenk Star Model 8101 pyranometers with tilt from the horizontal. Source: Andersson et al. (1981).

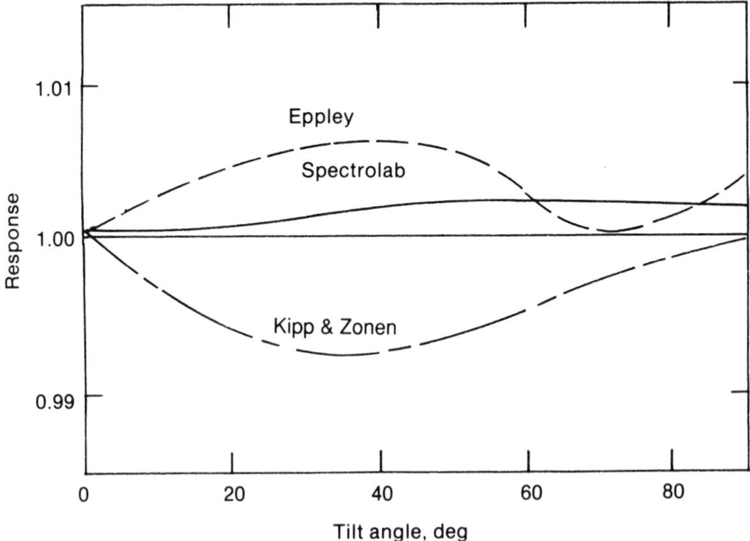

Figure 5.61
Variations in response of three pyranometers as a function of tilt from the horizontal. Source: Flowers (1984b).

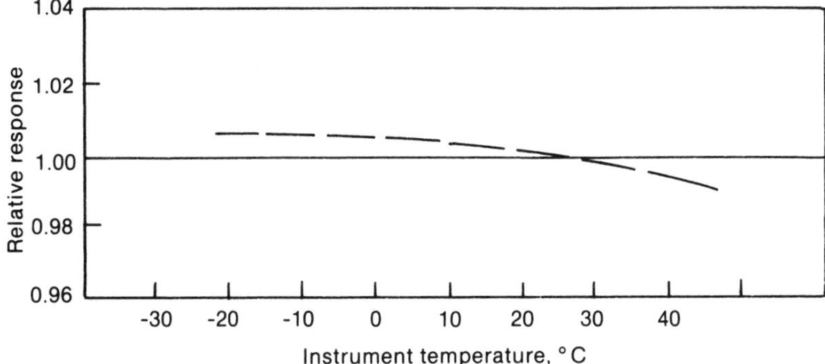

Figure 5.62
Temperature dependence plot of the instrument constant of Eppley Model PSP SN 19129 pyranometer. Source: Eppley Laboratory (n.d.3).

5.10.1.3 Temperature Response Problems in Pyranometry For a number of years, the Eppley Laboratory has furnished temperature response curves with every Model NIP pyrheliometer and PSP pyranometer sold. We at DSET Laboratories, Inc., regularly use these data in our solar collector performance testing procedures by adjusting pyranometer calibration values throughout each day to account for changing ambient temperature. A typical sensitivity plot is presented in figure 5.62 for DSET's working standard Eppley Model PSP, SN 19129. Although we have learned that few laboratories adjust the sensitivity of their field instruments during collector testing, we use the temperature sensitivity function of all pyranometers in precision irradiance measurements whether temperature is compensated or not. Both Andersson et al. (1981) and Flowers (1981, 1984b) have measured the response changes in pyranometers as a function of temperature by heating the constantly irradiated instrument in a temperature-controlled chamber. Their data are in excellent agreement for the Eppley Model PSP, Kipp and Zonen Model CM5, and Schenk Star pyranometers studied and are typified by the plots presented in figure 5.63.

5.10.1.4 Linearity of Response Most researchers using pyranometers assume they are dealing with radiometers whose instrument constants are linear with an irradiance level in the range of interest. This is not necessarily true, as shown by Andersson et al. (1981). Although the maximum variation in the sensitivity observed in Model PSP pyranometers by the authors cited was on the order of only a couple of tenths of a percent between 200

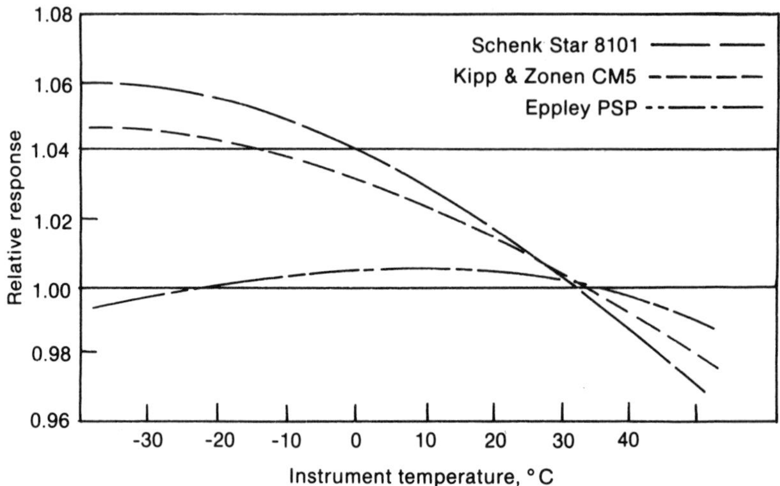

Figure 5.63
Temperature dependence plot of the instrument constant of three different pyranometers. Sources: Andersson (1981) and Flowers (1981, 1984b).

and 1,000 W/m^2, sensitivity decreased by about 5% in the same range in the two Schenk Star and two Kipp and Zonen Model CM5 pyranometers.

Significant errors can be introduced in determining the incident angle modifier $k_{\tau,\alpha}$ [equation (5.1)] required in thermal performance testing of solar collectors as a result of linearity effects combined with other factors. These include potential cosine deviation at the higher incident angles required by ASHRAE Standard 93-77 (1977) and the combination of disproportionate convective cooling at low irradiance levels and tilt errors at concomitant high tilts from horizontal.

5.10.2 Calibration of Primary Reference Pyranometers

5.10.2.1 General Considerations Until an absolute cavity or a self-calibrating pyranometer becomes available that has a 180° solid angle field of view and a cosine response approaching the theoretical at high angles of incidence (that is, low solar elevations), the only outdoor method for calibrating primary reference pyranometers involves comparison to absolute cavity pyrheliometers (see section 5.4.1), or normal incident pyrheliometers. Transfer of calibration from pyrheliometers to pyranometers can only be accomplished by limiting the field of view of the pyranometer—

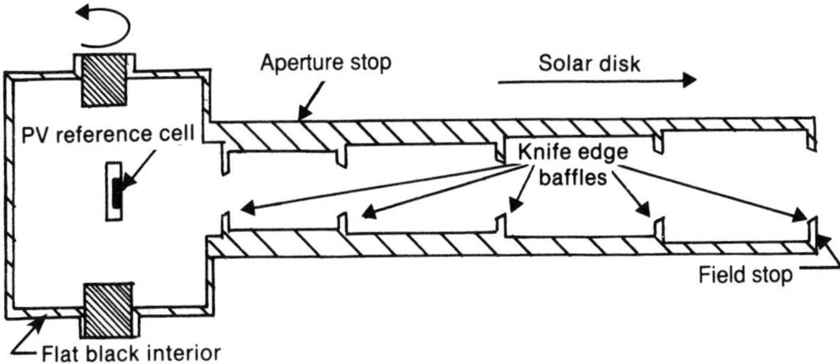

Figure 5.64
Pyrheliometer-comparison tube for calibrating photovoltaic reference cells to direct normal irradiance.

either by a pyrheliometer-comparison tube that occludes the sky or by one of two shade-disk methods that occludes the sun. In the first case, transfer is direct as long as the slope and opening angles of the collimator closely agree with the primary reference pyrheliometer. In the more traditional of the shade methods, the comparison is indirect and is made by dividing the difference between the pyranometer's voltage response for the alternately unshaded and shaded conditions by the irradiance measured with the primary reference pyrheliometer. In the newer shade technique, commonly termed the component summation method, the comparison is direct and is achieved by dividing the signal of the unshaded test pyranometer by the sum of the measured direct irradiance and diffuse (sky) irradiance measured with a continuously shaded reference pyranometer.

Although there are pros and cons to each method, the pyranometer-comparison tube method has fallen into some disfavor, and the shading disk approach is the more or less official technique prescribed by WMO and its constituent participants. However, the component summation version of the shade method is finding increasing favor throughout the world.

5.10.2.2 Occulting-Tube Approach A typical pyrheliometer-comparison tube consists of a blackened enclosure for mounting and optically aligning the pyranometer exactly beneath a baffled occulting tube. Figure 5.64 shows the apparatus used at DSET for calibrating photovoltaic standard reference cells.[10]

Although this method permits direct transfer of instrument constants from the primary reference pyrheliometers, it has several disadvantages. Chief among these disadvantages are solar heating-induced drift of pyranometer response, optical misalignment between the occulting tube and pyranometer receiver, and optical interaction (for example, stray light) between the domes and pyranometer case and the view-limiting aperture stop and enclosure. (The DSET pyrheliometer-comparison tube used in photovoltaic reference cell calibrations has a rotating axle that permits precise determination of the cell's cosine response to beam radiation. Such a device is impractical for domed pyranometers because of the above-mentioned alignment and interaction problems.)

Although a ventilated occulting-tube calibration instrument could most likely be constructed to transfer calibration accurately to a specific pyranometer design such as the Eppley Model PSP, residual optical alignment and vignetting effects most likely will produce errors much greater than experienced with the component-summation shading disk techniques.

5.10.2.3 Shade Disk Approach (Intermittent Shade) The shading disk technique for transferring calibration to pyranometers from pyrheliometers has been used throughout the world for many years; however, no standard methods of calibration were promulgated by user nations until 1982 when two methods were prepared in ASTM Subcommittee E-44.02. These methods deal with shading disk calibrations of pyranometers with the axis vertical (horizontal receiver) and the axis tilted (ASTM, n.d.3, n.d.4). These methods require calibration to the WRR scale. (Traceability to IPS 1956 is not permitted.)

In both methods, two types of calibrations are specified: Type I, checked against a self-calibrating absolute cavity pyrheliometer (a primary standard pyrheliometer), and Type II, checked against either a secondary standard pyrheliometer or a first class pyrheliometer that must be calibrated by transfer from an absolute cavity pyrheliometer in accordance with ASTM Standard E 816 (n.d.). In both methods, direct traceability is required to either IPC IV or V, to NRIP I–NRIP VI, to any future intercomparisons of comparable reference quality, or to any absolute cavity pyrheliometer tested in those intercomparisons.

The methods were substantially taken from an informal consensus developed during discussions with experts in the United States and other countries. The methods require mounting the pyrheliometer on either an equa-

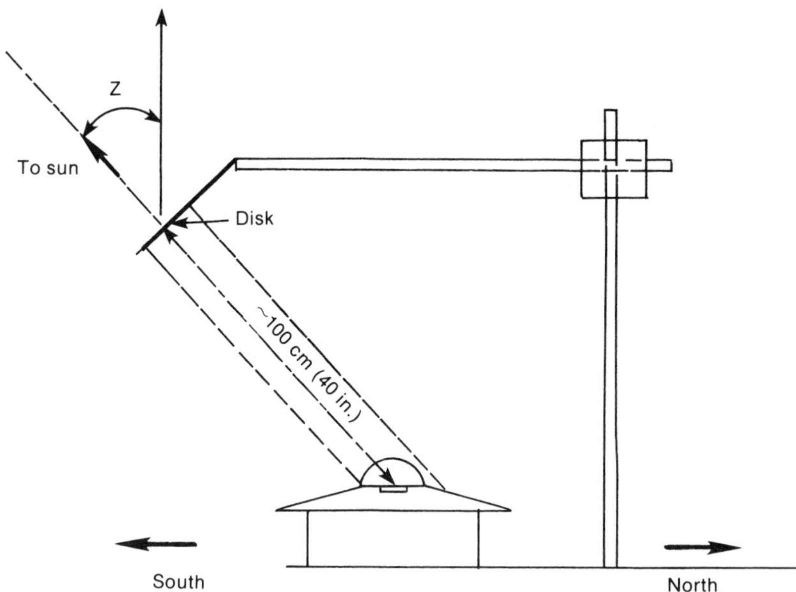

Figure 5.65
Shading disk apparatus for calibrating pyranometers with axis vertical, that is, with the receiver in the horizontal plane. Source: ASTM (n.d.3).

torial or altazimuth sun-following tracker and mounting the pyranometer on the appropriate test stand equipped with a shade disk. The geometrical relation is shown in figures 5.65 and 5.66 for axis vertical and tilted, respectively. (Although the shading disk technique can be performed at any incident angle on the tilted pyranometer receiver, figure 5.66 shows the specific case of normal incidence [0°] at the tilt angle defined by time of day.) Essentially, requirements in both methods are that the shadow cast by the disk completely cover the dome of the pyranometer, and that the field of view occulted by the shade disk equal that of the pyrheliometer used as the reference radiometer. Because Eppley Model NIP pyrheliometers have traditionally been used as reference pyrheliometers, especially before the wide availability of absolute cavity radiometers, the 5.7° field of view used is twice the opening angle. Practical considerations have dictated that a 10-cm-diameter disk located 100cm from the center of the pyranometer's receiver adequately shadows the domes and provides a field of view of 5.7°checked against the center of the receiver.

It is important to understand that such a shadow only poorly simulates the point-source nature of the sun, and although the opening angles (fields

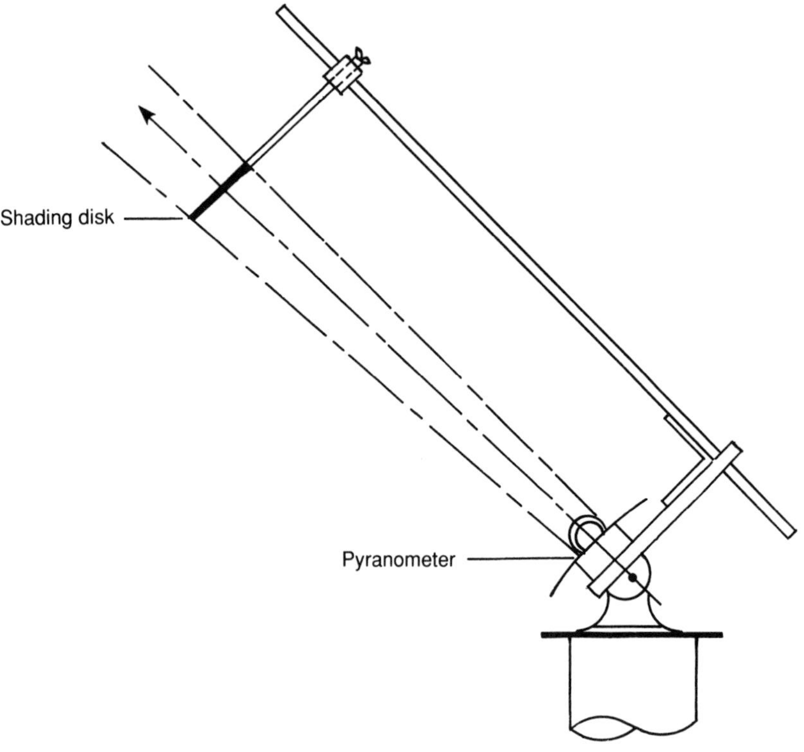

Figure 5.66
Shading disk apparatus for calibrating pyranometers with axis tilted from the horizontal—normal incidence mode shown. Source: ASTM (n.d.4).

of view) can be matched to the reference pyrheliometer, the slops angle cannot [see equations (5.3) and (5.4) and figure 5.10 in section 5.3.3.1]. As a result, large amounts of circumsolar radiation can result in small but significant errors in transfer of calibration by this method.[11] Furthermore, no clouds should be within a 30° solid angle of the sun, and calibrations should be done only under minimum acceptable sky conditions represented by a direct beam component of 0.80 or greater.[12]

The methods require that the pyranometer be alternately shaded and unshaded and that the difference between the two response signals be divided by the product of the beam irradiance (determined by the reference pyrheliometer) and the cosine of the incident angle of the direct beam component:

$$k = \frac{V_u - V_s}{I_d \cos\theta}, \tag{5.11}$$

where k is the instrument constant in $\mu V\, W^{-1}\, cm^2$; V_u and V_s are the voltage signals of the pyranometer when unshaded and shaded, respectively; I_d is the direct beam irradiance; and θ is the angle of incidence of the direct beam to the plane of the pyranometer's receiver.

For ASTM Standard E-913 with axis vertical, the sun's zenith angle z and the cosine angle are identical. In this case,

$$\cos\theta = \cos z = \sin\phi \sin\delta + \cos\phi \cos\delta \cos\omega, \tag{5.12}$$

where z is the sun's zenith angle; ϕ is the station latitude; δ is the solar declination, $\phi = 23.45 \sin[9863(n \pm 283.4)]$; and ω is the hour angle from solar noon, determined at $\pm 15°$ per hour on each side of solar noon.

For ASTM Standard E-941 with axis tilted, the angle of incidence is computed as

$$\begin{aligned}\cos\theta = {} & (\sin\delta \sin\phi \cos\beta) - (\sin\delta \cos\phi \sin\beta \cos\gamma) \\ & + (\cos\delta \cos\phi \cos\beta \cos\omega) \\ & + (\cos\delta \sin\phi \sin\beta \cos\gamma \cos\omega) \\ & + (\cos\delta \sin\beta \sin\gamma \sin\omega),\end{aligned} \tag{5.13}$$

where ϕ is the latitude, δ is the declination (defined above), β is the tilt angle from the horizontal, γ is the azimuth angle with 0 being due south (east negative and west positive), and ω is the hour angle (defined above). For the special case of shading disk calibrations at normal incidence, equation (5.13) reduces to equation (5.14):

$$k = \frac{V_u - V_s}{I_d}. \tag{5.14}$$

Historically, the shade/unshade timing sequence has varied from laboratory to laboratory. In the United States, the Eppley Laboratory, Inc., NOAA/SRF at Boulder, and DSET Laboratories, Inc., have all used different sequences. These include 6 min unshaded/5 min shaded, 1 min unshaded/1 min shaded, and 30 s unshaded/30 s shaded. The formula specified in ASTM Standards E-913 and E-941 result in a shade/unshade sequence of 30 s each. However, as discussed in the following sections, these ASTM standards are being rewritten first to indicate a preference for the

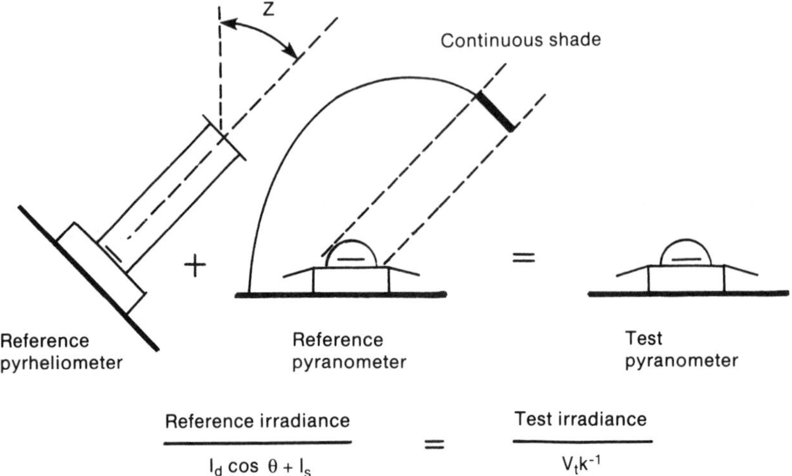

Figure 5.67
New shading disk method for calibrating pyranometers: the component-summation (two-reference-radiometer) method. Source: Zerlaut (1986).

component summation method and second to suggest that the shade/unshade time sequence should consider the thermodynamics of the particular pyranometer being calibrated.

5.10.2.4 Component-Summation Method The major concerns with the intermittent-shade calibration method are (1) the relationship between the amount and distribution of diffuse radiation and the cosine response of the pyranometer and (2) thermodynamic interactions such as dome heating and cooling in relation to the pyranometer's time constant. These concerns are critical to properly selecting the shade/unshade timing.

Concerns such as these surrounding the intermittent-shade method have prompted the adoption of a two-reference-radiometer method that simply precludes the generation of significant thermal transients. This method is more correctly termed the component-summation method and requires the use of a reference pyrheliometer (preferably an absolute cavity radiometer) to measure the direct beam component and a well-characterized, continuously shaded pyranometer for the diffuse (sky) component. Figure 5.67 is a schematic of the component-summation method.

In this case, the calibration is derived from the relationship

$$k = \frac{V_t}{I_d \cos \theta + I_s} (V\, W^{-1}\, m^2), \tag{5.15}$$

where V_t is the signal of the unshaded test instrument, I_s is the irradiance measured with the continuously shaded pyranometer, I_d is the direct irradiance measured with the pyrheliometer, and θ is the incident angle (the solar zenith angle for a horizontally mounted pyranometer).

The uncertainties in the component-summation method are significantly less than in the intermittent-shade method. Because these calibrations should be performed only when the diffuse irradiance is less than 20% of total irradiance, the combined transfer uncertainties are much lower than the transfer uncertainties in the intermittent-shade method. For example, a 2% uncertainty in the measurement of the diffuse component by the continuously shaded reference pyranometer will contribute an uncertainty of only 0.3% to the measurement of the source irradiance when the diffuse component is 15%. Since direct irradiance can be measured to an accuracy of 0.3% by using an absolute cavity pyrheliometer, the combined uncertainties in the component-summation method are thus only about 0.6%, and are a major advantage in using the method. Indeed, Flowers (1985a) has adopted the component-summation method at NOAA/SRF and earlier reached similar conclusions regarding the advantages of this method, citing a combined accuracy for this method of 99.3% (or an uncertainty of 0.7%).

A second major advantage of the component-summation method is the ease of operation and the ability to calibrate numerous pyranometers at one time (compared with one-at-a-time calibration using the intermittent-shade method).

The principal disadvantage of the component-summation method is the necessity for ensuring the linearity and the accurate characterization of the reference pyranometer used to measure the diffuse component. Although Forgan (1983) has provided an iterative method for assuring the accuracy of the measured diffuse component using two reference pyranometers, that is, three reference radiometers, some experts believe that the proper technique is to use the intermittent-shade method carefully to calibrate the reference pyranometer for subsequent component-summation calibrations. However, I believe that additional studies are required to develop a shade/unshade duty cycle that precludes secondary time constants from affecting calibration results.

5.10.2.5 The Concept of Duty Cycle in Shade Calibrations The problem of providing an accurate instrument constant, or sensitivity function, for the continuously shaded pyranometer in the two-radiometer, component-summation calibration method is important in obtaining optimal accuracy in pyranometer calibrations.

During NRIP V in May 1982, considerable discussion centered upon the need for a detailed study of the concept of duty cycles, or illuminated-versus-shade-time durations. Traditional methods involved essentially 50% duty cycles at the several laboratories represented by participants in NRIP V; these ranged from 30 s/30 s to 5 min/6 min cycles. As a result of this discussion, the proposed ASTM standards dealing with the shade calibration of pyranometers were changed to reflect a 50% duty cycle based on multiples of the pyranometer's primary time constant. This resulted in shade/ illumination durations of only 30 s to 1 min for WMO first class pyranometers, such as the Eppley Model PSP and the Kipp and Zonen Model CM-11. The result was that this timing sequence became the operative duty cycle in the now-published ASTM Standards E-913 and E-941 (n.d.3, n.d.4).

The rationale for the ASTM timing sequence was based on a simple energy balance presented at the IEA Pyranometer Experts Meeting in Norrköping, Sweden, in January 1984 (Zerlaut and Maybee, 1984). Here the ASTM method received considerable criticism, some of which was entirely justified. In that paper, we correctly addressed the concept of energy balance but made some important errors: (1) the Eppley Model PSP pyranometer's first time constant is between 1.5 and 2 s (not 1 s) for the purposes of the duty cycle, (2) the pyranometer's signal growth curve is essentially identical to its decay curve and the growth curves do not quickly reach steady state as reported earlier, and (3) 20 s ($\sim$10 time constants) is insufficient for initiating data taking after either shading or unshading.

Subsequent analyses by Flowers (1985a) have elucidated the problem created by assuming an incorrect signal-growth curve. Additionally, Flowers correctly argues, backed by experimental evidence from his cited reference (reproduced in figure 5.68), that failure to reach steady state in both the unshaded (illuminated) and shaded cycles can lead to a quickly decreasing signal difference that indeed does not represent the true direct beam irradiance—resulting in instrument constants that are too low.

5.10.2.6 A Pyranometer Energy Balance I am continually plagued by the concern that equal steady-state duty cycles also cause an error—one that the following analysis indicates may be of essentially equal magnitude to those discussed by Flowers (1985a) but of opposite sign.

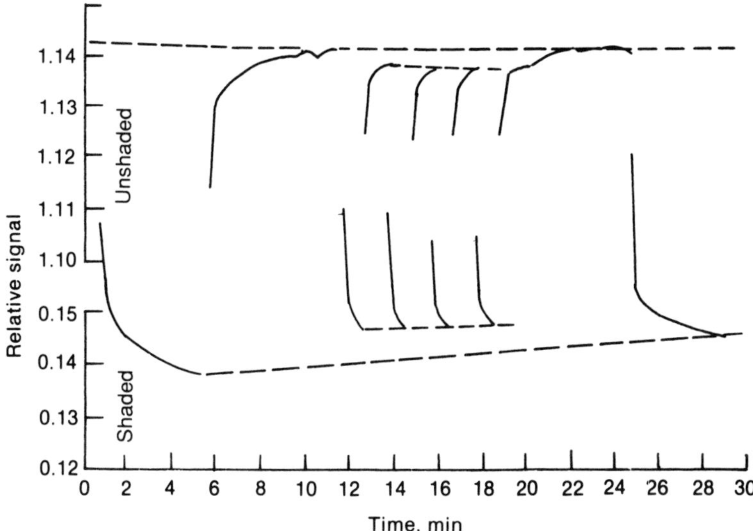

Figure 5.68
Duty cycle graph showing signal decay and growth curves during shade calibration of a pyranometer. Source: Flowers (1984b).

Consider the energy balance depicted in figure 5.69 and the signal growth/decay curve presented in figure 5.70. First, let the ordinate of figure 5.70 represent the typical signal of a Model PSP (shown in millivolts), and then let us accept the condition that a pyranometer has several time responses (the most responsive one being associated with the illumination of the receiver and slower time constants associated with the thermal irradiance of the receiver from heated domes and case surrounds).

Under steady-state illumination, it is therefore logical to assume that signal V_a is proportional to the sum of all contributions:

$$V_a = k[I_D + I_S + \sum Q], \tag{5.16}$$

where k is a proportionality constant we term the instrument sensitivity function, or the instrument constant. It is also easy to accept that in the steady-state shaded condition, V_a is simply proportional only to I_S, the sky irradiance—the contributions of the direct beam and thermal irradiance (due to dome and surrounds heating) having decayed to zero.

It then follows that there is some unknown point V_c where thermal

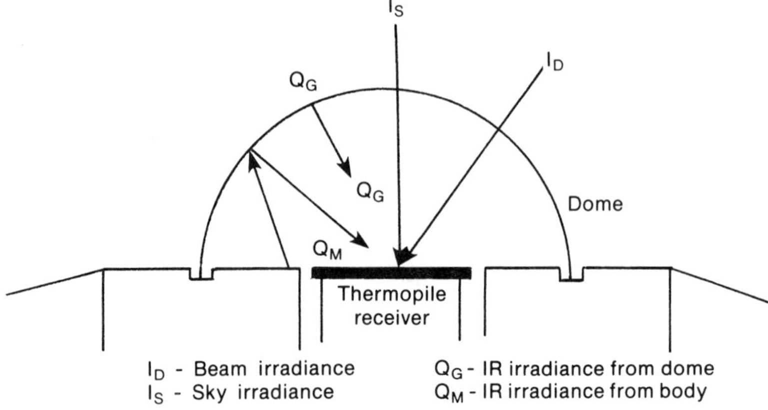

Figure 5.69
Schematic showing a suggested energy balance for an illuminated pyranometer (neglecting receiver losses).

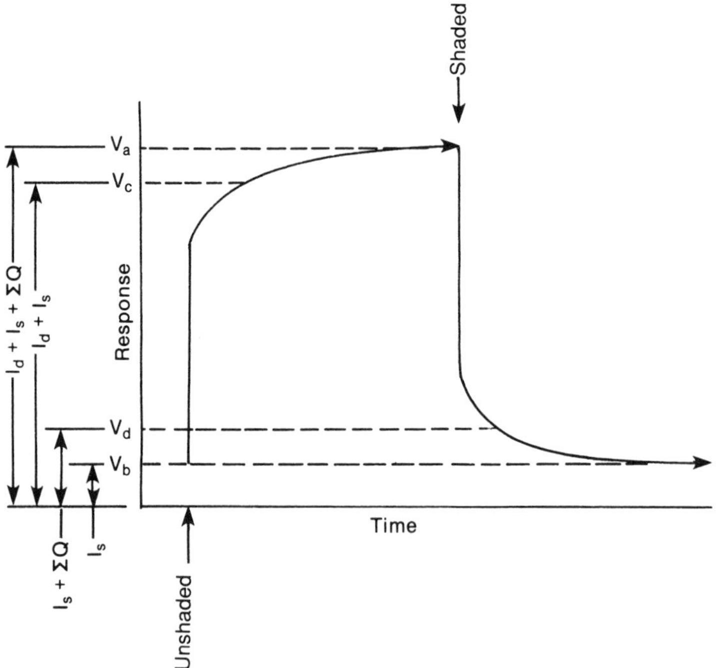

Figure 5.70
Duty cycle schematic of the intermittent shade calibration of a pyranometer.

heating is minimal and the contribution by the components I_D and I_S are saturated:

$$V_c = k[I_D + I_S]\,. \tag{5.17}$$

Likewise, the reverse is true; namely, that there is also some point where the copper receiver has lost its heat much faster than the glass and surrounds, and the signal V_d therefore represents the sum of the sky and thermal irradiance:

$$V_d = k[I_S + \textstyle\sum Q]\,. \tag{5.18}$$

Hence, the identities created by setting the measured beam irradiance I_B[13] equal to the difference of either $V_a - V_b$ (I) or $V_c - V_d$ (II) in figure 5.70 are incorrect, and either $V_a - V_d$ or $V_c - V_b$ are more apt to be correct. Also, to make a more precise shade calibration, the ratio of instrument sensitivity between the shaded and unshaded conditions must be known (if not unity). Letting

$$I_d = I_D - (I_D^r + I_D^a) \tag{5.19}$$

and

$$I_s = I_S - (I_S^r + I_S^a) \tag{5.20}$$

from figure 5.69, we can write the following identity for the steady-state, unshaded condition:

$$V_u = V_a = k_u[I_D - (I_D^r + I_D^a) + I_S - (I_r^S + I_S^a) + Q_B + Q_{G_d} + Q_{G_S}]\,. \tag{5.21}$$

For the nonsteady-state, shaded condition defined by ASTM standards E-913 and E-914, we have

$$V_s = V_d = k_s[I_S - (I_S^r + I_S^a) + Q_B^* + Q_{G_d}^* + Q_{G_S}]\,, \tag{5.22}$$

where k_u and k_s are the instrument sensitivities at the thermopile temperatures reached under the two conditions, and Q_B^* and $Q_{G_d}^*$ are decayed values of the thermal contributions from the domes and body (and where it is assumed that the thermal contribution to dome heating from sky radiation has not decayed). Since

$$kI_B = V_u - V_S \tag{5.23}$$

in the general case, we can substitute equations (5.21) and (5.22) into equation (5.23). Also, if we set $k_u/k_s = n$, it can be shown that the otherwise very complicated identity reduces to

$$I_B = I_D - (I_D^r + I_D^a) + I_S(1-n) - I_S^r(1-n) - I_S^a(1-n)$$
$$+ (Q_B - nQ_B^*) + (Q_{G_d} - nQ_{G_d}^*) - Q_G(1-n)\,. \qquad (5.24)$$

If $n = 1$ ($k_u = k_s$) and readings are taken at V_c in figure 5.70 (such that $\sum Q = 0 = Q_B + Q_{G_d} + Q_{G_S}$) and at V_d (such that $\sum Q = Q_B^* + Q_{G_d}^* + Q_{G_S}$), then

$$I_B = I_D - (I_D^r + I_D^a) - Q_B^* - Q_{G_d}^*, \qquad (5.25)$$

and the instrument constant derived from calibrations performed in this manner is too low. We agree with Flowers that this often occurs in using the ASTM methods where the signal is recorded too quickly in the illuminated, unshaded sequence. Conversely, if the identity is derived from equal duty cycles that represent steady-state conditions, where Q_B, Q_G, and $Q_{G,S}$ are finite (V_a for the illuminated condition) and $Q_B = Q_G = 0$ for the steady state, shaded condition (represented by V_b), then

$$I_B = I_D - (I_D^r + I_D^a) + Q_B^* + Q^*_{G_d}\,, \qquad (5.26)$$

and an equal but positive error in the instrument constant occurs.

However, if the unshaded, illuminated signal is taken at V_a under steady-state conditions and the shaded signal is taken at V_d before significant decay of the thermally-generated signal, then

$$I_B = I_D - (I_D^r + I_D^a)\,, \qquad (5.27)$$

and a correct identity is achieved. This expression externalizes the inferred beam component that is transmitted through the domes to reach the receiver (as is also the case for Model NIP pyrheliometers but not for absolute cavities). Equation (5.27) states that any change in the reflected component of the direct beam (during use) that is caused by the beam impinging on glass disparities, or on areas with deviations from sphericity, or any changes in the dome's solar absorptance early and late in the day, will create the errors indicated in the measurement of total irradiance, namely,

$$|Q_B^* + Q_{G_d}^*|\,.$$

Also, it is instructive to observe that an inverse duty cycle in which the non-steady-state, illuminated signal is taken at V_c and the shaded signal is

Table 5.16
Instrument constants of SN 19129 as a function of illuminated duty cycle (normalized to 25°C)

Cycle	Percent unshaded	Timing, unshaded/shaded	Instrument constant (μV/W m^{-2})	Ratio B/X
A	50	5 min/5 min	10.292	0.994
B	91	5 min/30 s	10.228	1.000
C	50	30 s/30 s	10.178	1.005

recorded at V_b under steady-state conditions (in which Q_B and Q_{G_d} are finite and $Q_B^* = Q_{G_d}^* = 0$) also results in the identity defined by equation (5.26).

Obviously, to apply this concept precisely, one needs to know much more about the influence of temperature on thermopile cold junctions and the resulting signals (even for temperature-compensated thermopiles). Furthermore, research is needed to evaluate the separate pyranometer time constants and the heat transfer mechanisms associated with generating a signal in real pyranometers.

Although further work is necessary, we have performed a set of measurements to define whether or not use of a 91% illuminated duty cycle (as shown in figure 5.70) results in an instrument constant that is intermediate between the two 50% duty cycles of 5 min/5 min and 30 s/30 s. Using DSET's reference pyranometer SN 19129 and the SN 17142 Eppley Model H-F absolute cavity pyrheliometer, we did normal incidence shade calibrations usng the proposed new 91% duty cycle and the two 50% duty cycles described above. Results of this experiment, presented in table 5.16, are in agreement with the thesis presented by the analysis of equations (5.16)–(5.27) and figure 5.70. Furthermore, we observe a 0.5% difference between A and B and a 1% difference between A and C, which is consistent in sign and magnitude with field experience (the difference being greater in magnitude than normal field experience).

Appendix: List of World and Regional Radiation Centers

World Radiation Centers

Davos (Switzerland)

Leningrad (USSR)

Regional Radiation Centers

• Region I (Africa)	Cairo (Egypt) Khartoum (Sudan) Kinshasa (Zaire) Lagos (Nigeria) Tamanrasset (Algeria) Tunis (Tunisia)
• Region II (Asia)	Poona (India) Tokyo (Japan)
• Region III (South America)	Buenos Aires (Argentina)
• Region IV (North and Central America	Toronto (Canada) Washington (U.S.)
• Region V (Southwest Pacific)	Aspendale
• Region VI (Europe)	Bracknell (UK) Budapest (Hungary) Davos (Switzerland) Leningrad (USSR) Norrköping (Sweden) Trappes/ Carpentras (France) Uccle (Belgium)

Notes

1. WMO changed the classification of instruments in 1983 (see section 5.5). For the most part, Class I pyranometers are now known as First Class instruments, and certain selected pyranometers from the class meet the new requirements of a Secondary Standard Pyranometer.

2. While the optical efficiency defined in equation (5.1) has no thermodynamic importance for solar collectors other than swimming pool collectors that may operate at near ambient temperatures, it is a highly useful diagnostic tool for assessing agreement between theory and observation in a laboratory.

3. The wavelength at which the transmittance is 50% of maximum value.

4. A 30% nickel stainless steel with low thermal expansion coefficient.

5. No longer manufactured.

6. Lupolen-H.

7. TUVRs are now furnished with translucent Teflon® diffusers.

8. The International Pyrheliometric Conference is held every 5 years at Davos, Switzerland.

9. NRIP procedures were adopted at IPC VI, held in October 1985 (Davos).

10. Because the diffuse portion of the sky is "blue" and spectrally quite different from the "yellow" beam radiation, photovoltaic (PV) reference cells designed for PV concentrator testing must be calibrated in a sky-occulting tube only.

11. Most experts agree, however, that these errors are smaller than those in the occulting tube method.

12. Several experts believe that the value should be at least 0.85.

13. As measured with an absolute cavity pyrheliometer.

References

Abbott, C. G., and L. B. Aldrich. 1908. "Pyrheliometric Observations." *Annals of the Astrophysics Observatory of the Smithsonian Institution.* 2:34–49.

Abbott, C. G., and L. B. Aldrich. 1932. "An Improved Water Flow Pyrheliometer and the Standard Scale of Solar Radiation." *Smithsonian Miscellaneous Collections* 87(15):1–8.

Ambrosetti, P., H. E. B. Andersson, L. Liedquist, C. Froelich, C. Wehrli, and H. C. Talarek. 1984. "Results of Outdoor and Indoor Pyranometer Comparisons." IEA Document No. 111.A.3. Jülich, Federal Republic of Germany: Kernforschungsanlage-Jülich, GmbH.

American Society for Testing and Materials (ASTM). n.d.1. "Calibration of Secondary Reference Pyrheliometers and Pyrheliometers for Field Use," E-816. *Annual Book of ASTM Standards.* Volume 12.02. Philadelphia, PA: ASTM.

American Society for Testing and Materials (ASTM). n.d.2. "Standard Method for Transfer of Calibration from Reference to Field Pyranometers," E-824. *Annual Book of ASTM Standards.* Volume 12.02. Philadelphia, PA: ASTM.

American Society for Testing and Materials (ASTM). n.d.3. "Standard Test Method for Calibration of Reference Pyranometers with Axis Vertical by the Shading Method," E-913. *Annual Book of ASTM Standards.* Volume 12.02. Philadelphia, PA: ASTM.

American Society for Testing and Materials (ASTM). n.d.4. "Standard Test Method for Calibration of Reference Pyranometers with Axis Tilted by the Shading Method," E-941. *Annual Book of ASTM Standards.* Volume 12.02. Philadelphia, PA: ASTM.

American Society of Heating, Refrigeration and Air Conditioning Engineers (ASHRAE), 1977. *ASHRAE Standard 93-77.*

Andersson, H. E. B., L. Liedquist, J. Lindblad, and L. A. Norsten. 1981. *Calibration of Pyranometers.* Report SP-PAPP 1981:7. Statens Provningsanstalt (National Testing Institute), Borås, Sweden.

Ångström, A. K., and A. J. Drummond. 1961. "Basic Concepts Concerning Cutoff Glass Filters Used in Radiation Measurements." *Journal of Meteorology* 18:360–367.

Anon. 1977. *Pyrheliometer Comparisons 1975, Results and Symposium*, IPC IV, Working Report WR No. 581, World Radiation Center, Davos, Switzerland.

Anon. 1981. *Pyrheliometer Comparisons 1980, Results and Symposium*, IPC V, Working Report WR No. 94, World Radiation Center, Davos, Switzerland.

Anson, D. 1980. "Instrumentation of Solar Radiation Measurements." Chapter 7. *Solar Energy Technology Handbook, Part A: Engineering Fundamentals*, W. C. Dickinson and P. M. Cheremisinoff, eds., New York: Marcel Dekker, Inc.

Bahm, R. J., and J. C. Nakos. 1979. *The Calibration of Solar Radiation Measuring Instruments.* Report No. BER-1(79)DOE-684-1. DOE Contract No. EM-78-S-04-5336. Albuquerque, NM: University of New Mexico.

Bird, R. E., and R. L. Hulstrom. 1981. *Solar Spectral Measurements and Modeling*. SERI/TR-642-1013. Golden, CO: Solar Energy Research Institute.

Budde, W. 1983. "Physical Detectors of Optical Radiation." Vol. 4 of *Optical Radiation Measurements*. New York: Academic Press.

Compagnie Industrielle Radioelectrique (CIR). 1980. *PMO-6 Absolute Radiometer* (brochure). Doc. T586. Bern, Switzerland.

Crommelynck, D., and J. R. Latimer. 1980. "Spectral Irradiance Measurements of Natural Sources." Chapter 5. *An Introduction to Meteorological Measurements and Data Handling for Solar Energy Applications*. DOE/ER-0084, IEA Task IV. Washington, DC: U S. Department of Energy.

Dirmhirn, I. 1958. "Untersuchungen an Stempyranometer." *Anch. Met. Geopheys*. Bioklin, Vienna, B.9 (2): 124–148.

Drummond, A. J. 1966. *Pyrheliometric Calibrations at the Eppley Laboratory*. Eppley memorandum, Newport, RI: Eppley Laboratory, Inc.

Drummond, A. J., ad J. R. Hickey. 1968. "The Eppley-JPL Solar Constant Measurement Program." *Solar Energy* 12: 217–232.

Drummond, A. J., and J. J. Roche. 1965. "Corrections to Be Applied to Measurements Made with Eppley (and Other) Spectral Radiometers When Used with Schott Colored Glasses." *Journal of Applied Meteorology* 7: 741–744.

Dunkelman, L., and R. Scolnik. 1959. "Solar Spectral Irradiance and Vertical Atmospheric Attenuation in the Visible and Ultraviolet." *Journal of the Optical Society of America* 49(4).

Eko Instruments Trading Co. n.d. Technical Literature. Tokyo, Japan.

Eppley Laboratory, Inc. 1977. "The Self-Calibrating Sensor of the Electric Satellite Pyrheliometer Program." *Proceedings of the 1977 Annual Meeting, American Section, ISES*, Vol 1., Sec. 15, International Solar Energy Society.

Eppley Laboratory, Inc. n.d.1. *The Eppley-Ångström Compensation Pyrheliometer and Associated D.C. Electrical Instrumentation*. Undated instruction manual. Newport, RI: Eppley Laboratory, Inc.

Eppley Laboratory, Inc. n.d.2. *Model NIP*. Sales and Technical Bulletin. Newport, RI.

Eppley Laboratory, Inc. n.d.3. *The Eppley Model PSP*. Sales and Technical Bulletin, Newport, RI.

Eppley Laboratory, Inc. n.d.4. *Eppley Precision Infrared Radiometer (Pyrgeometer)*. Newport, RI.

Estey, R. S., and C. H. Seaman. 1981. *Four Absolute Cavity Radiometer (Pyrheliometer) Intercomparisons at New River, Arizona*. JPL Publication 81-60. Jet Propulsion Laboratory.

Flowers, Edwin C. 1974. "The 'So-Called' Parson's Black Problem with Old-Style Eppley Pyranometers." *Solar Energy Data Workshop*. C. Turner, ed. NSF-RA-N-062. Washington, DC: U.S. Government Printing Office.

Flowers, Edwin C. 1981. "Calibration and Comparison of Pyranometers and Pyrheliometers." *Summary Proceedings, Workshop on Accurate Measurement of Solar Radiation, 1981 Annual Meeting of American Section of ISES (Philadelphia)*. Trinity University, San Antonio, TX.

Flowers, Edwin C. 1982. Personal communication, Boulder, CO: National Oceanic and Atomspheric Administration.

Flowers, Edwin C. 1984a. Personal communication, Boulder, CO: National Oceanic and Atomspheric Administration.

Flowers, Edwin C. 1984b. "Solar Radiation Measurements." *Proceedings of the Recent Advances in Pyranometry, International Energy Agency IEA Task IX Solar Radiation and Pyranometer Studies.* Downsview, Ontario, Canada: Atmospheric Environment Service.

Flowers, Edwin C. 1985a. *Tests of IEA Pyranometers at the Solar Radiation Facility, January–June 1984.* Boulder, CO: National Oceanic and Atmospheric Administration.

Flowers, Edwin C. 1985b. Personal communication, Boulder, CO: National Oceanic and Atmospheric Administration, December 18, 1985.

Forgan, B. 1983. "A Direct-Diffuse Calibration Method for Pyranometers." *Conference Proceedings, International Solar Energy Society Solar World Congress.* S. V. Szokolay, ed. Perth, Australia.

Grether, D. F., D. Evans, A. Hunt, and M. Wahlig. 1980. "Measurement and Analysis of Circumsolar Radiation." *Proceedings of the Annual DOE Insolation Reference Assessment Program Review.* SERI/CP-642-977. Golden, CO: Solar Energy Research Institute.

Grether, D. F., D. Evans, A. Hunt, and M. Wahlig. 1981. "The Effect of Circumsolar Radiation on the Accuracy of Pyrheliometer Measurements of the Direct Solar Radiation." *Summary Proceedings, Accurate Measurement of Solar Radiation, 1981 Annual Meeting of American Section of ISES (Philadelphia).* Trinity University, San Antonio, TX.

Hanson, Kirby J. 1974. "Comments on the Quality of the NWS Pyranometer Network from 1954 to the Present." *Solar Energy Data Workshop.* C. Turner, ed. NSF-RA-N-062. Washington, DC: U.S. Government Printing Office.

Hickey, J. 1981. "Report of Tests and Abbreviated Description of Calibration Techniques." *Proceedings of the International Energy Agency Conference on Pyranometer Measurements.* SERI/TR-642-1156R. Golden, CO: Solar Energy Research Institute, pp. 235–245.

Hickey, J. 1984. Personal communication, The Eppley Laboratory, Inc., August 24, 1984.

Hollis Geosystems Corp. 1978. Technical Bulletins MR-3 (A) and MR-5. Nashua, NH.

IGY Instruction Manual. 1958. *Part VI, Radiation Measurements and Instruments.* London, England: Pergamon Press.

Inn, E. C. Y., and Y. Tanaka. 1953. "Absorption Coefficient of Ozone in the Ultra-Violet and Visible Regions." *Journal of the Optical Society of America* 43:870–873.

Iqbal, M. 1983. *An Introduction to Solar Radiation.* New York: Academic Press.

Johnson, N. 1984. Personal communication. International Division, Rockville, MD: National Oceanic and Atmospheric Administration.

Karoli, A. R., J. R. Hickey, and R. G. Frieden. 1984. "Self Calibrating Cavity Radiometers at the Eppley Laboratory: Capabilities and Applications." *Proceedings of the Recent Advances in Pyranometry,* International Energy Agency IEA Task IX Solar Radiation and Pyranometer Studies. Downsview, Ontario, Canada: Atmospheric Environmental Services.

Kendall, J. M., Sr. 1969.*Primary Absolute Cavity Radiometer.* Technical Report 32-1396. Pasedena, CA: Jet Propulsion Laboratory.

Kendall, J. M., Sr., and C. W. Berdahl. 1970. "Two Blackbodies of High Accuracy." *Applied Optics* 9(5):1082–1091.

Kipp and Zonen. n.d.1. *Sales and Technical Literature.* Delft, Holland.

Kipp and Zonen. n.d.2. *The CM11 Pyranometer.* Sales and Technical Bulletin 8102. Delft, Holland.

Kliman, A. W., and H. G. Eldering. 1981. "Design and Development of a Solar Spectroradiometer." *Electro-Optical Systems Design,* Nov.:53–61.

Labs, D., and H. Neckel. 1962a. "Die absolute Strahlungenintensitaet der Sonnenmitte im Spektralbereich $4010 \leq \lambda \leq 6569$ Å." *Z. Ap.* 55:269.

Labs, D., and H. Neckel. 1962b. "Die absolute Strahlungenintensitaet der Mitte der Sonnenscheibe im Spektralbereich 6389 $\leq \lambda \leq$ 12,480 Å." *Z. Ap.* 57:283.

Labs, D., and H. Neckel. 1967. "Die absolute Strahlungenintensitaet der Mitte der Sonnenscheibe im Spektralbereich 3288 $\leq \lambda \leq$ 12,480 Å." *Z. Ap.* 65:133.

Labs, D., and H. Neckel. 1968. "The Radiation of the Solar Photosphere from 2000 Å to 100 μ." *Z. Ap.* 69:1.

Labs, D., and H. Neckel. 1970. "Transformations of the Absolute Solar Radiation Data into the 'International Practical Temperature Scale of 1968.'" *Solar Phys.* 15:79.

Ley, W. 1979. "Survey of Solar Simulator Test Facilities and Initial Results of IEA Round Robin Tests Using Solar Simulators." *IEA Task III Report on Performance Testing of Solar Collectors.* Cologne, Germany: Deutsche Forschungs- und Versuchsanstalt für Luft- und Raumfahrt ev. (DFVLR).

LiCor. 1982a. *Radiation Measurements and Instrumentation.* Technical Literature Publication No. 8208-LM. Lincoln, NE.

LiCor. 1982b. *LI-1800 Portable Spectroradiometer Specification.* No. 8208-LM. Lincoln, NE: LiCor.

Loxom, F., and W. C. Hogan. 1981. "Measurement of Irradiation of Tilted Surfaces." *Summary Proceedings, Accurate Measurement of Solar Radiation, 1981 Annual Meeting of American Section of ISES (Philadelphia).* Trinity University, San Antonio, TX.

Lufkin, D. H. 1980. "New U. S. Network for Solar Radiation Measurements." Chapter 6. *Solar Energy Technology Handbook, Part A: Engineering Fundamentals.* W. C. Dickinson and P. M. Cheremisinoff, eds New York: Marcel Dekker, Inc., pp. 167–172.

Martin, M. R., and R. D. Anson. 1980. "The Measurements of Infrared Radiation." Chapter 6. *An Introduction to Meteorological Measurements and Data Handling for Solar Energy Applications.* DOE/ER-0084, IEA Task IV. Washington, DC: U.S. Department of Energy.

McGreggor, J. 1982. Personal communication. Cardiff, Wales, UK: Cardiff University.

Neckel, H., and D. Labs. 1973. *Proceedings of the Symposium on Solar Radiation.* Washington, DC: Smithsonian Institution, p. 326.

Nelson, D., Haas, R., DeLuisi, J., and Zerlaut, G. 1987. "Results of the NRIP 7 Intercomparison November 18–21, 1985," NOAA Technical Memorandum REL ARL-161, National Oceanic and Atmospheric Administration (Air Resources Laboratory), Silver Spring, MD.

Norris, D. J. 1974. "Calibration of Pyranometers in Inclined and Inverted Positions." *Solar Energy* 16:53–55.

Pergamon Press. *Radiation Instruments and Measurements.* Part IV. 1958. *IGY Instruction Manual.* New York: Pergamon Press.

Plamondon, J. A. 1969. *JPL Space Program Summary for 1969.* III:37–59.

Proctor, D., and E. S. Trickett. 1982. *Solar Energy* 29(3):189–194.

Reed, K. A. 1984. Personal communication. Washington, DC: National Bureau of Standards, April 10, 1984.

Riches, Michael R., ed. 1980. *An Introduction to Meteorological Measurements and Data Handling for Solar Energy Applications.* Report DOE/ER-0084. International Energy Agency Task IV, Washington, DC: U.S. Department of Energy.

Riches, Michael R., T. L. Stoffel, and C. V. Wells. 1982. *Proceedings of the International Energy Agency Conference on Pyranometer Measurements.* SERI/TR-642-1156R. Golden, CO: Solar Energy Research Institute.

Schenk, P. n.d. Data Sheet B901/1. Vienna, Austria.

Shaw, G. E., J. A. Regan, and B. A. Herman. 1973. *Journal of Applied Meteorology* 12(2): 374–380.

Stair, R. 1951. "Ultraviolet Spectral Distribution of Radiant Energy from the Sun." *Journal of Research of the National Bureau of Standards.* 46(5).

Stair, R. 1952. "Ultraviolet Radiant Energy from the Sun Observed at 11,190 Feet." *Journal of Research of the National Bureau of Standards.* 49(3).

Stair, R. 1953. "Filter Radiometry and Some of Its Applications." *Journal of the Optical Society of America* 43(11):971.

Stair, R., and R. G. Johnston. 1956. "Preliminary Spectroradiometric Measurements of the Solar Constant." *Journal of Research of the National Bureau of Standards.* 57:205.

Stair, R., R. G. Johnston, and T. C. Bagg. 1954. "Spectral Distribution of Energy from the Sun." *Journal of Research of the National Bureau of Standards.* 53(2).

Streed, E. R., et al. 1978. *Results and Analysis of a Round Robin Test Program for Liquid-Heating Flat-Plate Solar Collectors.* NBS-975. Gaithersburg, MD: U.S. National Bureau of Standards.

Swissteco. 1983. Technical Data Sheet 55-25. Oberriet, Switzerland.

Talarek, H. D. 1979. *Testing of Liquid-Heating Flat-Plate Collectors Based on Standard Procedures.* KFA-IKP-113-4/79 (Kernforschungsanlage-Jülich). Presented by E. Streed to ASTM Committee E-44 as an IEA progress report.

Talarek, H. D., and C. Froelich. 1980. "Results of a Pyranometer Comparison." *International Energy Agency IEA/PMOD (WRC) Report.*

Thekaehara, M. P. 1966. "Solar Radiation Measurements: Techniques and Instrumentation." *Solar Energy* 18(4):309–325.

Wardle, D. I., and D. C. McKay, eds. 1984. *Symposium Proceedings, Recent Advances in Pyranometry.* International Energy Agency Task IX Solar Radiation and Pyranometer Studies (Norrköping, Sweden). Downsview, Ontario, Canada: Atmospheric Environment Service.

Wells, C. V. 1981. "An Experiment to Compare Data Acquisition Methods Employed in the IPC V and New River Comparisons." *Pyrheliometer Comparisons 1980, Results and Symposium IPC V.* WR No. 94. Davos, Switzerland: World Radiation Center.

Willson, R. C. 1973. "Active Cavity Radiometer." *Applied Optics* 12:810–817.

World Meteorological Organization (WMO). 1971. "Measurement of Radiation and Sunshine." Chapter 9. *Guide to Meteorological Instrument and Observing Practices*, fourth ed. WMO-No. 8, TP. 3. Geneva: WMO.

World Meteorological Organization (WMO). 1983. "Measurement of Radiation." *Guide to Meteorological Instruments and Methods of Observation*, fifth ed. WMO-No. 8. Geneva: WMO.

World Radiation Center (WRC). 1977. *Pyrheliometer Comparisons 1975, Results and Symposium.* IPC IV, WR No. 581. Davos, Switzerland: World Radiation Center.

World Radiation Center (WRC). 1981.*Pyrheliometer Comparisons 1980, Results and Symposium.* IPC V, WR No. 94. Davos, Switzerland: World Radiation Center.

Zerlaut, G. A. 1981a. "The Challenge of Solar Energy." ASTM Committee E44. *ASTM, Standardization News.*

Zerlaut, G. A. 1981b. "Scanning Solar Spectroradiometer: Facilities Design, Calibration and Significant Data." B. H. Glenn and G. E. Franta, eds. *Proceedings of the 1981 Annual Meeting, the American Section, International Solar Energy Society (Philadelphia).* Newark, DE: University of Delaware.

Zerlaut, G. A. 1981c. "Why Standard Pyranometer Calibrations Are Inappropriate for Solar Collector Testing." B. H. Glenn and G. E. Franta, eds. *Proceedings of the Annual Meeting, the American Section, International Solar Energy Society (Philadelphia).* Newark, DE: University of Delaware.

Zerlaut, G. A. 1983. "The Calibration of Pyrheliometers and Pyranometers for Testing Photovoltaic Devices." *Solar Cells* 7:119–129.

Zerlaut, G. A. 1984. "The New River Intercomparisons of Absolute Cavity Pyrheliometers (NRIP I–VI)." D. I. Wardle and D. C. McKay, eds. *Symposium Proceedings, Recent Advances in Pyranometry, International Energy Agency Task IX Solar Radiation and Pyranometry Studies (Norrköping, Sweden).* Downsview, Ontario, Canada: Atmospheric Environmental Service.

Zerlaut, G. A. 1986. "Solar Radiometry Instrumentation, Calibration Techniques, and Standards." *Solar Cells* 18:189–203.

Zerlaut, G. A., and B. Garner. 1983. "International Standardization in the Plastics Industry —a Viewpoint." *ASTM Standard News* No. 4.

Zerlaut, G. A. and G. D. Maybee. 1983. "Spectroradiometer Measurements in Support of Photovoltaic Device Testing." *Solar Cells* 7:97–106.

Zerlaut, G. A. and G. D. Maybee. 1984. "The Comparison of Pyranometers at Horizontal, 45° Tilt, and Normal Incidence." *Proceedings of the Recent Advances of Pyranometry, International Energy Agency IEA Task IX Solar Radiation and Pyranometer Studies.* Downsview, Ontario, Canada: Atomspheric Environment Service.

Zerlaut, G. A., and J. S. Robbins, Jr. 1984. *Accelerated Outdoor Exposure Testing of Coil Coatings by the EMMA(QUA)r Test Method.* Brussels, Belgium: European Coil Coaters Association. 1984 Annual Congress. (Preprint to be published.)

Zerlaut, G. A., et al. 1980. *Operational and Calibration Procedures of the DSET Spectroradiometer* (Vol. 1). DSET Contractor Report R22075-1, JPL Contract BQ-713137. Pasadena, CA: Jet Propulsion Laboratory.

6 Spectral Terrestrial Solar Radiation

Richard Bird

6.1 Introduction

The intensity of solar radiation varies with the wavelength of the radiation, and the functional relationship between intensity and wavelength is called the solar spectral distribution. This distribution is important in solar applications because the performance of many solar collectors is spectrally dependent. For example, the reflectivity of the front surface of concentrating collectors may be spectrally dependent.

The spectral distribution of solar radiation outside the earth's atmosphere, called the extraterrestrial or air mass zero (AMO) spectrum, is well characterized. It resembles the spectrum of a blackbody at 5,900K with the exception of absorption lines caused by attenuation of radiation in the medium surrounding the sun (see figure 6.1). When this solar radiation passes through the earth's atmosphere, the spectral distribution is modified by absorption and scattering of the radiation by atmospheric constituents, such as aerosols, ozone, and water vapor. Figure 6.2 shows a spectral distribution at the earth's surface and identifies regions in the spectrum where absorption by constituents in the sun's and earth's atmospheres modifies the solar spectral distribution.

The exact spectral distribution at the earth's surface at any time depends on local atmospheric conditions and the path length of solar radiation through the atmosphere (or air mass). Air mass depends on the sun angle, which varies with location, time of day, and day of year. Since both air mass and atmospheric conditions vary throughout the day and year for different locations, solar spectral distributions at the earth's surface vary geographically and temporally.

Two methods are used to determine spectral solar radiation distributions for particular locations and atmospheric conditions: modeling and measurements. Both methods are necessary because it is impractical to measure the spectrum for all conditions, and it is difficult (if not impossible) to separate atmospheric effects experimentally. This chapter discusses how each method is used to determine these spectral distributions.

6.2 Spectral Solar Radiation Modeling

A rigorous model of radiation transport through the atmosphere uses numerical methods to solve the integral form of the radiative transfer

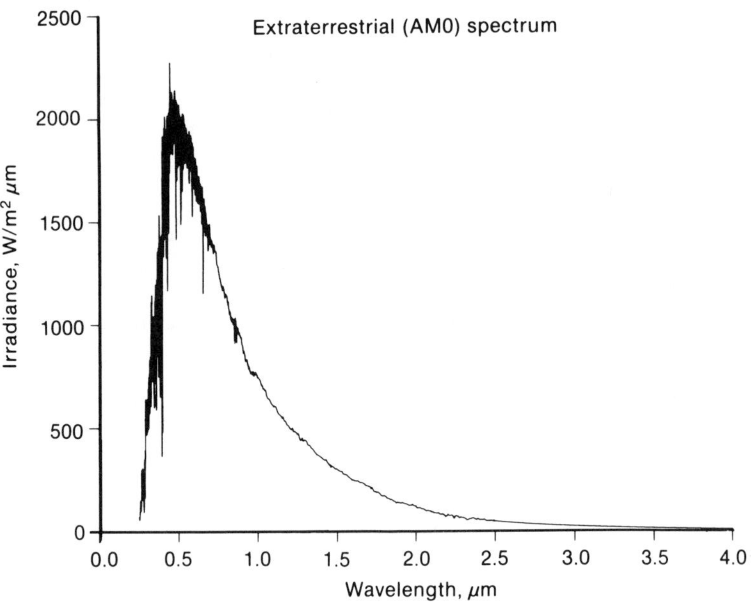

Figure 6.1
Revised and extended Neckel and Labs (1981) extraterrestrial spectrum, plotted to 4 μm. Source: Fröhlich and Wehrli (1981).

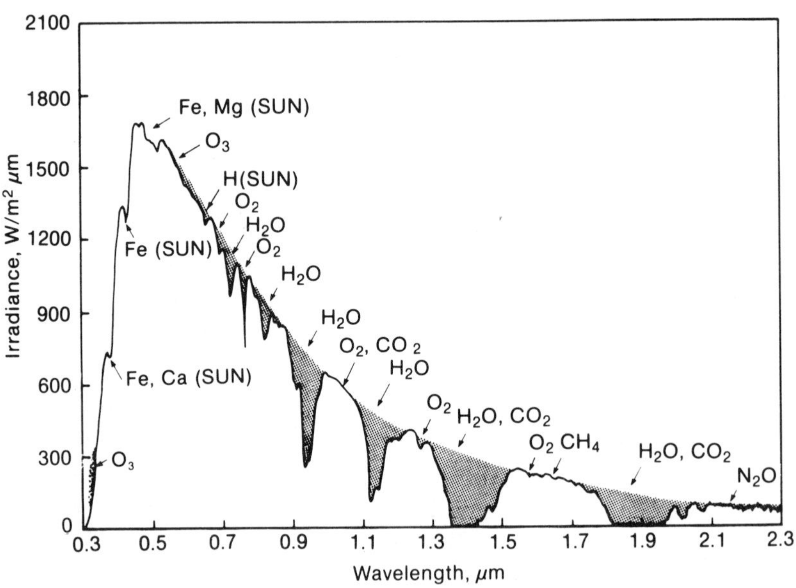

Figure 6.2
Global (180° FOV—field of view) spectrum measured on a horizontal surface 18 July 1980 12:36 Local Standard Time in Bedford, MA, showing absorption regions (shaded) (latitude 42.5°N, longitude 71.25°W).

equation (Chandrasekhar, 1950). Several numerical methods have been devised to solve this equation; one of them is the Monte Carlo method illustrated in section 6.2.3. These solutions consider absorption, scattering, and transmittance of the extraterrestrial solar radiation by the earth's atmosphere. The solar radiation transmitted to the earth's surface can be separated into several distinct components, each having a different spectral distribution. For solar collector applications, spectral solar radiation models are used to predict the spectral distribution of the solar radiation components that particular collectors use. This section discusses the fundamentals of atmospheric transmittance, solar radiation components, and spectral solar radiation models and their application.

6.2.1 Fundamentals of Atmospheric Transmittance

A simple expression for the transmittance T_λ of monochromatic radiation through the atmosphere is

$$T_\lambda = \frac{I_\lambda}{I_{0\lambda}} = \exp - \gamma_\lambda XM, \tag{6.1}$$

where

I_λ = transmitted radiation at wavelength λ,

$I_{0\lambda}$ = initial radiation at wavelength λ,

γ_λ = extinction coefficient at wavelength λ,

X = length of the vertical path considered, and

M = relative optical mass.

This expression, called Beer's law, Lambert's law, or Bouguer's law, assumes that the atmosphere is homogeneous over the path length X. If the atmosphere is not homogeneous (γ_λ varies along the path), equation (6.1) should be written as

$$T_\lambda = \exp - \left(\int_{s2}^{s1} \gamma_{\lambda s}\, ds \right), \tag{6.2}$$

where s is the path of the radiation. Equation (6.1) strictly applies only to monochromatic radiation, but it can be used quite accurately to model finite spectral bandwidths in spectral regions where rapid changes caused by molecular absorption do not occur. In molecular absorption bands with structure, one of many band absorption models should be used.

The relative optical mass M accounts for the path length through the atmosphere when the sun is not directly overhead. It is the ratio of the path length along a nonvertical path to the path length in the vertical direction. The following expression formally defines relative air mass:

$$M = \int_0^\infty \rho \, ds \Big/ \int_0^\infty \rho \, dz, \tag{6.3}$$

where

ρ = density of the medium,

ds = increment of length along the actual path of the radiation, and

dz = increment of length in the vertical direction.

For greatest accuracy, equation (6.3) should be solved separately for different attenuators in the atmosphere, such as air molecules, water vapor, and ozone.

The absolute, or pressure-corrected, optical mass for air or the uniformly mixed gases (CO_2, N_2O, CH_4, CO, N_2, and O_2), is

$$M' = \frac{P}{P_0} M, \tag{6.4}$$

where

P = measured surface pressure and

P_0 = surfaced pressure at sea level (nominally 1,013 mb).

For solar zenith angles (Z) less than 60°, we can obtain sufficient accuracy for optical air mass using

$$M = \sec Z, \tag{6.5}$$

where Z is measured from the local vertical to the position of the sun. For $60° < Z < 80°$, the earth's curvature becomes important. Geometric considerations give

$$M = [(r/H)^2 \cos^2 Z + 2(r/H) + 1]^{1/2} - (r/H) \cos Z, \tag{6.6}$$

where r is the earth's radius and H is the scale height defined by

$$H = \int_0^\infty \rho/\rho_o \, dz. \tag{6.7}$$

The constant ρ_o is the surface level density. For air we see that $H = P_o/(\rho_o g)$, where g is the acceleration due to gravity. For $Z > 80°$, atmospheric refraction becomes important. The correct value for the relative air mass at all solar zenith angles is available in tables (Kasten, 1966; Kondratyev, 1969) and can be calculated with approximate expressions. An expression that is correct to within 1% for $Z < 89°$ is

$$M = [\cos Z + 0.15(93.885 - Z)^{-1.253}]^{-1}. \tag{6.8}$$

For greatest accuracy, Z in equation (6.8) should be the apparent solar zenith angle (see Zimmerman [1981] and Walraven [1978] for the calculation of Z).

The atmospheric extinction coefficient γ_λ can be subdivided into absorption and scattering mechanisms in the atmosphere as follows:

$$\gamma_\lambda = \sigma_{a\lambda} + \sigma_{m\lambda} + k_{a\lambda} + k_{m\lambda}. \tag{6.9}$$

Scattering and absorption are represented by σ and k, respectively, and a and m indicate aerosol or molecular processes, respectively. Molecular scattering is usually referred to as Rayleigh scattering, and aerosol scattering is often called Mie scattering (Kerker, 1969). Both $\sigma_{a\lambda}$ and $k_{a\lambda}$ are usually determined using the Mie theory, $\sigma_{m\lambda}$ is calculated using the Rayleigh scattering theory, and $k_{m\lambda}$ is determined from absorption measurements or a band absorption model.

Beer's law can be expressed as total optical depth, given by

$$\tau_\lambda = \gamma_\lambda X M. \tag{6.10}$$

From equations (6.1) and (6.10), it is evident that τ_λ is the negative logarithm of transmittance. The total optical depth can be separated into components similar to those in equation (6.9):

$$\tau_\lambda = (\tau_{a\lambda} + \tau_{r\lambda} + \tau_{ma\lambda})M, \tag{6.11}$$

where

$\tau_{a\lambda}$ = aerosol optical depth in a vertical path (also called turbidity),

$\tau_{r\lambda}$ = Rayleigh scattering optical depth in a vertical path, and

$\tau_{ma\lambda}$ = molecular absorption optical depth in a vertical path.

Molecular absorption includes effects of constituents such as water vapor, ozone, and uniformly mixed gases.

Turbidity $\tau_{a\lambda}$ includes both aerosol scattering and absorption effects. To arrive at equation (6.11), we assume that the relative optical mass M is the same for molecules and aerosols. An examination of equation (6.2) shows that this is not strictly true. Since aerosols and some molecular absorbers have a density profile different from air, the relative mass for each can be slightly different. These differences are considered in transmittance expressions for absorption and scattering processes in simple solar radiation models by developing different relative mass expressions.

6.2.2 Solar Radiation Components

When solar radiation enters the earth's atmosphere, it is transmitted directly to the earth's surface or absorbed or scattered by molecules and aerosols; a fraction of the scattered radiation reaches the earth's surface. Solar radiation transmitted directly to the earth's surface is called direct solar radiation. Scattered solar radiation that reaches the earth's surface is called diffuse solar radiation. The sum of the direct and diffuse solar radiation is called total, or global, solar radiation.

Solar collectors that focus the sun's energy (concentrators) are designed to track the sun and use direct solar radiation plus a small component of scattered radiation around the sun called circumsolar radiation. Since concentrators are oriented with the sun's rays perpendicular to the collectors, the direct solar radiation is called direct normal. Nonfocusing, flat-plate collectors use direct and diffuse solar radiation. If the flat plates track the sun, solar radiation received at the collector surface is referred to as global-normal solar radiation, which is the sum of the direct normal component and the diffuse component in the collector field of view (FOV). Solar radiation received by a flat-plate collector that is not oriented normally to the sun is the sum of the direct normal component projected onto the surface and the diffuse component in the collector FOV. Flat-plate collectors tilted from the horizontal also receive a small amount of solar radiation reflected from the ground.

Each of these solar radiation components has a different spectral distribution, and each distribution changes with the sun's angle and atmospheric conditions. Therefore, in solar radiation modeling for solar applications, we must consider the direct and diffuse components, the collector FOV, sun angle, and atmospheric conditions.

6.2.3 Spectral Solar Radiation Models

Some of the first spectral solar radiation modelers (Moon, 1940) used a simple Beer's law approach to determine spectral solar radiation at the earth's surface. This method is accurate for calculating the direct solar radiation; however, the diffuse (scattered) component is much more difficult to calculate accurately. Over the years, solar radiation modelers have continued to use methods similar to those Moon used (Thomas and Thekaekara, 1976; Leckner, 1978) to calculate direct solar radiation and have used approximate methods to obtain the diffuse component.

Outside of the solar community, several rigorous methods of solving the equation of atmospheric radiative transfer were developed (Lenoble, 1977) that can be used to accurately estimate the diffuse component. Some of these methods were applied to solar radiation modeling (Dave, 1978; Bird, 1982). The standard approach when using rigorous models is to divide the atmosphere into several layers and to define key atmospheric parameters at each layer height. The parameters normally defined are temperature, pressure, the density of important molecular species, and the aerosol density. The Mie scattering theory is used to model the effects of aerosols, the Rayleigh scattering theory is used to model molecular scattering effects, and a molecular band model is used to model for molecular absorption.

One such rigorous model implemented at the Solar Energy Research Institute (SERI) is the BRITE Monte Carlo radiative transfer model (Bird, 1982). Examples of spectra generated with BRITE are shown in figure 6.3. Here, a 5-nm resolution version of the revised and extended Neckel and Labs AMO spectrum (Fröhlich and Wehrli, 1981; Neckel and Labs, 1981) is shown with AM1, AM1.5, and AM2 direct-normal spectra. A global spectrum is also shown for a flat collector surface facing the sun and tilted 37° from the horizontal. Atmospheric parameters used in the BRITE model to generate the spectra at the earth's surface were

ground albedo = 0.2,

turbidity at 0.5 μm = 0.27,

water vapor = 1.42 cm, and

ozone = 0.344 atm-cm.

The air mass values depend on solar zenith angle. A U.S. standard atmosphere (McClatchey et al., 1972) and a rural aerosol model (Shettle and Fenn, 1975) were used to define attributes and constituents of the

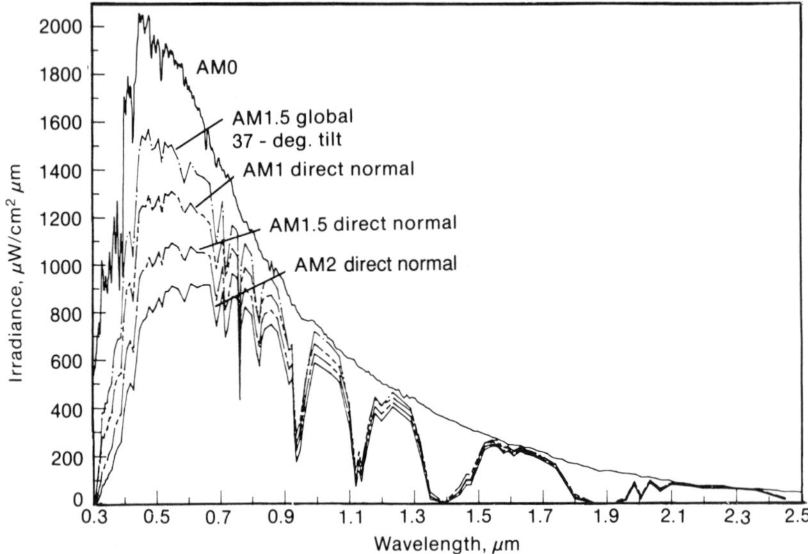

Figure 6.3
Extraterrestrial (AMO) spectrum and four examples of spectra generated with the SERI rigorous radiative transfer model (BRITE).

atmospheric layers. These spectra provide a quick comparison between the extraterrestrial spectrum and several spectra at the earth's surface. Section 6.3 gives a comparison of model calculations with measured data.

The rigorous models require large computer codes and computing capability and are relatively time-consuming and, consequently, expensive to use. They often require someone with an atmospheric and radiative-transfer background to operate them correctly. To simplify the complex computer codes, SERI researchers used the BRITE code to evaluate the use of a single, homogeneous atmospheric layer and deterministic (rather than statistical) methods to solve the radiative transfer equation. The one-layer atmosphere applies only to calculations of solar radiation at the top or bottom of the atmosphere. Results showed that identical solar radiation values could be obtained with a single-layer model for the direct component in spectral regions where strong molecular absorption structure is not present. This was demonstrated for zenith angles between 0° and 80°. The requirement imposed is that the total optical depth be the same in the single-layer and multilayer models.

For the strong absorption bands, nearly identical multilayer and single-layer results could be obtained by finding an equivalent temperature and pressure to use in the band absorption model. The equivalent temperature and pressure for the homogeneous layer for oxygen was different from that for water vapor, for example. This is a reasonable result because oxygen and water vapor are distributed differently with altitude and therefore are concentrated at different temperatures and pressures in the real atmosphere.

Further results at SERI showed that the diffuse component in the single-layer atmosphere was calculated to within 15% of the multilayer result in the strong absorption bands. Some of these differences are probably due to statistical inaccuracies in the BRITE (Monte Carlo) calculations; i.e., large photon numbers are required for meaningful statistical results in a strong absorption region of the spectrum. The diffuse component was calculated to within 7% of the multilayer result for $0^\circ < Z < 80^\circ$ in spectral regions where strong absorption is absent.

As a result of these analyses, SERI researchers concluded that a single-layer model using deterministic methods (Stamnes and Swanson, 1981) could be constructed that would be very simple and efficient and would give accurate results. Recent publications (Brine and Iqbal, 1983; Justus and Paris, 1985; Bird, 1984) demonstrated limited success in accurately calculating the scattered component using very simple spectral solar radiation models. SERI researchers examined these simple methods extensively by comparing the results from rigorous models (BRITE) with limited outdoor measurements (section 6.3). A new SERI simple spectral model based on previous work at SERI and models of Justus and Paris (1985) and Hay and Davies (1978) can be implemented on a microcomputer to calculate clear-sky direct and diffuse solar radiation on horizontal and tilted surfaces. Research is continuing at SERI to verify the model using outdoor spectral measurements and to add a cloud-cover modifier to the clear-sky model.

In the SERI simple spectral model, direct-normal solar radiation is calculated using

$$I_{d\lambda} = I_{o\lambda} D T_{r\lambda} T_{a\lambda} T_{w\lambda} T_{o\lambda} T_{u\lambda}, \tag{6.12}$$

where

$I_{o\lambda}$ = extraterrestrial solar radiation at the mean earth-sun distance,

D = the earth-sun distance correction factor, and

$T_{r\lambda}$, $T_{a\lambda}$,
$T_{w\lambda}$, $T_{o\lambda}$, $T_{u\lambda}$ = the transmittance functions of the atmosphere at wavelength λ for molecular (Rayleigh) scattering, aerosol attenuation, water vapor absorption, ozone absorption, and uniformly mixed gas absorption, respectively.

Direct solar radiation on a surface that is not normal to the sun's rays is obtained by multiplying $I_{d\lambda}$ by the cosine of the incidence angle. Bird and Riordan (1986) give expressions for each of the transmittance functions in the direct-normal equation.

Diffuse solar radiation on a horizontal surface is modeled by

$$I_{s\lambda} = I_{r\lambda} + I_{a\lambda} + I_{g\lambda}, \tag{6.13}$$

where

$I_{r\lambda}$ = the Rayleigh scattering component,

$I_{a\lambda}$ = the aerosol scattering component, and

$I_{g\lambda}$ = the component that accounts for multiple reflection between the ground and air.

Equations for the diffuse component are also given in Bird and Riordan (1986). These equations are based on previous work at SERI and on the Justus and Paris model (1985). An algorithm developed by Hay and Davies (1978) for broadband solar radiation is used in the SERI simple spectral model to convert diffuse solar radiation on a horizontal surface to diffuse solar radiation on a tilted surface. This conversion algorithm accounts for reflection of solar radiation from the ground onto the surface and for sky brightening around the sun.

The only inputs required to use the SERI simple spectral model are atmospheric ozone, water vapor, turbidity at 0.5 μm, solar zenith angle, surface pressure, ground albedo, and collector tilt angle. These parameters are obtained as follows. Ozone is calculated using the method of Van Heuklon (1797). Water vapor can be measured (section 6.8) or obtained from the local National Weather Service (NWS). Turbidity can be measured (section 6.8) or extrapolated from data in Flowers, McCormick, and Kurfis (1969). Surface pressure can be calculated using elevation data, measured or obtained from NWS. Solar zenith angle is calculated from site longitude and latitude and time of day and year. Ground albedo is estimated

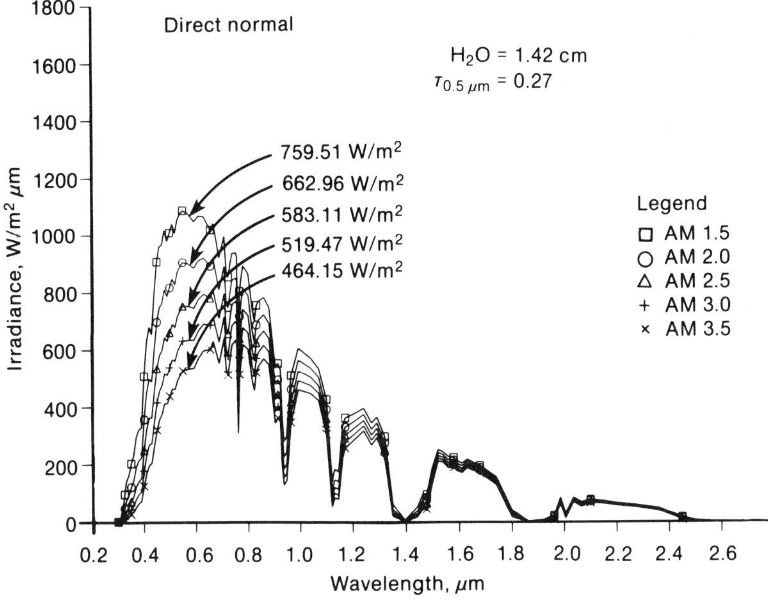

Figure 6.4
Direct-normal spectral solar radiation as a function of air mass.

(usually about 0.2 for dry soil to 0.80 for snow) and can be held constant across all wavelengths if the wavelength-dependent albedo is not known. Finally, the modeler can vary the collector tilt angle.

6.2.4 Application of Spectral Solar Radiation Models

The rigorous and simple spectral solar radiation models are used to study the sensitivity of spectral distributions to atmospheric conditions and to develop spectral solar radiation standards and data sets that can be used to study the effects of solar spectral distributions on the performance of solar collectors. Examples of these applications follow.

6.2.4.1 Sensitivity Studies Figures 6.4–6.9 show specific examples of how we can use the SERI simple spectral model to study the effect of air mass, turbidity, and water vapor on the direct normal and global spectra. These data allow us to examine the variation in spectral distribution for a range of conditions that might occur at a particular location. Figures 6.4 and 6.5 show the effect of changes in air mass on the direct- and global-

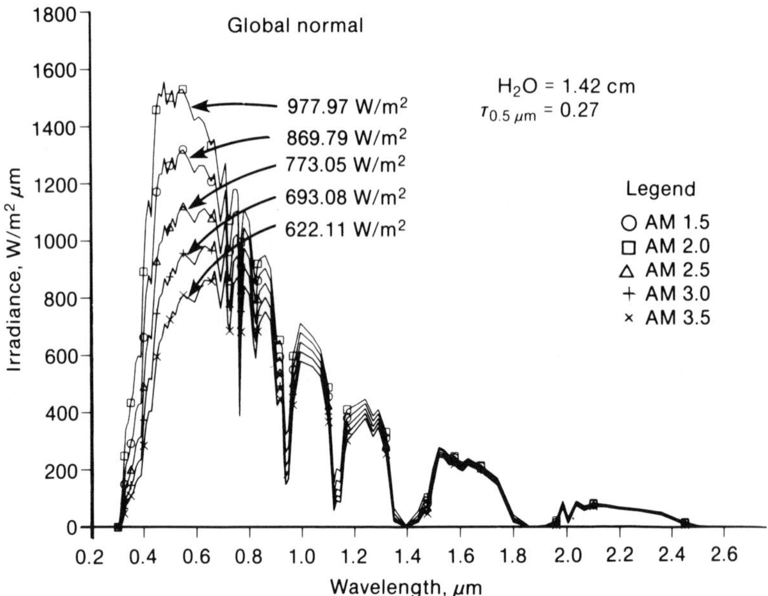

Figure 6.5
Global-normal spectral solar radiation as a function of air mass.

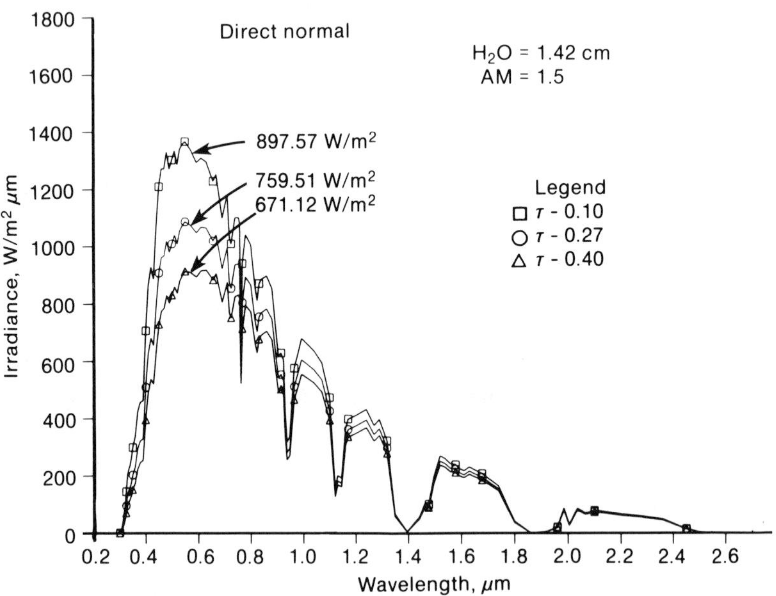

Figure 6.6
Direct-normal spectral solar radiation as a function of turbidity [$\tau(0.5\ \mu m)$].

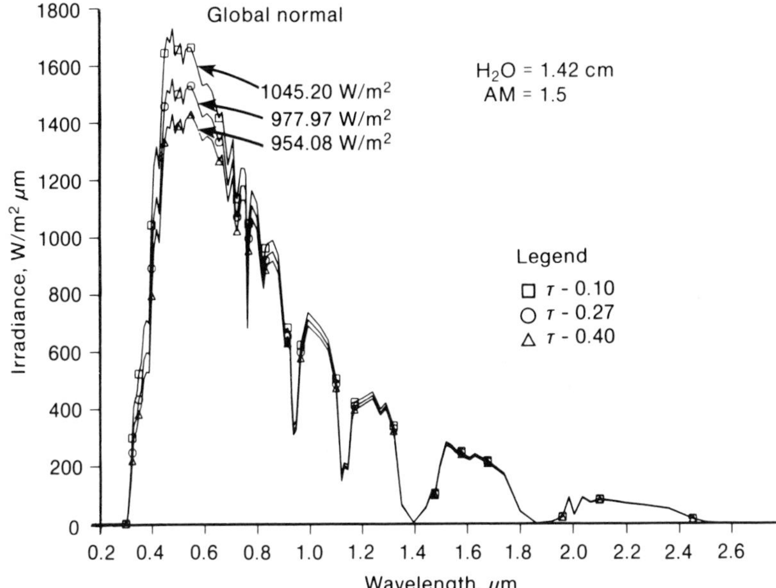

Figure 6.7
Global-normal spectral solar radiation as a function of turbidity [$\tau(0.5\ \mu m)$].

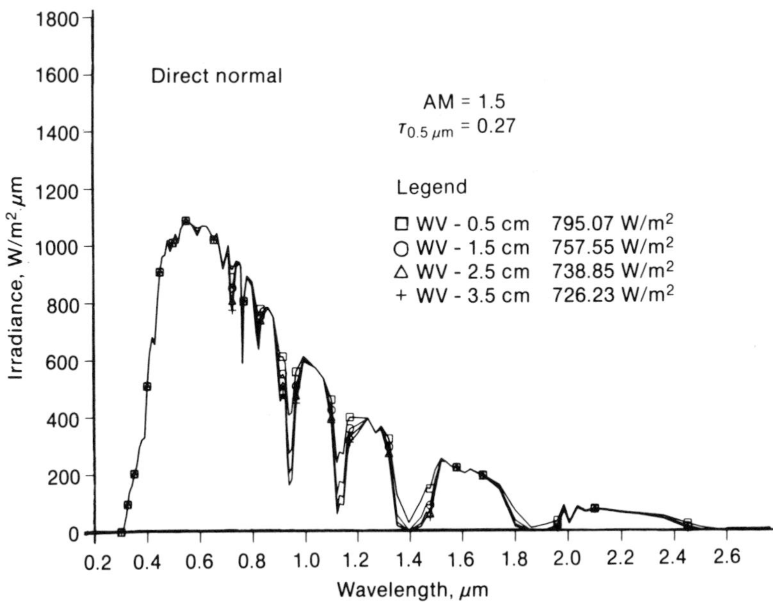

Figure 6.8
Direct-normal spectral solar radiation as a function of water vapor.

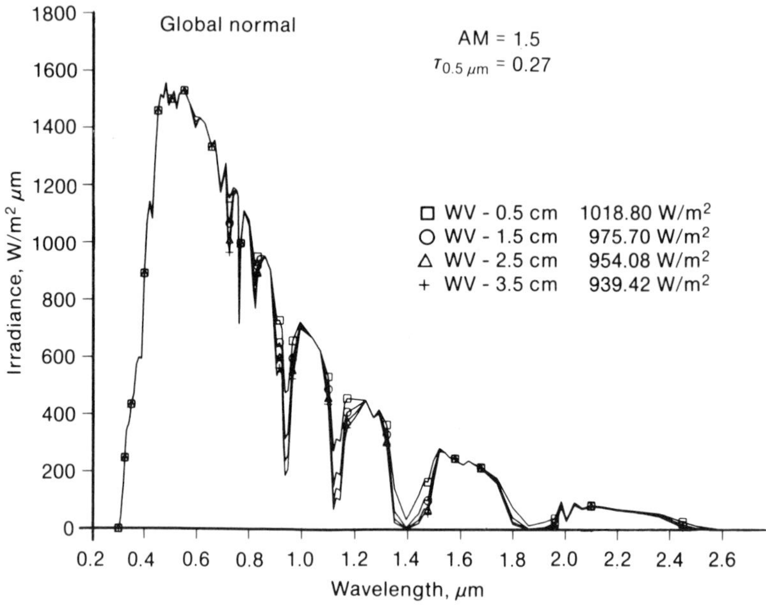

Figure 6.9
Global-normal spectral solar radiation as a function of water vapor.

normal spectra, respectively, when water vapor and turbidity are held constant. Figures 6.6 and 6.7 show the effect of turbidity on direct- and global-normal spectra, respectively, when water vapor and air mass are held constant. Finally, figures 6.8 and 6.9 show the effect of water vapor on direct- and global-normal spectra when turbidity and air mass are held constant.

6.2.4.2 Spectral Solar Radiation Standards and Data Sets Spectral solar radiation data sets generated by SERI, using SERI's rigorous radiative transfer code (BRITE), were adopted by the American Society for Testing and Measurements as standards in 1982 (ASTM, 1982). These data are for direct-normal spectral solar radiation at AM1.5 and for global spectral solar radiation on a surface tilted 37° from the horizontal at AM1.5, from 0.305 to 2.45μm. The two data sets were improved and updated by SERI (Bird, Hulstrom, and Lewis, 1983; Hulstrom, Bird, and Riordan, 1985) and are being considered for updates to the 1982 ASTM standard spectra and for international use as reference spectra to compare solar conversion device performance.

Other spectral solar radiation data sets were generated by SERI (Bird and Hulstrom, 1983a) or can be generated using SERI's simple spectral solar radiation model (Bird and Riordan, 1986). These data can be combined with data on spectrally dependent solar collector materials to design and predict the performance of the collectors.

6.3 Spectral Solar Radiation Measurements

Various instruments to measure solar radiation can be used to take measurements at selected wavelengths with narrow bandwidth (wavelength interval) filters, to take broadband spectral measurements over selected wavelength intervals, and to take high-resolution spectral measurements over a portion of the solar spectrum. (High resolution as it is used here is a relative term since we could resolve individual absorption lines within absorption bands with a truly high-resolution instrument.) Examples of each of these measurements and their applications follow.

6.3.1 Spectral Solar Radiation Measurements at Selected Wavelengths

A sun photometer is an instrument used to measure solar radiation at selected wavelengths. This instrument is usually constructed with a limited FOV, interference filters, and detectors. Sun photometers measure the radiation from the sun within the bandwidth of the optical filter. Typical filter bandwidths are 5–10 nm at half-maximum, and the FOV is usually less than 3°. Measurements at selected wavelengths can be used to calculate the total optical depth of the atmosphere at these wavelengths. For this purpose, the World Meteorological Organization recommends that the central wavelengths of the filters be at 0.368 μm, 0.5 μm, 0.778 μm, and 0.862 μm.

Sun photometers can be calibrated in the laboratory, but most of the calibrations are performed outdoors using the Langley plot method. This is accomplished by plotting the logarithm of the voltage response at each wavelength channel versus the relative air mass. If the atmospheric conditions are stable and the atmosphere is horizontally homogeneous during the measurements, this plot will form a straight line. By extrapolating the straight line to AMO, we can determine the response of the instrument to extraterrestrial radiation in that channel. Assuming that Beer's law holds, the total optical depth becomes

$$\tau_\lambda = -\ln(V_\lambda/V_{o\lambda}), \tag{6.14}$$

where V_λ is the terrestrial voltage reading for the channel and $V_{o\lambda}$ is the extraterrestrial voltage reading taken from the Langley plot intercept. Equation (6.14) can then be written as

$$\tau_{a\lambda} = \tau_{\lambda/M} - \tau_{r\lambda} - \tau_{ma\lambda}. \tag{6.15}$$

The total optical depth τ_λ is obtained from the sun photometer measurement (Shaw, 1983). We can calculate the Rayleigh optical depth $\tau_{r\lambda}$ (Shaw, 1983; Young, 1981) and $\tau_{ma\lambda}$ if molecular absorption is present. The end result is the turbidity for each wavelength channel.

The turbidity is often measured only at a 0.5-μm wavelength. To obtain the turbidity at other wavelengths, we can use the following equation:

$$\tau_{a\lambda} = \beta\lambda^{-\alpha}, \tag{6.16}$$

where β is called the Ångström turbidity coefficient (Ångström, 1961), and α is a parameter that depends on the aerosol particle size distribution. Equation (6.16) assumes that a log-log plot of turbidity versus wavelength produces a straight line, which is not always the case (King and Herman, 1979). A reasonable value to use for α is 1.2, but it can vary between -0.3 and 4.0.

A sun photometer can also be used to calculate precipitable water vapor if the wavelength interval of one of the filters corresponds to a water vapor absorption band. We can use the ratio of the voltage reading at this wavelength to the reading at a wavelength where no water vapor absorption occurs in equations to calculate precipitable water vapor (Bird and Hulstrom, 1982, 1983b). Typical wavelengths for this method are 0.862 μm (no water vapor absorption) and 0.942 μm (strong water vapor absorption).

As previously described in section 6.2.3, turbidity and water vapor are two important parameters needed to calculate spectral solar radiation using simple models. These two parameters are also essential to analyze the wavelength characteristics of measured spectra.

6.3.2 Broadband Spectral Solar Radiation Measurements

Solar radiation measurement devices equipped with special filters can be used to measure solar radiation in relatively broad portions of the spectrum for particular applications (Rao, 1984). For example, filters that select a portion of the spectrum from 0.4 μm to 0.7 μm are used to measure solar radiation in the photosynthetically active region of the spectrum for plant studies. Another example is a sensor with a spectral responsivity curve equal to the response of the average human eye (photopic response). This sensor can be used for studies that require illuminance data, such as building systems studies on daylighting.

6.3.3 High-Resolution Spectral Solar Radiation Measurements

Relatively complex instruments (spectroradiometers) are available to make high-resolution spectral solar radiation measurements over a broad wavelength range. One of the more sophisticated spectroradiometers was built under SERI's direction. This instrument can complete a spectral scan from 0.3 μm to 2.5 μm with less than 1-nm resolution in 2.5 min. The instrument uses visible and infrared wavelength channels that simultaneously view an integrating sphere. The unique features of this instrument are continuous spectral solar radiation calibration, continuous wavelength calibration, continuous monitoring of broadband (0.3–1.1 μm) solar radiation stability, and self-correction for changes in the response at each wavelength. Details of the instrument are given in Kliman and Eldering (1981).

High-resolution spectral measurements are needed to study the fine structure in spectral distributions. It is important to record ancillary meteorological data, such as water vapor, with spectral measurements to isolate specific atmospheric effects on the spectra. Many of the published spectral measurements (Henderson and Hodgkiss, 1963; Gates and Harrop, 1963; Randerson, 1970; Laue, 1970; McCartney and Unsworth, 1979; Kok, Chalmers, and Turner, 1979; Arndt, Bloss, and Hewig, 1979; Burtin et al., 1980) can be used to evaluate spectrally sensitive solar devices directly, but the atmospheric data needed to extend the analyses to other locations and to verify models are lacking.

Examples of spectra measured with the SERI spectroradiometer are shown in figures 6.10–6.13. These plots show the global-normal, direct-normal, global-horizontal, and diffuse-horizontal spectral solar radiation at approximately 0.7-nm resolution in 0.1-μm sections of the spectrum from 0.35 μm to 0.75 μm. The spectra were measured on 24 October 1980 in Golden, CO (latitude 39.75°N, longitude 105.15°W), within a 1.75-h period in the morning. This was a cloudless day with a measured atmospheric turbidity at 0.5 μm of 0.05. Several Fraunhofer lines from the sun and a few absorption bands of the earth's atmosphere are identified. These data illustrate the relative importance of scattered light in the UV portion of the spectrum (although the effect is minimized here because of the very clear sky conditions). The effect of receiver FOV and orientation, indicated by the different components measured, is very evident in these data.

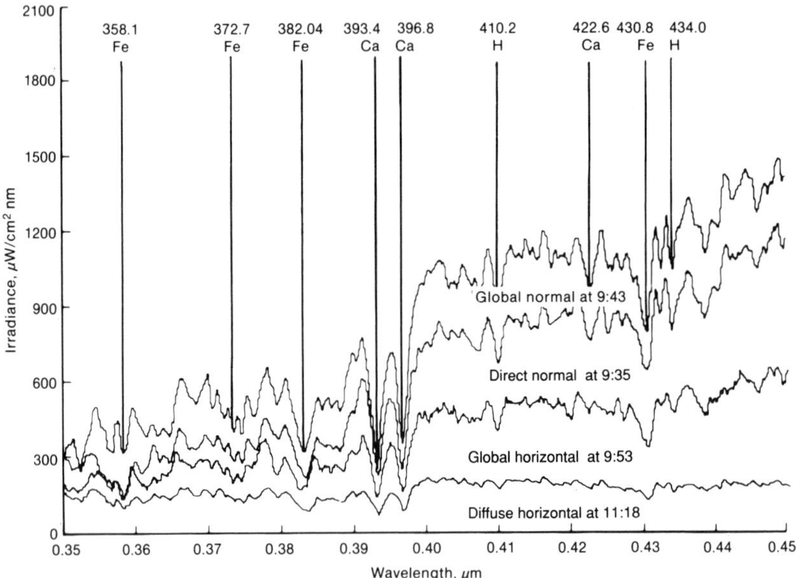

Figure 6.10
Global-normal, direct-normal, global-horizontal, and diffuse-horizontal spectral solar radiation from 0.35 μm to 0.45 μm measured with the SERI spectroradiometer, 24 October 1980, at Golden, CO.

Examples of measured spectra that are used to study atmospheric effects on spectral distributions are also shown in figures 6.14 and 6.15. Figure 6.14 shows the effect of air mass on the direct-normal spectrum. These data show that the greatest change in the spectrum due to air mass occurs in the visible portion of the spectrum and diminishes with increasing wavelength. Figure 6.15 shows the effects of clouds on the spectral distribution. The 9:37 spectrum consists almost entirely of diffuse light. The 12:24 spectrum is also mostly diffuse but has a smaller cosine effect and includes reflection from cloud edges. The 12:56 spectrum includes the direct horizontal component, diffuse light scattered by the air, and diffuse light reflected from cloud edges. The fractional difference between cloudy and clear spectra, defined as $F = (I_{\text{cloud}} - I_{\text{clear}})/I_{\text{clear}}$, is plotted as a function of wavelength in figure 6.16 using three spectra from figure 6.15. I_{clear} is the 12:59 spectrum; the 12:56 spectrum under cloud cover with the solar disk unobscured and the 12:24 spectrum under cloud cover with the solar disk obscured are used for I_{cloud}. Analyses of these data indicate that clouds

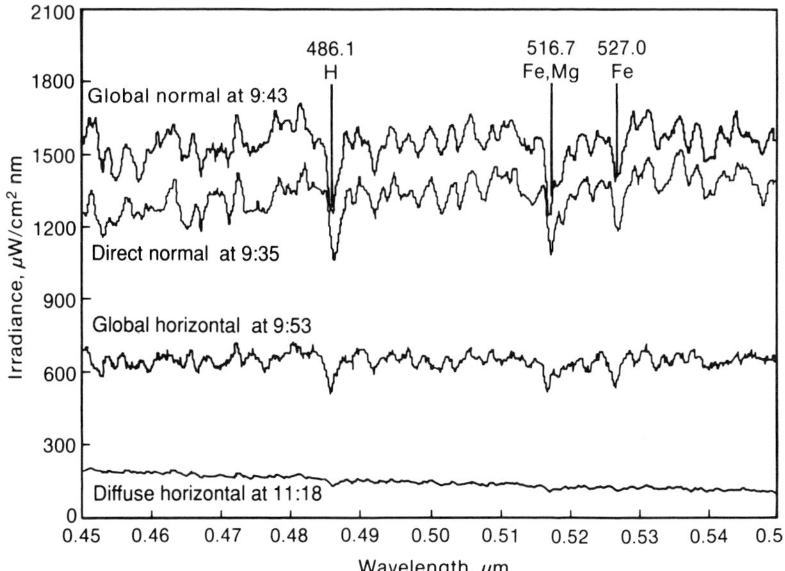

Figure 6.11
Global-normal, direct-normal, global-horizontal, and diffuse-horizontal spectral solar radiation from 0.45 μm to 0.55 μm measured with the SERI spectroradiometer, 24 October 1980, at Golden, CO.

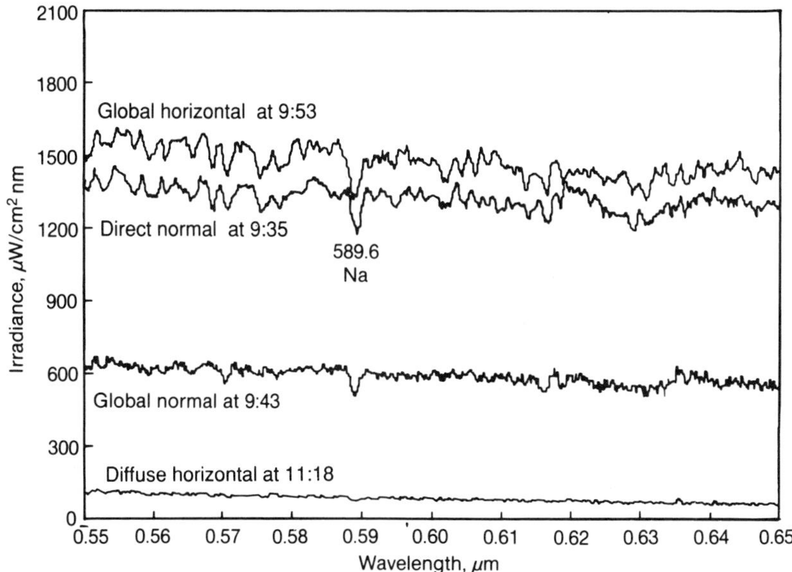

Figure 6.12
Global-normal, direct-normal, global-horizontal, and diffuse-horizontal spectral solar radiation from 0.55 μm to 0.65 μm measured with the SERI spectroradiometer, 24 October 1980, at Golden, CO.

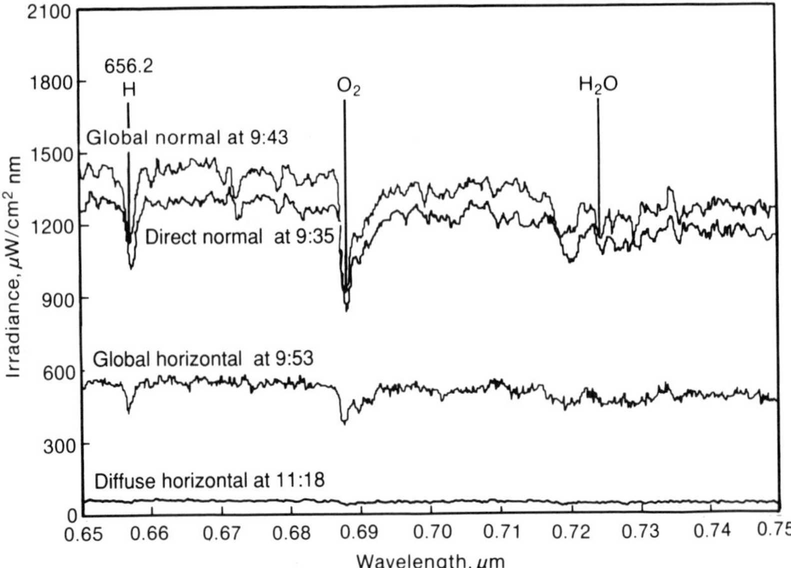

Figure 6.13
Global-normal, direct-normal, global-horizontal, and diffuse-horizontal spectral solar radiation from 0.65 μm to 0.75 μm measured with the SERI spectroradiometer, 24 October 1980, at Golden, CO.

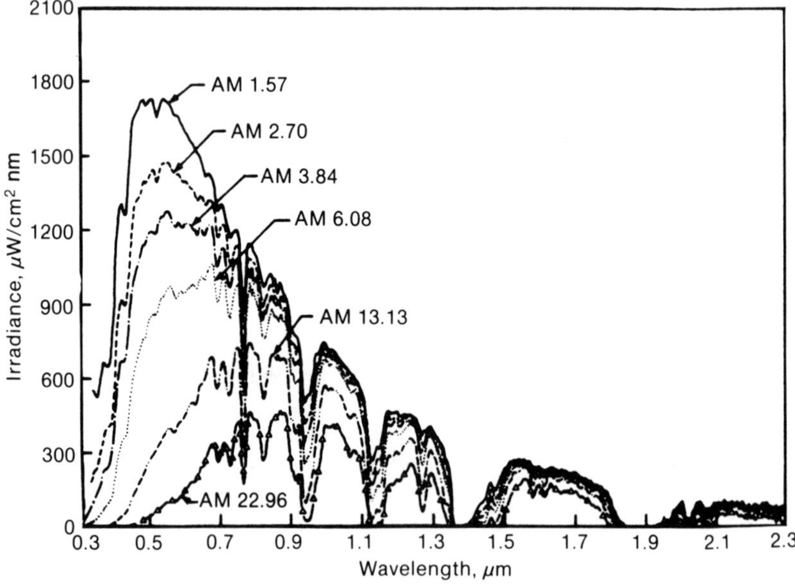

Figure 6.14
Spectral solar radiation measured 21 October 1980 at Golden, CO, showing the effects of air mass on the direct-normal spectrum (turbidity at 0.5 μm ranged 0.04 to 0.05 during these measurements and with no clouds).

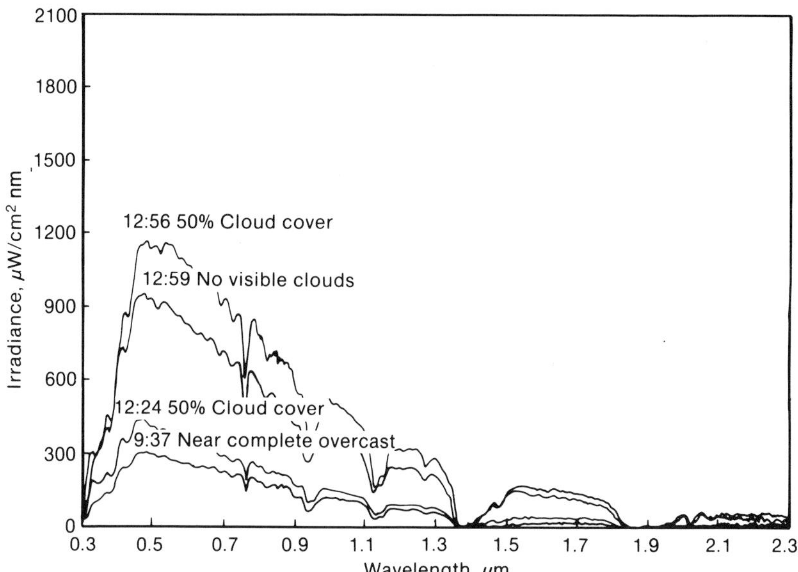

Figure 6.15
Spectral solar radiation measured 29–30 January 1981 in Golden, CO, under cloudy and clear conditions.

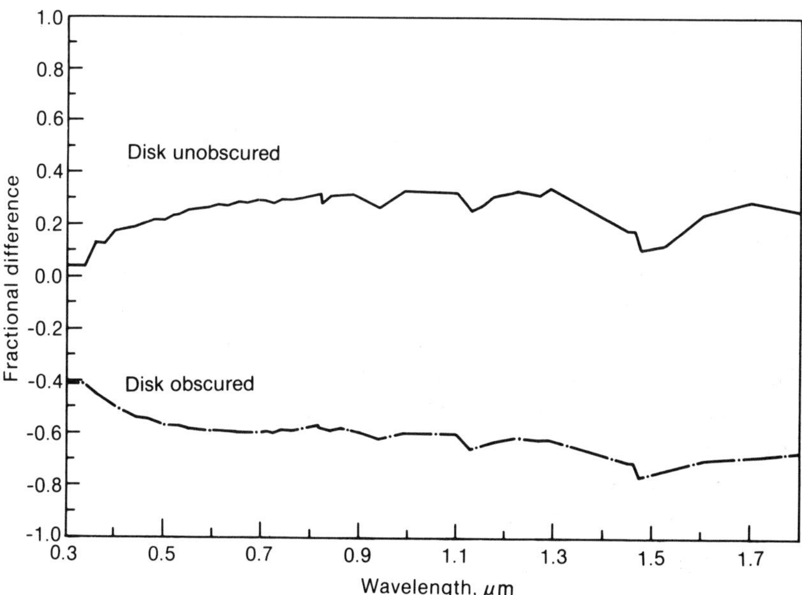

Figure 6.16
Fractional difference between cloudy-sky and clear-sky spectra shown in figure 6.15.

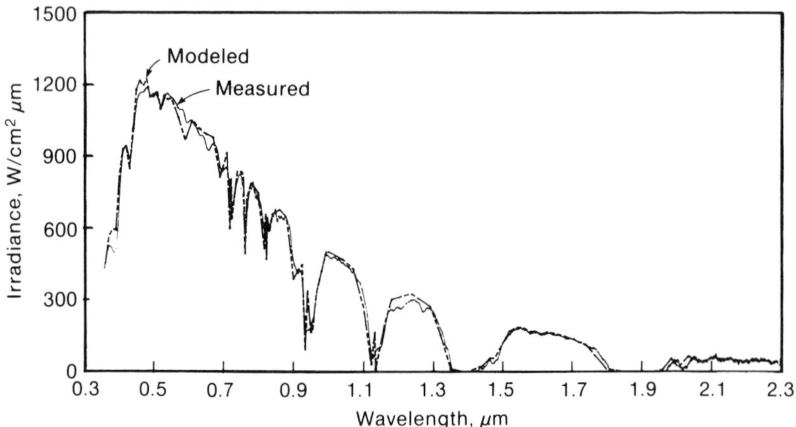

Figure 6.17
Comparison of spectral solar radiation measured at 15:09 MST on 5 August 1981 at Golden, CO, modeled using the SERI rigorous spectral solar radiation model (BRITE).

greatly affect the amplitude of solar radiation and have a relatively minor effect on the global horizontal spectral distribution beyond a 0.5 μm wavelength with the possible exception of additional structure in the water vapor absorption bands. Measured spectra similar to those described here are used to study the effects of other atmospheric parameters such as turbidity and water vapor on spectral distributions and to verify spectral models.

An example of model verification using spectroradiometer data is given in figure 6.17. This plot shows a global horizontal spectrum measured at 10-nm resolution with the SERI spectroradiometer and a spectrum calculated using the SERI BRITE model. The measured spectral data were recorded on 5 August 1981 in Golden, CO, at 15:09 MST with turbidity and water vapor calculated from sun photometer measurements. The following atmospheric conditions recorded at the time of the spectral measurements were used in the model calculations:

solar zenith angle = 44.8°,
turbidity at 0.368 μm = 0.39,
turbidity at 0.500 μm = 0.28,
turbidity at 0.862 μm = 0.13,
water vapor = 2.25 cm, and
surface pressure = 829 mb.

The ground albedo was assumed to be 0.2 for all wavelengths. The amount of ozone was calculated to be 0.33 atm-cm (Van Heuklon, 1979), which produced an ozone optical depth of 0.009 in the vertical (Inn and Tanaka, 1953) at 0.5 μm.

6.4 Summary

This chapter is not intended to be a thorough treatment of spectral modeling and measurements. It gives the reader a quantitative measure of the solar spectrum under varying conditions and presents state-of-the-art modeling and measurement methods. A much greater understanding of the subjects discussed here can be obtained by examining the references (for example, Iqbal, 1983). The data presented here should help the reader to understand the intricate nature of the solar spectrum, which becomes even more intricate at higher resolutions than are shown here.

References

American Society for Testing and Materials (ASTM). 1982. Standard terrestrial direct normal solar spectral irradiance tables for air mass 1.5. E-891-82. Standard for solar spectral irradiance tables at air mass 1.5 for a 37-deg tilted surface. E-892-82. *Annual Book of ASTM Standards*. Philadelphia, PA: ASTM.

Ångström, A. K. 1961. Technique of determining the turbidity of the atmosphere. *Tellus* 13:214–231.

Arndt, W., W. H. Bloss, and G. H. Hewig. 1979. Determination of the spectral distributions of global radiation with a rapid spectral radiometer and its correlation with solar cell efficiency. In *Proc. 2nd E. C. Photovoltaic Solar Energy Conference*. Boston, MA: D. Reidel Publishing Co.

Bird, R. E. 1982. Terrestrial solar spectral modeling. *Solar Cells* 7:107–111.

Bird, R. E. 1984. A simple spectral model for direct normal and diffuse horizontal irradiance. *Solar Energy* 20:461–471.

Bird, R. E., and R. L. Hulstrom. 1982. Precipitable water measurements with sun photometers. *Journal of Applied Meteorology* 21:1196–1201.

Bird, R. E., and R. L. Hulstrom, 1983a. Additional solar spectral data sets. *Solar Cells* 8:85–95.

Bird, R. E., and R. L. Hulstrom. 1983b. Reply. *Journal of Climate and Applied Meteorology* 22:1969–1970.

Bird, R. E., and C. Riordan. 1986. Simple solar spectral model for direct and diffuse irradiance on horizontal and tilted planes at the earth's surface for cloudless atmospheres. *Journal of Climate and Applied Meteorology* 25:87–97.

Bird, R. E., R. L. Hulstrom, and L. J. Lewis, 1983. Terrestrial solar spectral data sets. *Solar Energy* 30:563–573.

Brine, D. T., and M. Iqbal. 1983. Solar spectral diffuse irradiance under cloudless skies. *Solar Energy* 30:447–453.

Burtin, B., R. Carels, D. Crommelynk, and R. Dogniaux. 1980. *Development of Discrete and Continuous Spectral Measurements of Global Solar Radiation*. Royal Meteorological Institute of Belgium.

Chandrasekhar, S. 1950. *Radiative Transfer*. Oxford: Clarendon Press.

Dave, J. V. 1978. Extensive data sets of the diffuse radiation in realistic atmospheric models with aerosols and common absorbing gases. *Solar Energy* 21:361–369.

Flowers, E. C., G. A. McCormick, and K. R. Kurfis. 1969. Atmospheric turbidity over the United States. *Journal of Applied Meteorology* 8:955–962.

Fröhlich, C., and C. Wehrli, 1981. Data provided on computer tape. Davos, Switzerland: World Radiation Center.

Gates, D. M., and W. J. Harrop. 1963. Infrared transmission of the atmosphere to solar radiation. *Applied Optics* 2:887–898.

Hay, J. E., and J. A. Davies. 1978. Calculation of solar radiation incident on an inclined surface. In *Proc. First Canadian Solar Radiation Data Workshop*. J. E. Hay and T. K. Won, eds., Toronto, Ontario, Canada: Minister of Supply and Services.

Henderson, S. T., and D. Hodgkiss. 1963. The spectral energy distribution of daylight. *British Journal of Applied Physics* 14:125–131.

Hulstrom, R. L., R. E. Bird, and C. Riordan, 1985. Spectral solar irradiance data sets for selected terrestrial conditions. *Solar Cells* 15:365–391.

Inn, E. C. Y., and Y. Tanaka. 1953. Absorption coefficient of ozone in the ultra-violet and visible regions. *Journal of the Optical Society of America* 43:870–873.

Iqbal, M. 1983. *An Introduction to Solar Radiation*. New York: Academic Press.

Justus, C. G., and M. V. Paris. 1985. A model for solar spectral irradiance at the bottom and top of a cloudless atmosphere. *Journal of Climate and Applied Meteorology* 24:193–205.

Kasten, F. 1966. A new table and approximate formula for relative optical air mass. *Arch. Meteorol. Geophys. Bioklimatol.*, Ser. B14., pp. 216–223.

Kerker, M. 1969. *The Scattering of Light and Other Electromagnetic Radiation*. New York: Academic Press.

King, M. O., and B. M. Herman. 1979. Determination of the ground albedo and the index of absorption of atmospheric particulates by remote sensing, part I: theory. *Journal of the Atmospheric Sciences* 36:163–173.

Kliman, A. W., and H. G. Eldering. 1981. Design and development of a solar spectroradiometer. *Electro Opt. Sys. Design*, pp. 53–61.

Kok, C. J., A. N. Chalmers, and R. Turner. 1979. Spectroradiometry of daylight at sea level in the southern hemisphere: Durban. *S. Afr. Tydskr. Fis.* 2:47–53.

Kondratyev, K. Ya. 1969. *Radiation in the Atmosphere*. New York; Academic Press.

Laue, E. G. 1970. The measurement of solar spectral irradiance at different terrestrial elevations. *Solar Energy* 13:43–57.

Leckner, B. 1978. The spectral distribution of solar radiation at the earth's surface—the elements of a model. *Solar Energy* 29:143–150.

Lenoble, J. 1977. *Standard Procedures to Compute Atmospheric Radiative Transfer in a Scattering Atmosphere*. Boulder, CO: National Center for Atmospheric Research.

McCartney, H. A., and M. H. Unsworth. 1979. Spectral distribution of solar radiation. I: direct radiation. *Quarterly Journal of the Royal Meteorological Society* 104:699–718.

McClatchey, R. A., R. A. Fenn, J. E. A. Selby, F. E. Volz, and J. S. Garing. 1972. *Optical Properties of the Atmosphere*, third edition. AFCRL-72-0497. Bedford, MA: Air Force Cambridge Research Laboratories.

Moon, P. 1940. Proposed standard solar-radiation curves for engineering use. *Journal of the Franklin Institute* 20:583–617.

Neckel, H., and D. Labs. 1981. Improved data of solar spectral irradiance from 0.33 to 1.25 μm. *Solar Physics* 74:231–249.

Randerson, D. 1970. A comparison of the spectral distribution of solar radiation in a polluted and a clear air mass. *Journal of the Air Pollution Control Association* 20:546–548.

Rao, C. R. Nagaraja. 1984. Photosynthetically active components of global solar radiation: measurements and model computations. *Arch. Met. Geoph. Biocl.* B34:353–364.

Shaw, G. E. 1983. Sun photometry. *Bulletin American Meteorological Society* 64:4–10.

Shettle, E. P., and R. W. Fenn. 1975. Models of atmospheric aerosols and their optical properties. In *Proc. AGARD Conference No. 183 Optical Propagation in the Atmosphere*, pp. 2.1–2.16.

Stamnes, K., and R. A. Swanson. 1981. A new look at the discrete ordinate method for radiative transfer calculations in anisotropically scattering atmospheres. *Journal of the Atmospheric Sciences* 3:387–399.

Thomas, A. P., and M. P. Thekaekara. 1976. Experimental and theoretical studies on solar energy for energy conversion. In *International and U. S. Programs Solar Flux*. K. W. Boer, ed., Vol. 1, pp. 338–355.

Van Heuklon, T. K. 1979. Estimating atmospheric ozone for solar radiation models. *Solar Energy* 22:63–68.

Walraven, B. R. 1978. Calculating the position of the sun. *Solar Energy* 20:393–397.

Young, A. T. 1981. On Rayleigh-scattering optical depth of the atmosphere. *Journal of Applied Meteorology* 20:328–380.

Zimmerman, J. C. 1981. *Sun-Pointing Programs and Their Accuracy*. SAND-81-0761. Albuquerque, NM: Sandia Laboratories.

7 Insolation Forecasting

John Jensenius

7.1 Introduction

The use of solar energy will increase as it gains recognition as a valuable resource, thus increasing the need to predict solar radiation. Predictions of solar radiation are based on past climatic data and day-to-day climatic data. Past climatic data, for example, might be used to determine the probability of 3, 4, or 5 consecutive days of cloudy weather. The probabilities could then be incorporated into cost-benefit studies for determining the optimal heat storage and backup heating systems for a particular location. Climatic data might also be used to estimate the payback period for various solar energy systems. In general, predictions based only on past climatic data assume that weather patterns are somewhat regular. This is a safe assumption because although the weather can fluctuate considerably from the normal during a season or year, weather conditions averaged over the lifetime of a solar energy system will usually deviate only slightly from the normal. Consequently, the climatic type of forecast is more useful than the day-to-day forecast in designing solar energy systems, energy storage facilities, and backup systems.

Once a solar energy system is operational, however, a day-to-day forecast of the expected incoming energy is more useful than the climatic estimates (Hulstrom, 1981). For example, a homeowner with a solar water heater may wish to schedule household chores requiring hot water on days when the amount of incoming solar radiation is predicted to be high. Where peak and off-peak power prices differ, homeowners with photovoltaic-electric systems may wish to buy and store energy during the night when the cost is lower if they know that their system will not be able to generate the needed energy for the following day (Aronson and Caskey, 1981). In areas where a significant number of homes use solar heating or photovoltaic systems, utilities may find that a forecast of the incoming solar radiation allows them to anticipate and plan for the future energy demand. This may be especially important on cloudy days during the winter when the incoming solar radiation is low. During these times, utilities can expect an increased load because of the use of backup systems for domestic space or water heating. The load will be even greater after several days of cloudy weather when most stored energy (solar) supplies have been exhausted. Utilities using the sun as a primary source of energy also will need predictions of the expected incoming energy to complement solar power efficiently with electricity generated from other sources.

Solar radiation forecasts are important to the agricultural community, which uses them to schedule various farm activities (Brown and Lambert, 1981; Cox, 1981). Insolation forecasts must be included in crop models to predict critical stages of plant development accurately. Forecasts should also be considered before applying pesticides, fungicides, and herbicides since the sun's radiation affects many of these chemical compounds, as well as the microclimate both above and below the surface of the soil. High amounts of incoming solar radiation aid in the evaporation of vegetative wetness due to precipitation or condensation. Hay should be cut when several days of good drying are forecast. Predictions of insolation should also be considered when determining irrigation schedules since both evaporation and evapotranspiration are strongly related to the incoming solar energy. In summary, many farm activities are either directly or indirectly affected by the sun's energy and can be better planned if the expected insolation is known.

The insolation forecasts discussed in the remainder of this chapter are of the total solar radiation received on a horizontal surface. Ma and Iqbal (1983a) discuss methods for relating the radiation received on a horizontal surface to direct and diffuse radiation. However, little or no effort has been made to predict the direct beam or diffuse (sky) radiation components. Similarly, no work has been done to predict the insolation received by inclined or vertical surfaces. Ma and Iqbal (1983b) have also examined techniques for estimating insolation on these surfaces.

This chapter presents only those methods developed and tested to predict incoming radiation. Methods for estimating solar radiation based on coincident weather observations are not included here, but are discussed in chapter 3 of this book. Techniques used to forecast sunshine (Cope and Bosart, 1982; National Weather Service, 1983), although not specifically covered by this chapter, are similar to those used to forecast insolation.

7.2 Solar Forecast Research and Development

Although considerable effort has been made to utilize solar energy efficiently, only minimal resources have been directed toward forecasting the incoming energy. In most cases, those attempting to forecast insolation limited their analysis to a small number of stations. Several techniques were used to develop methods to forecast insolation. One common method uses statistical regression equations that relate the observed insolation to other meteorological variables. The Model Output Statistics (MOS) technique (Glahn and Lowry, 1972), which is an example of this method, is currently used by the National Weather Service (NWS) to provide forecasters with guidance for many meteorological variables. In developing MOS equations to forecast insolation, scientists use linear regression to determine the relationship between the observed insolation (predictand) and the output of numerical weather prediction models (predictors). The output usually consists of meteorological variables such as temperature, wind, pressure, and humidity, for various times and levels in the atmosphere. Often, climatic variables are included as predictors. Nonlinear relationships can be accounted for through transformations of the basic predictors.

In 1979, the NWS (Jensenius and Carter, 1979) used the MOS technique to develop and test insolation forecasts for three agricultural stations in South Carolina. Equations were developed relating the observed daily insolation to forecast output produced by one of the NWS numerical weather prediction models. When tested with independent data for one growing season, the equations produced forecasts with mean absolute errors of 0.74, 0.80, and 0.89 kWh/m^2 for advance periods of 1, 2, and 3 days, respectively. The mean absolute error for forecasts based on the expected climatic value was 1.04 kWh/m^2. These same equations are currently used during the growing season to produce insolation forecasts for the three testing locations in South Carolina.

Baker and Casper (1981) developed and tested three types of regression equations for objectively forecasting the percent of extraterrestrial radiation (ETR) received at Bismarck, ND, Madison, WI, and Chicago, IL. For the first type of equation, the relationship between MOS forecasts of the probability of precipitation and the observed insolation was determined. The resulting equation and a forecast of the probability of precipitation were then used daily to predict the insolation. For the second type of equation, a MOS approach was used to relate the observed insolation to relative humidity forecasts from a numerical weather prediction model. The third type of equation was similar to the second except that the observed relative humidity was related to the insolation rather than to the predicted relative humidity. Of course, to forecast insolation with the third type of equation, the model predictions of humidity were substituted into the equation. Verification results for a 24-h forecast indicated that this third technique was best. In terms of the percent of ETR, the root mean square (rms) error for this method was 18.0. Using the objective forecasts as guidance, several students and faculty at the University of Michigan made subjective predictions of insolation. Forecasts were made daily and a consensus of the forecasts was achieved. When verified, the consensus forecast had an rms error (in terms of percent of ETR) of 17.1.

Using a combination of theoretical, empirical, and statistical functions, Falconer (1981) developed a method for forecasting the expected insolation for clear or partly cloudy conditions at Albany, NY. First he estimated the average precipitation expected during the day. He then plotted this value and the day of the year to determine the expected clear-sky insolation. Finally, he reduced the clear sky insolation using a cloud effect factor, which he determined to be 0.85 ± 0.15 for days with scattered clouds and 0.55 ± 0.25 for days with broken clouds. Falconer did not determine factors for cloudy days or for days with rapidly changing cloud conditions. He also did not present any independent test verifications of his method in his report.

In 1979, the National Oceanic and Atmospheric Administration (NOAA) proposed a new system that would provide solar energy forecasts for the conterminous United States. Using the MOS technique and data for the period from February through June of 1977 and 1978 (excluding April 1978), Jensenius and Cotton (1981) developed and tested equations for predicting the daily insolation 1 and 2 days in advance. They used data from 34 sites in the NOAA Solar Radiation Network, and developed and tested equations only for the spring season (March–May). They used a forward stepwise selection procedure to derive the multiple linear regression equations. Two basic approaches for producing the insolation forecasts were tested. For the first, the "indirect" approach, they derived equations to predict the percent of ETR received. They then multiplied the percent values by the ETR to obtain the insolation amount forecasts. For the second, the "direct" approach, they derived equations to predict the insolation amount directly. In addition, two types of equations were developed and tested for each approach. For the "single-station" equation type, they developed a separate equation for each of the 34 stations. For the "regionalized" equation type, they divided the United States into six regions and combined the data for all stations within a region. They then derived a separate equation for each region. One advantage of a regionalized equation is that, once developed, the equation can be applied to other locations within the region if the new locations are climatically similar to the stations for which the equation was derived. In contrast, the single-station equation accounts for the specific characteristics of the station for which it was derived and usually should not be applied to other locations.

For each equation type, Jensenius and Cotton generated and verified forecasts for a 1-month independent sample (April 1978). In addition, they made and verified forecasts based solely on the expected climatic value and forecasts based solely on the previous day's observation (persistence). The verification results showed that regionalized equations derived with the indirect approach gave the best results. The mean absolute error, in terms of percent of ETR, ranged from 10% for daylight periods ending 24 h after the initial forecast time to 13% for daylight periods ending approximately 60 h after the initial forecast time. Although the MOS regionalized indirect approach verified best, all of the MOS methods tested were better than the persistence or climatic forecasts.

7.3 The NWS Operational Solar Insolation Forecast System

Although the MOS approach worked well for the 34 stations used to develop the equations, there was no way to determine how well the equations would do at other locations where insolation data were unavailable. As an alternative to the strict MOS approach, the National Weather Service (Jensenius, 1983) also developed and tested equations using a second-order MOS approach. In this development, existing MOS forecasts of cloud cover (National Weather Service, 1981) and dew point (National Weather Service, 1985) were used as predictors instead of the raw model output. About 30 of the stations in the NOAA solar radiation network used available MOS cloud and dew point forecasts. A single equation was derived for all stations and seasons. This equation was then applied to the more than 230 stations for which MOS cloud and surface dew point forecasts are made. The MOS cloud forecast data that are used in the equations are expressed in terms of the probability (a number ranging from 0.0 to 1.0) of the sky being clear (0–5% of the sky covered by clouds), scattered (5–55%) of the sky covered by clouds), broken (55–95% of the sky covered by clouds), or overcast (95–100% of the sky covered by clouds). The dew point forecasts are given in degrees Fahrenheit. Forecasts of both cloud cover and dew point valid at 1200 Greenwich Mean Time (GMT) (7 AM EST), 1800 GMT (1 PM EST), and 0000 GMT (7 PM EST) were included in the development. Global (total on a horizontal surface) solar energy observations and MOS forecasts for the period from January 1977 through December 1978 were originally available for the development. Forecasts from these equations were disseminated to NWS field offices from October 1981 through June 1982. Since June 1982, the forecasts have been produced by revised equations developed from 4 years (1977–1980) of data. A series of statistical regressions were used to develop the final forecast equation, which has the form

$$\text{PETR} = \sum_{h=1}^{3} \text{COEF}_h[0.69 - 0.25(\text{BKN}_h) - 0.54(\text{OVC}_h) - 0.0025(\text{DP}_h) + 0.015(\text{ETR})],$$

where

h = 1, 2, and 3 for the 1200, 1800, and 0000 GMT values, respectively,
PETR = percent of ETR,
BKN = probability of a broken cloud cover,
OVC = probability of an overcast sky,
DP = surface dew point (°F),
ETR = daily extraterrestrial radiation (in kWh/m^2), and
COEF = weighting coefficient.

The equations for the weighting coefficients (COEF) are as follows:

$$\text{COEF}_1 = 0.84 - 0.0041(\text{LONG}) - 0.022(\text{DAY LENGTH}),$$
$$\text{COEF}_2 = 0.84 - 0.0041(\text{LONG}) - 0.002(\text{DAY LENGTH}),$$
$$\text{COEF}_3 = -0.68 + 0.0082(\text{LONG}) + 0.024(\text{DAY LENGTH}),$$

where LONG = station longitude and DAY LENGTH = hours from sunrise to sunset.

In the PETR forecast equation, the weighting coefficients for 1200, 1800, and 0000 GMT account for longitudinal and seasonal differences in the importance of the forecasts for the three different times. For example, the 1200 GMT forecast would have more weight along the East Coast, where the local standard time (LST) is 7 AM, than along the West Coast, where the LST is only 4 AM. Similarly, the 0000 GMT forecast has more weight along the West Coast, where the LST is 4 PM, than along the East

Table 7.1
Solar energy forecast calculations for the two examples for Washington, DC, in April

Equation terms	Example 1 term contribution	Example 2 term contribution
0.69	0.69	0.69
−0.25 (BKN)	−0.00	−0.00
−0.54 (OVC)	−0.54	−0.00
−0.0025 (DP)	−0.15	−0.05
+0.015 (ETR)	+0.15	+0.15
Forecast PETR	0.15	0.79

Coast, where the LST is 7 PM. Neither the clear nor scattered cloud cover categories contributed significantly to the equation and, therefore, were not included. The regression constant, however, does account for the combined effect of these two predictors.

To understand better the forecast equation, consider the following two examples for Washington, DC, in late April. At that time, the daily ETR is about 10 kWh/m^2. For the first example, assume that the skies are expected to be overcast all day with virtually no chance of any clearing; thus, the probability of overcast skies would remain at 1.0 all day. Also, assume that the dew point is about 60°F (16°C). The calculated values for each of the terms in the equation are given in table 7.1. Note that, for simplicity, the dew point and probability of cloudiness are assumed to be constant throughout the day; therefore, each term in the forecast equation can be multiplied by the sum of the weighting coefficients, which is 1. In this example, the equation predicts that only 15% of the ETR, or about 1.5 kWh/m^2, will reach the earth's surface during the day. In contrast, consider a second example with clear skies for which the probabilities of broken or overcast sky conditions are equal to zero. In this example, suppose that the dew point remains at about 20°F (−7°C) during the day. Table 7.1 also gives the calculations for this example. In this case, the equation predicts that 79% of the ETR, or 7.9 kWh/m^2, will reach the ground during the day.

7.3.1 A Sample Forecast

A sample of the two maps that compose the solar forecast chart is shown in figures 7.1 and 7.2. These maps contain the predicted total daily energy for

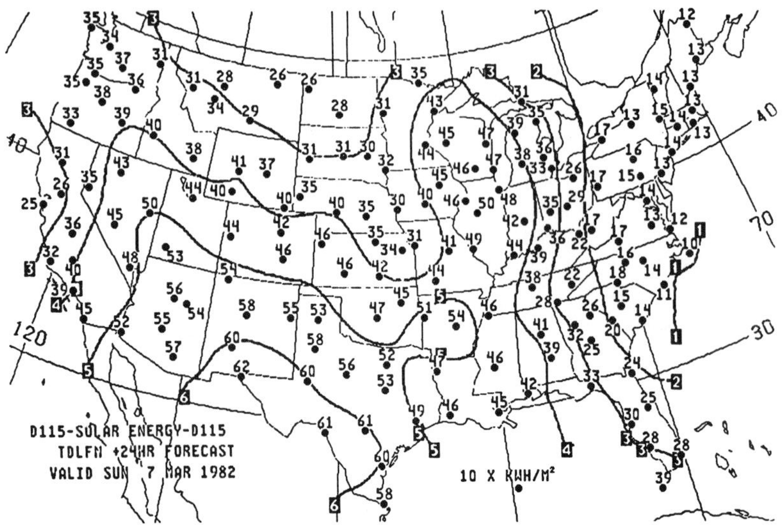

Figure 7.1
NWS solar energy forecast for 7 March 1982. The values plotted are the expected number of kilowatt-hours per square meter multiplied by 10. The contour intervals are every kilowatt-hour per square meter.

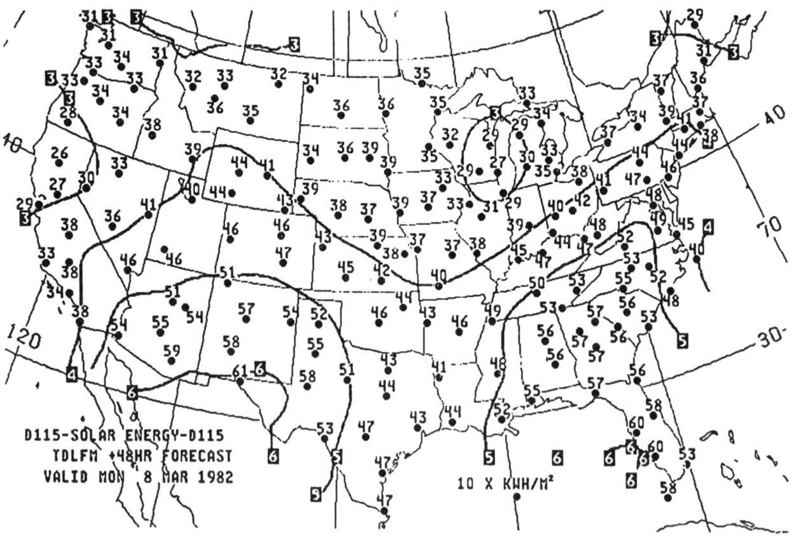

Figure 7.2
NWS solar energy forecast for 8 March 1982. The values plotted are the expected number of kilowatt-hours per square meter multiplied by 10. The contour intervals are every kilowatt-hour per square meter.

Figure 7.3
Satellite picture taken at 1 PM EST, 7 March 1982.

7 and 8 March 1982 and are based on weather data observed at 7 PM EST on 6 March. NWS field offices received the maps about 1 AM EST, 7 March. The values plotted at cities on the chart are the number of kilowatt-hours of energy expected per square meter multiplied by 10. Contour lines are drawn at intervals of every 1 kWh/m^2. Satellite pictures taken at 1 PM EST on these days are shown in figures 7.3 and 7.4 for comparison. Keep in mind that the satellite pictures show only the instantaneous cloud cover, whereas the solar forecasts are for the total energy received over the entire day. Also, the solar forecasts are for specific locations and should not be interpolated to locations where local geographical features cause significant differences in cloudiness; e.g., the values for Denver and Grand Junction, CO, should not be averaged to obtain a value for the intermediate mountains.

The solar forecast for Sunday, 7 March (figure 7.1), shows a large area of low-energy values over most of the eastern part of the country. High values are predicted from Arizona and New Mexico eastward to Alabama and northward through the Midwest to Lake Superior. Lower values are predicted for Montana southeastward to northwestern Missouri and also in northern and central California. By Monday, 8 March, the forecast (figure 7.2) indicates that the incoming energy should increase substantially over the entire East Coast area, with particularly high values in the Southeast.

Figure 7.4
Satellite picture taken at 1 PM EST, 8 March 1982.

High values are also forecast for most of the South and Southwest. Low values are predicted in northern California and Oregon and in the area around Lake Michigan. The satellite pictures taken at 1800 GMT (1 PM EST) on 7 March (figure 7.3) and 8 March (figure 7.4) show that the areas of cloudiness coincide fairly well with areas forecast to have low incoming energy. In the East, several storms moving along the Atlantic Coast caused cloudiness over much of the area on Sunday; however, by Monday, this cloudiness had shifted eastward off the coast. Note that the brighter area from West Virginia and Pennsylvania northward on the 8 March picture is caused by snow on the ground, not by cloudiness, except in northern New England, where some clouds are present over the snow-covered terrain. In the northern Great Plains, Upper Midwest, and Great Lakes regions, the snow cover also makes it difficult to distinguish cloudy and clear areas; however, a close examination of the pictures reveals ground features such as rivers, lakes, and forests in areas with clear skies. In the Midwest, a storm causing cloudiness from Montana southeastward to eastern Nebraska and western Missouri on 7 March moved to the Great Lakes area by 8 March. Another storm approaching the California coast from the Pacific was responsible for clouds over most of central and northern California and parts of Nevada on 7 March and most of northern California and Oregon on 8 March. Some cloudiness is also observed in the Rocky Mountains on both days, although most of this is limited to the mountainous areas between stations.

7.3.2 Daily Operational Forecast Procedure

Since the process used to produce the solar forecasts is complex, consider step-by-step how the solar forecasts are generated. The procedure begins twice a day at 1200 and 0000 GMT, i.e., 7 AM and 7 PM EST. At these times, weather observations are taken at the ground and at numerous levels in the atmosphere at locations throughout the world. These obsevations are fed into the NOAA computers in Suitland, MD, for analysis of the current weather conditions. These analyzed fields are then used as the starting conditions for a mathematical computer model of the atmosphere. From the initial data, this weather prediction model progresses forward in time in much the same way as the real atmosphere would. Both the conditions in the model atmosphere and in the real atmosphere are controlled by Newton's equations of motion, laws of thermodynamics, and gravity. At given computer-simulated time intervals, usually every 6 to 12 h, the predictions of temperature, humidity, pressure, wind, etc., produced by the numerical model are saved. These quantities are then fed into the MOS system, which transforms the atmospheric model results into forecasts of specific weather elements. As discussed earlier, of particular importance for predicting the solar energy are the MOS forecasts of cloud cover and of surface dew point. The cloud cover and dew point predictions, along with climatic factors, are then used in the appropriate statistical equation to produce forecasts of the percentage of daily ETR expected on a horizontal plane at the earth's surface. This percentage is multiplied by the actual daily ETR to obtain the solar energy amount forecasts. The forecasts are then transmitted to NWS field offices in map form via facsimile circuit and the Automation of Field Operations and Services (AFOS) system.

7.3.3 Performance of the NWS Solar Forecast System

Before the solar energy guidance was first issued, the NWS conducted a limited amount of testing to determine its accuracy. As mentioned earlier, the original operational equations were developed from 2 years of data (1977–1978). The equations for the 0000 GMT cycle were then tested on independent data for the period March through May 1979. Table 7.2 gives the results of these tests. Note that the forecasts were made and verified in

Table 7.2
NWS solar forecast verification statistics for the March–May 1979 test period (forecast, observed, and error values are given in terms of percent of daily ETR)

Statistic	24-h statistical forecast	24-h persistence	Climatic estimate	48-h statistical forecast	48-h persistence
Mean forecast value	0.53	0.54	0.54	0.53	0.54
Mean observed value	0.54	0.54	0.54	0.54	0.54
Mean absolute error	0.09	0.15	0.15	0.11	0.17
Root mean square error	0.12	0.20	0.18	0.14	0.24
Correlation of the forecasts and observations	0.79	0.48	0.38	0.70	0.28

terms of percent of the daily ETR. Forecasts based on expected climatic values and forecasts set equal to the previous day's observation (persistence) were also verified, and these results are given for comparison. About 1,500 forecasts were verified for each forecast type and projection. The errors for the statistical forecasts were considerably smaller than those for the forecasts based on persistence or climate. The correlation between the observed and forecast values was also much higher for the statistical forecasts.

Although the operational NWS forecasts have not been verified on a nationwide scale, several independent studies have been conducted for specific locations. Justus and Tarpley (1983) verified the NWS insolation forecasts for November 1981 through October 1982 for Atlanta, GA. They found the NWS forecasts to be more reliable than either persistence or climatic forecasts for all projections. Riordan (1984) compared the solar energy forecasts for Denver with observations taken in Boulder and Golden, CO. She found the mean absolute errors of the forecasts (in terms of percent of ETR) to range from 9% at the 24-h projection to 13% at the 60-h projection.

The most important consideration, however, in judging the overall performance of a solar forecast system is whether the forecasts are sufficiently accurate to be of value to specific users. Although the accuracy of the NWS solar forecasts is known to some extent, the accuracy needed for specific applications is not known. Because of the variety of possible uses for the NWS solar forecast product, each individual application will need to be evaluated separately (for example, Picard and Winn, 1985). Cost-benefit studies may be necessary to determine whether or not decisions should be made based on the forecasts of solar energy.

7.3.4 Availability of the NWS Solar Forecasts

The statistical solar forecasts are currently transmitted daily from Suitland, MD, to NWS forecast offices across the country via facsimile circuit and on the AFOS system. At these offices, the statistical predictions can be used as guidance to help prepare forecasts of the incoming solar energy. These predictions can then be disseminated to the media in general public or agricultural forecast messages. At present, little demand exists for the solar predictions in most areas of the country (possibly because the solar community is unaware of the product); consequently, few forecast offices are disseminating the solar forecast. Also, because of the need for higher-priority weather information, many NWS forecast offices currently may not be receiving the solar energy forecasts. However, in areas where the need for, or interest in, the forecasts has been sufficient to justify dissemination of this information, forecast offices are making the solar forecasts available to the public. Inquiries concerning the local availability of the NWS solar energy forecasts should be directed to the nearest NWS office.

7.4 Summary

The need for and possible uses of a forecast of the incoming solar radiation have been discussed. Currently, the National Weather Service is producing objective forecasts of global solar radiation for more than 200 stations throughout the conterminous United States. These solar forecasts have been shown to be reliable for projections of 1 and 2 days. At present, however, because little demand exists for the solar forecasts, most NWS offices do not disseminate this information to the public. Public demand for the solar forecasts is not expected to increase until solar energy is more widely utilized and until the specific benefits of the solar forecast are demonstrated.

References

Aronson, E. A., and D. L. Caskey, 1981. The value of prediction to grid connected photovoltaic systems with storage. In *Proc. First Workshop on Terrestrial Solar Resource Forecasting and on the Use of Satellites for Terrestrial Solar Resource Assessment*. Newark, DE, Boulder, CO: American Solar Energy Society, pp. 3–7.

Baker, D. G., and M. A. Casper, 1981. Subjective forecasting of received solar radiation. In *Proc. First Workshop on Terrestrial Solar Resource Forecasting and on the Use of Satellites for Terrestrial Solar Resource Assessment*. Newark, DE, Boulder, CO: American Solar Energy Society, pp. 8–11.

Brown, M. E., and J. R. Lambert, 1981. Daily solar radiation measurements and forecasts for application of agricultural crop simulation models. In *Proc. First Workshop on Terrestrial Solar Resource Forecasting and on the Use of Satellites for Terrestrial Solar Resource Assessment*. Newark, DE, Boulder, CO: American Solar Energy Society, pp. 114–118.

Cope, A. M., and L. F. Bosart, 1982. A percentage of possible sunshine forecasting experiment at Albany, New York. *J. of Applied Meteorology* 21(9): 1217–1227.

Cox, W. T., 1981. Agricultural uses for solar radiation forecasts. In *Proc. First Workshop on Terrestrial Solar Resource Forecasting and on the Use of Satellites for Terrestrial Solar Resource Assessment*. Newark, DE, Boulder, CO: American Solar Energy Society, pp. 128–130.

Falconer, P., 1981. *A Methodology for Forecasting Morning, Afternoon, and Daily Totals of Normal Incidence Solar Radiation at Ground Level under Clear or Partly Cloudy Skies*. ASRC Publication 805. Albany, NY: State University of New York, 31 pp.

Glahn, H. R., and D. A. Lowry, 1972. The use of model output statistics (MOS) in objective weather forecasting. *J. of Applied Meteorology* 11(8): 1203–1211.

Hulstrom, R. L., 1981. Solar user needs for forecasts and satellite mapping of insolation. In *Proc. First Workshop on Terrestrial Solar Resource Forecasting and on the Use of Satellites for Terrestrial Solar Resource Assessment*. Newark, DE, Boulder, CO: American Solar Energy Society, pp. 131–133.

Jensenius, J. S., Jr., 1983. Automated forecasts of daily global solar energy. *Progress in Solar Energy*, Newark, DE: American Solar Energy Society, pp. 859–864. Presented at the 1983 Annual Meeting/Solar Technologies Conference of the American Solar Energy Society, Minneapolis, MN, 1–3 June 1983, Boulder, CO: American Solar Energy Society.

Jensenius, J. S., Jr., and G. M. Carter, 1979. *Specialized Agricultural Weather Guidance for South Carolina*. TDL Office Note 79-15, Silver Spring, MD: National Weather Service, National Oceanic and Atmospheric Administration, U.S. Department of Commerce, 16 pp.

Jensenius, J. S., Jr., and G. F. Cotton, 1981. The development and testing of automated solar energy forecasts based on the model output statistics (MOS) technique. In *Proc. First Workshop on Terrestrial Solar Resource Forecasting and on the Use of Satellites for Terrestrial Solar Resource Assessment*, Newark, DE, Boulder, CO: American Solar Energy Society, pp. 22–29.

Justus, C. G., and J. D. Tarpley, 1983. Accuracy and availability of solar radiation data from satellites and from forecast estimates. *Preprints, Fifth Conference on Atmospheric Radiation*, Boston, MA: American Meteorological Society, pp. 243–245.

Ma, C. C. Y., and M. Iqbal, 1983a. Statistical comparison of solar radiation correlations. *Progress in Solar Energy*, Newark, DE: American Solar Energy Society, pp. 865–870. Presented at the 1983 Annual Meeting/Solar Technologies Conference on the American Solar Energy Society, Minneapolis, MN, 1–3 June 1983.

Ma, C. C. Y., and M. Iqbal, 1983b. Statistical comparison of models for estimating solar radiation on inclined surfaces. *Progress in Solar Energy*, Newark, DE, American Solar Energy Society, pp. 871–875. Presented at the 1983 Annual Meeting/Solar Technologies Conference of the American Solar Energy Society, Minneapolis, MN, 1–3 June 1983.

National Weather Service, 1981. *The Use of Model Output Statistics for Predicting Ceiling, Visibility, Cloud Amount, and Obstructions to Vision*. NWS Technical Procedures Bulletin 303, Silver Spring, MD: National Oceanic and Atmospheric Administration, U.S. Department of Commerce, 11 pp.

National Weather Service, 1983. *Sunshine and Solar Energy Guidance*. NWS Technical Procedures Bulletin 334, Silver Spring, MD: National Oceanic and Atmospheric Administration, U.S. Department of Commerce, 8 pp.

National Weather Service, 1983. *Automated Daytime Maximum, Nighttime Minimum, 3-Hourly Surface Temperature, and 3-Hourly Surface Dew Point Guidance.* NWS Technical Procedures Bulletin 356, Silver Spring, MD: National Oceanic and Atmospheric Administration, U.S. Department of Commerce, 14 pp.

Picard, P., and C. B. Winn, 1985. Optimal control of a direct gain system having load managed storage. In *Proc. Joint ASME-ASES Solar Energy Conference.* New York: American Society of Mechanical Engineers. Presented at the 7th Annual Meeting of the Solar Energy Division of ASME, Knoxville, TN, 25–28 March 1985.

Riordan, C. J., 1984. *A Preliminary Comparison of Insolation Measurements, Forecasts and Estimates from Satellite Imagery.* SERI/TR-215-2046, Golden, CO: Solar Energy Research Institute, 39 pp.

8 Illuminance Models and Resources in the United States

Claude Robbins

8.1 Introduction

This chapter presents the background and current research conducted in the daylight or exterior illuminance resource of the United States. This spectral area of radiation has a history and research significantly different from the radiation measurement and modeling that has occurred in this country. In fact, the United States has historically taken a predominant role in the measurement and modeling of radiation, and several European countries have taken a similar role in the measurement and modeling of illuminance.

Light is defined in terms of the electromagnetic radiation falling in the spectral range of the human eye. Since the terms radiant flux, irradiance, and radiance include radiation outside of this range, these quantities are not used to describe light. Instead, analogous terms—luminous flux, illuminance, and luminance—are used. These terms are defined based on the spectral response of the human eye with respect to the photopic luminous efficiency function K_λ. Human visual sensitivity peaks at about 555 nm (yellow light) and declines to zero at about 380 nm (purple) and 760 nm (red), respectively. The luminous efficiency function of the human eye within this range is illustrated in table 8.1 (IES, 1981).

Light emanating from point (usually man-made) sources of light is described differently from that produced by extended (usually natural) sources of light. The primary difference is that for the former, the rays of light diverge radially outward from a single point, but from an extended light source such as the sun and sky, the rays of light propagate in a variety of directions from every point in the source. For the reader interested in a more detailed treatise on the physics of daylight, two good books on the subject are available: *The Science of Daylight* (Walsh, 1961) and *Principles of Natural Lighting* (Lynes, 1969).

The key photometric quantities are defined in terms similar to the corresponding radiant terms. These quantities are also defined in terms of human visual response and not the directional aspects of the light; therefore, both point and extended light sources can be defined as follows:

luminous flux: $$\mathbf{\Phi}_v = 683 \int_{380}^{760} K_\lambda \mathbf{\Phi}_\lambda \, d\lambda \qquad \text{(lm)}; \qquad (8.1)$$

Table 8.1
Photopic spectral luminous efficiency, $V(\lambda)$ (unity at wavelength of maximum luminous efficacy)

Wave-length λ (nm)	Standard values	Values interpolated at intervals of 1 nm								
		1	2	3	4	5	6	7	8	9
380	.00004	.000045	.000049	.000054	.000059	.000064	.000071	.000080	.000090	.000104
390	.00012	.000138	.000155	.000173	.000193	.000215	.000241	.000272	.000308	.000350
400	.0004	.00045	.00049	.00054	.00059	.00064	.00071	.00080	.00090	.00104
410	.0012	.00138	.00156	.00174	.00195	.00218	.00244	.00274	.00310	.00352
420	.0040	.00455	.00515	.00581	.00651	.00726	.00806	.00889	.00976	.01066
430	.0116	.01257	.01358	.01463	.01571	.01684	.01800	.01920	.02043	.02170
440	.023	.0243	.0257	.0270	.0284	.0298	.0313	.0329	.0345	.0362
450	.038	.0399	.0418	.0438	.0459	.0480	.0502	.0525	.0549	.0574
460	.060	.0627	.0654	.0681	.0709	.0739	.0769	.0802	.0836	.0872
470	.091	.0950	.0992	.1035	.1080	.1126	.1175	.1225	.1278	.1333
480	.139	.1448	.1507	.1567	.1629	.1693	.1761	.1833	.1909	.1991
490	.208	.2173	.2270	.2371	.2476	.2586	.2701	.2823	.2951	.3087
500	.323	.3382	.3544	.3714	.3890	.4073	.4259	.4450	.4642	.4836
510	.503	.5229	.5436	.5648	.5865	.6082	.6299	.6511	.6717	.6914
520	.710	.7277	.7449	.7615	.7776	.7932	.8082	.8225	.8363	.8495
530	.862	.8739	.8851	.8956	.9056	.9149	.9238	.9320	.9398	.9471
540	.954	.9604	.9661	.9713	.9760	.9803	.9840	.9873	.9902	.9928
550	.995	.9969	.9983	.9994	1.0000	1.0002	1.0001	.9995	.9984	.9969
560	.995	.9926	.9898	.9865	.9828	.9786	.9741	.9691	.9638	.9581
570	.952	.9455	.9386	.9312	.9235	.9154	.9069	.8981	.8890	.8796
580	.870	.8600	.8496	.8388	.8277	.8163	.8046	.7928	.7809	.7690
590	.757	.7449	.7327	.7202	.7076	.6949	.6822	.6694	.6565	.6437
600	.631	.6182	.6054	.5926	.5797	.5668	.5539	.5410	.5282	.5156

610	.503	.4905	.4781	.4658	.4535	.4412	.4291	.4170	.4049	.3929
620	.381	.3690	.3570	.3449	.3329	.3210	.3092	.2977	.2864	.2755
630	.265	.2548	.2450	.2354	.2261	.2170	.2082	.1996	.1912	.1830
640	.175	.1672	.1596	.1523	.1452	.1382	.1316	.1251	.1188	.1128
650	.107	.1014	.0961	.0910	.0862	.0816	.0771	.0729	.0688	.0648
660	.061	.0574	.0539	.0506	.0475	.0446	.0418	.0391	.0366	.0343
670	.032	.0299	.0280	.0263	.0247	.0232	.0219	.0206	.0194	.0182
680	.017	.01585	.01477	.01376	.01281	.01192	.01108	.01030	.00956	.00886
690	.0082	.00759	.00705	.00656	.00612	.00572	.00536	.00503	.00471	.00440
700	.0041	.00381	.00355	.00332	.00310	.00291	.00273	.00256	.00241	.00225
710	.0021	.001954	.001821	.001699	.001587	.001483	.001387	.001297	.001212	.001130
720	.00105	.000975	.000907	.000845	.000788	.000736	.000688	.000644	.000601	.000560
730	.00052	.000482	.000447	.000415	.000387	.000360	.000335	.000313	.000291	.000270
740	.00025	.000231	.000214	.000198	.000185	.000172	.000160	.000149	.000139	.000130
750	.00012	.000111	.000103	.000096	.000090	.000084	.000078	.000074	.000069	.000064
760	.00006	.000056	.000052	.000048	.000045	.000042	.000039	.000037	.000035	.000032

Source: IES (1981).

$$\text{illuminance:} \quad E_v = 683 \int_{380}^{760} K_\lambda E_\lambda d\lambda \quad (\text{lm/m}^2, \text{lux}); \tag{8.2}$$

$$\text{luminance:} \quad L_v = 683 \int_{380}^{760} K_\lambda L_\lambda d\lambda \quad (\text{lm/m}^2\ \text{sr}'). \tag{8.3}$$

The value 683 is the number of lumens of light stimulus produced by 1 W of electromagnetic radiation at the peak visual response (555 nm) as described by the Commission Internationale de L'Eclairage (CIE, 1970, 1957). The term K_λ is the photopic spectral luminous efficiency function as described by the Illumination Engineering Society of the United States (IES/US) (IES, 1981). The term Φ_λ is the spectral radiant flux in W/nm, E_λ is the spectral irradiance in W/m^2 nm, and L_λ is the spectral radiance in W/(m^2 sr nm).

A key in the understanding of light measurement and modeling is the concept of the steradian (sr). A steradian, or solid angle, is typically defined in terms of a closed curve about a point in space. The solid angle subtended by the curve at the point is defined to be the area of a radial projection of the curve onto a sphere of unit radius centered on the point. The unit of solid angle is the steradian. This definition is applicable to a point source of light in which a circle is used to define the solid angles. In daylighting, a rectangular curve is frequently used and the solid angle subtended by a rectangle at an arbitrary point in space may be represented by an integral around the boundary of the rectangle. The maximum solid angle is 4π sr, encompassing all directions about a point in space. The solid angle subtended by a hemisphere at its center is 2π sr.

The amount of exterior illuminance available for daylighting has the greatest potential for uncertainty of any value used in daylighting calculations. Further, without some knowledge of the daylight resource that can be expected at any given time for a given locale, it is virtually impossible to analyze daylighting systems in terms of lighting and building energy performance characteristics over time.

The traditional approach to analyzing the daylight resource has been to take detailed measurements for a given locale and use the measured data to establish a locale-specific mathematical model. Illuminance measurements used for this purpose have been made by numerous people in various countries over the past 85 years. The primary limitation of these models is that they apply only to one location, and they often deal only with the global illuminance on a horizontal surface E_{GH}.

In the past 5 years several general illuminance models have been developed. These models, which are considerably more complex than the locale-specific models, use a series of equations to describe the different components (direct, diffuse, and reflected light) of the daylight resource. The general models can be applied to a given locale by using recorded or calculated climate data to describe the local atmospheric conditions. These models are limited by the availability of detailed climate data and the number of cities for which such data has been recorded. In the United States, either typical meteorological year (TMY) or test reference year (TRY) climate data can be used. Typically TMY data are used, allowing most general illuminance models to be applied to the 244 cities throughout the TMY network in the United States (NCC, 1979).

Detailed illuminance measurements are needed to develop either the locale-specific or general illuminance models. These measurements are similar to irradiance measurements but use detectors that closely match the spectral range of the sensitivity of the human eye. In addition to illuminance measurements, the daylighting designer also needs some measurement of the amount of time the sky can be considered clear or overcast, since daylighting systems function differently under these conditions. Because daylighting is an instantaneous use of the available resource rather than a storing of energy for later use, a time metric such as half-day totals or monthly totals of illuminance is not useful in lighting design or lighting energy calculations.

8.1.1 Instrumentation and Measurement Error Sources

Detectors of radiation produce different responses to different wavelengths of radiation. Ideally, if one is trying to measure a relatively narrow band of radiation such as the visible spectrum, it would be desirable to have a detector that has a completely constant spectral sensitivity over all wavelengths contained in the bandwidth being measured. A silicon photovoltaic detector can be equipped with a filter that produces approximately flat response over the range from 400 to 900 nm. For illumination measurements, however, it is more desirable to have a filter that produces a spectral response that matches the photopic luminous efficiency function given in table 8.1. Silicon detectors corrected to have photopic response are the most common sensors for illumination measurements.

There are two basic types of light sensors: those that measure illuminance (lm/m^2) and those that measure luminance (lm/m^2 sr). Illuminance sensors receive visible radiation from within a specified solid angle, usually hemispherical, and provide an output signal, in voltage or current, that is proportional to the quantity of the flux per unit area received by the sensor. These is usually no distinction in the distribution characteristics of the light flux within the solid angle field of view of the sensor. Luminance sensors normally receive radiation from within a very small specified solid angle and provide an output signal that is proportional to the quantity of the flux per unit area and per unit of solid angle received by the detector.

Illuminance sensors are typically used for measuring the following: global horizontal, direct normal, diffuse, global vertical, and diffuse vertical illuminance, and scale model testing.

Luminance measurements are also significant in daylighting resource assessment because the directional distribution of luminance entering a fenestration aperture is important in determining the spatial distribution of luminance over the surfaces in the room. It is the reflected luminance that the eye sees. Luminance sensors are most often used to measure a fixed (small) portion of the sky by means of a Gershun tube to limit the field of view of the detector to the desired solid angle, usually 5° or 7.5°. Another approach to luminance measurement is to attempt to measure the entire 2π sr solid angle of the sky dome. One such piece of instrumentation developed especially to measure the luminance of the entire sky dome is the all-sky flux mapper developed by the Solar Energy Research Institute (SERI) for the U.S. Department of Energy (DOE). The flux mapper (shown in figure 8.1) is a video system that provides a rapid, real time processing of daylighting luminance data. The system uses on orthographic-projection fisheye (180°) lens to project the entire hemispheric field onto a vidicon target. The signal is then digitized and may be plotted as equal brightness contours or recorded on tape for later analysis (Robbins, Hunter, and Cannon, 1984).

The spectral response of illuminance and luminance sensors needs to be corrected to correspond with the IES luminous efficiency curve (table 8.1) for the human eye. This is accomplished by selecting filters that correct the spectral response of the specific detector chosen. Although the sandwiched filter glasses are fairly stable, they can vary by batch and can degrade with time. Photovoltaic detectors are usually less stable than their correction filters, necessitating recalibration on a regular basis. Many manufacturers of such detectors suggest recalibration every 6–12 months. For research

Figure 8.1A
SERI all-sky flux mapper. Source: Robbins (1986).

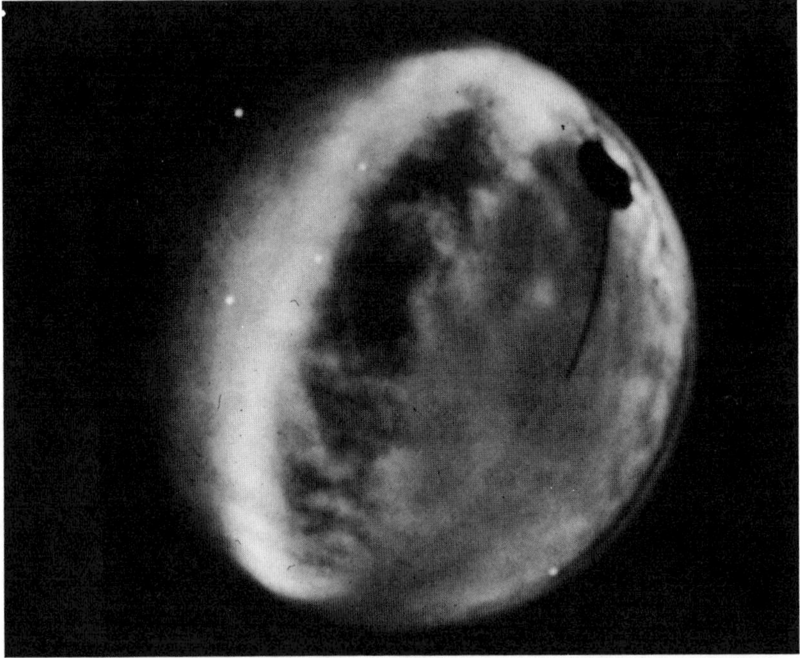

Figure 8.1B
View through the fluxmapper. Source: Robbins (1986).

purposes recalibration may be more frequent—as often as every 3 months (Hunter, Robbins, and Harr, 1983).

For typical daylight measurement purposes, there are three critical sources of error in the detectors. The first source of error is the light used to calibrate the detector. Ideally, one should use a light source whose spectral composition is identical or very similar to the light being measured. If the calibration source is significantly different, then the error can be substantial. This is a key problem in daylight measurement because many manufacturers do not calibrate sensors using light sources that spectrally match daylight and sunlight. Further, daylight sensors that only have a view of the sky dome should be calibrated differently from those that see the direct sun. Another part of this problem is the calibration of the source lamp itself. Typically, the source is National Bureau of Standards (NBS) traceable through two or three additional levels of calibration. This can create errors on the order of 5–8% that cannot easily be eliminated.

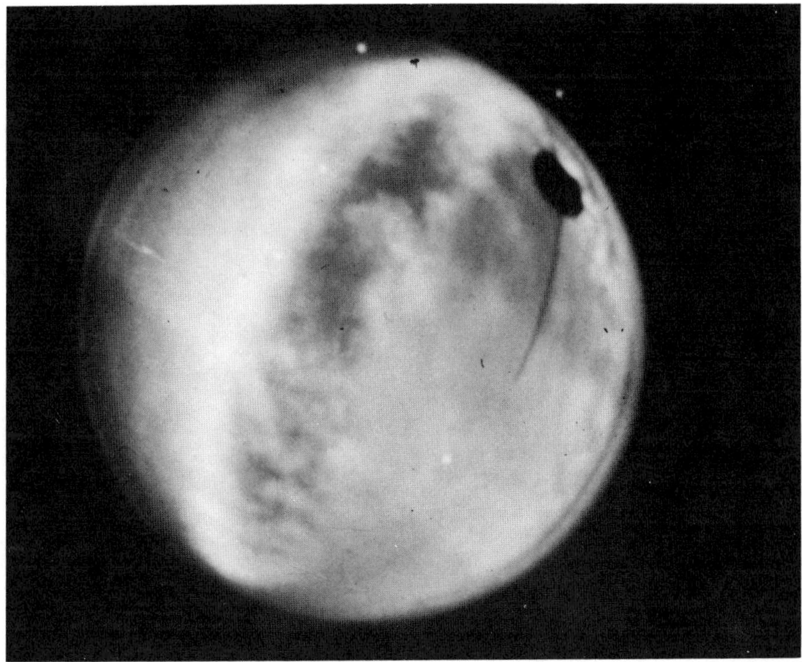

Figure 8.1C
Computer-enhanced view. Source: Robbins (1986).

A second major source of error is the cosine response, more correctly called the obliquity response, of the detector. A detector illuminated with a collimated beam will have a decrease in the flux incident to it that is directly proportional to the cosine of the angle of incidence. Typical deviations of a bare photodiode's output can vary as much as 5–10% from "true" cosine response. Optimal cosine correction may be achieved by manipulating the dimensions of the sensor until a desired measured angular response is achieved for the specific materials used in the detector and photopic filters (McCluney, 1983).

The third major source of error is the response of the silicon photodiode to temperature changes. Considering the fact that the response of illuminance sensors is in a very narrow waveband, temperature error is considerably less than for radiometers and other radiation sensors, with the exception of direct normal illuminance sensors. There are several thermisters available whose resistance is a function of temperature and can be used

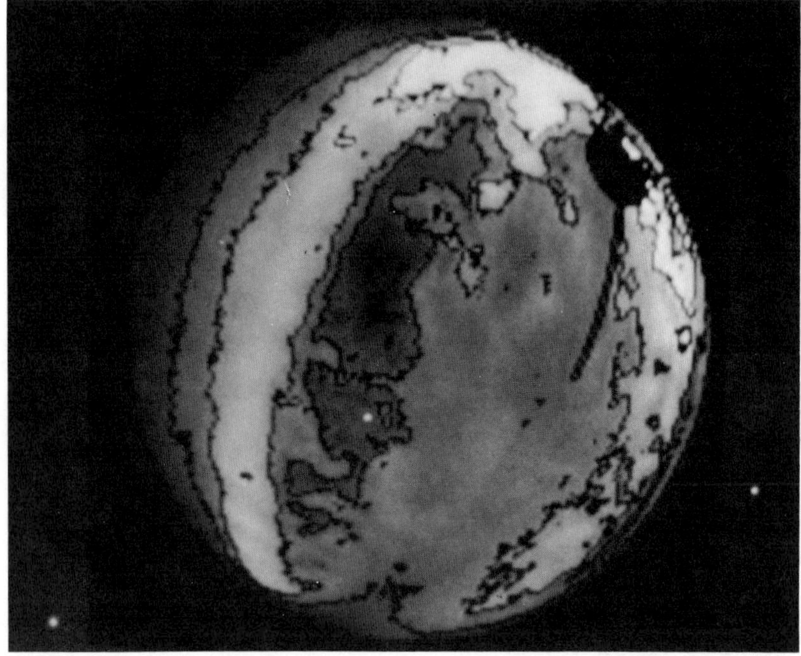

Figure 8.1D
Sky map showing lines of equal luminance. Source: Robbins (1986).

to compensate partially for thermal drifts in illuminance detectors. Any daylight sensor designed for outdoor measurement under direct sunlight should include some method for temperature compensation. Typical error due to thermal drift is on the order of 3–5%. When these three sources of error are added to data acquisition system error and other systematic error data, measurement accuracy is (at best) in the 13–18% error range (Roswell and Loudermilk, 1968).

Daylight measurements are currently being taken at the Lawrence Berkeley Laboratory. These measurements are under DOE's auspices. Other measurements are also being taken at NBS, with joint support for that effort from the National Fenestration Council, and at the Florida Solar Energy Center as part of a state-supported research effort. At all of these stations, measurements taken include zenith sky luminance, horizontal and vertical global illuminance, diffuse horizontal illuminance, and the corresponding solar radiation (irradiance) data.

8.1.2 Location-Specific Illuminance Models

Sky measurements have been made at different locales across the United States. The earliest recorded measurements were made by Basquin in 1897 in Chicago; however, these measurements were not published until 1906 (Basquin, 1906). Probably the best known measurements were those made by Kimball and Hand between 1919 and 1922 for Washington, DC, and Chicago, IL (Kimball, 1919; Kimball and Hand, 1921; Kimball et al., 1923). The majority of these measurements have been illuminance values on horizontal and vertical surfaces. The U.S. Weather Bureau (now the National Oceanic and Atmospheric Administration [NOAA]) took measurements in terms of a time metric (footcandle-hours) for Washington, DC, and Baltimore, MD, in 1954. All of the measurements for locale-specific models were of rather short duration, the longest being about 18 months. In most cases, measurements were taken for 1–6 months.

Europe has a substantial history of sky illuminance measurements. The Building Research Establishment (BRE) in Watford, England, has data for Teddington from 1923 to 1939 (McDermott and Gordon-Smith, 1951) and for Kew and Bracknell for 1964–1973 (Blackwell 1954). Krochmann, the German scientist, has collected short-term data for several countries including Germany, France, Finland, Austria, and the USSR (Krochmann and Seidl, 1974). Similar measurements have been made in Sweden (Elvegard and Sjostedt, 1940), the Netherlands (Dogniaux, 1968; Dogniaux et al., 1968), and Czechoslovakia (Kittler, 1962). Outside of Europe, measurements have been reported for Australia, Japan, India, and Nepal (Kendrick, 1984). Further, there have been extensive measurements, as yet unpublished, taken by the National Building Research Institute (NBRI) in Pretoria, South Africa (NBRI, 1984). For those interested in illuminance measurements, a substantial bibliography of papers and reports on the topic can be found in Robbins and Hunter (1983).

Because the majority of the sky measurements are similar in form, it is not surprising that the resulting illuminance models are also similar. Further, the clear and overcast skies are almost always considered separately, with most computer modelers presenting an algorithm for clear skies and a separate algorithm for overcast skies in their published work. The critical element in most equations for both clear and overcast sky is the solar altitude angle h.

Many of the models use, sometimes with slight variation, one of the following general forms (Robbins and Hunter, 1983):

$$E_H = A + B \sin h \qquad \text{(klx)}, \tag{8.4}$$

$$E_H = B \sin h \qquad \text{(klx)}, \tag{8.5}$$

$$E_H = (A \sin h) \exp(B/\sin h) \qquad \text{(klx)}, \tag{8.6}$$

$$E_H = A \sin h/(B + C \operatorname{cosec} h) \qquad \text{(klx)}, \tag{8.7}$$

where E_H is the illuminance on a horizontal surface, h is the solar altitude angle in degrees, and A, B, and C are coefficients to describe the atmospheric conditions of the specific locale being studied.

A brief description of several different clear sky and overcast sky models follows. These models represent some of the better known efforts by scientists from the United States, Europe, and Japan. No attempt was made to present all existing models, but common trends and differences in many of the exterior illuminance models are examined.

As mentioned previously, the work by Kimball and Hand (1919–1923) is probably the best known measurement effort in the United States. The results of that effort have been accepted by IES/US as the basis for daylighting design and analysis in their *IES Handbook*. This handbook is recognized as the standard lighting and illumination reference in the United States. The illuminance equation developed by Kimball and Hand took on the form (Kimball, 1919)

$$E_{dH,c} = B \sin h, \tag{8.8}$$

where $E_{dH,c}$ is the diffuse illuminance on a horizontal surface on a clear day in footcandles or lumens per square foot. The coefficients for B were developed for Washington, DC, and Chicago, IL, and compared with similar radiation calculations for other parts of the country that were made in 1914. Kimball then converted his calculations into a graphical form, which is used in the *IES Handbook*. Sample graphs are shown in figures 8.2, 8.3, and 8.4. Kimball used a ratio of the percentage of sunshine to the radiation intensity to develop similar charts for the overcast condition.

In England, work conducted by Moon and Spencer (1942) at the BRE was based on the measurements taken at Teddington from 1923 to 1939. The equations derived by Moon exhibit a linear relationship between the logarithm of the solar illuminance and the corresponding air mass. The Moon equations were only developed for clear-sky conditions. A second set of equations was developed to represent "less sunny" climates, such as Great Britain. Two formulas are given. The first applies to latitudes less than 23.45°, the second to latitudes greater than 23.45° such that

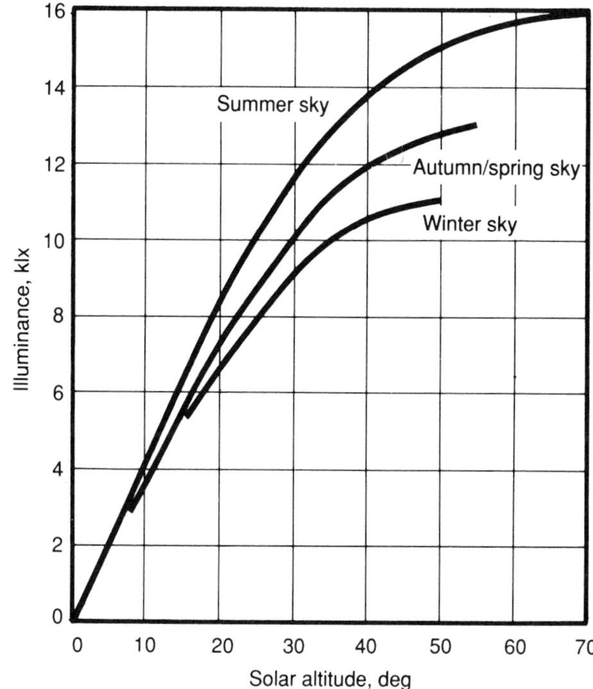

Figure 8.2
Clear-sky diffuse illuminance on a horizontal surface, $E_{\mathrm{dH,c}}$, according to the Kimball and Hand model. Source: Robbins (1986).

for latitude < 23.45°, sunny climate:

$$\log E_{\mathrm{GH,c}} = 4.1 - 0.1 \operatorname{cosec} h; \tag{8.9}$$

for latitude > 23.45°, sunny climate:

$$\log E_{\mathrm{GH,c}} = 3.95 - 0.1 \operatorname{cosec} h; \tag{8.10}$$

for latitude > 23.45°, less sunny climate:

$$\log E_{\mathrm{GH,c}} = 2.87 - 0.1 \operatorname{cosec} h; \tag{8.11}$$

for latitude > 23.45°, less sunny climate:

$$\log E_{\mathrm{GH,c}} = 2.76 - 0.1 \operatorname{cosec} h. \tag{8.12}$$

In Sweden, work conducted by Elvegard and Sjostedt (1940) was published in the United states in 1940. The model was based upon an analysis

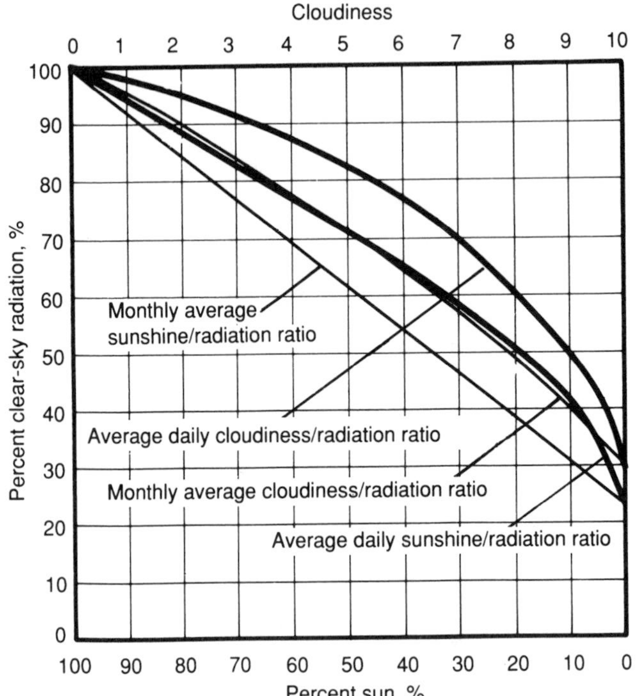

Figure 8.3
Ratio of percent sun to radiation, used in the Kimball and Hand model to determine overcast sky illuminance: (a) average daily cloudiness/radiation ratio; (b) average daily sunshine/radiation ratio; (c) monthly average cloudiness/radiation ratio; (d) monthly average sunshine/radiation ratio. Source: Robbins (1986).

of illuminance data from Stockholm. The model considers the direct $E_{DH,c}$ and diffuse $E_{dH,c}$ components of the global illuminance on a horizontal surface from the clear sky, such that

$$E_{GH,c} = E_{DH,c} + E_{dH,c} \quad \text{(klx)} \tag{8.13}$$

and

$$E_{DH,c} = 1.6 E_S (\sin h) 10^{-0.1m} \quad \text{(klx)}, \tag{8.14}$$

$$E_{dH,c} = 0.211 E_S \sin^{0.5} h \quad \text{(klx)}, \tag{8.15}$$

where E_S was defined as 77,000 lux, and m was the optical air mass.

For the overcast condition, the direct $E_{GH,o}$ and diffuse $E_{dH,o}$ components were defined in a manner similar to the clear sky components;

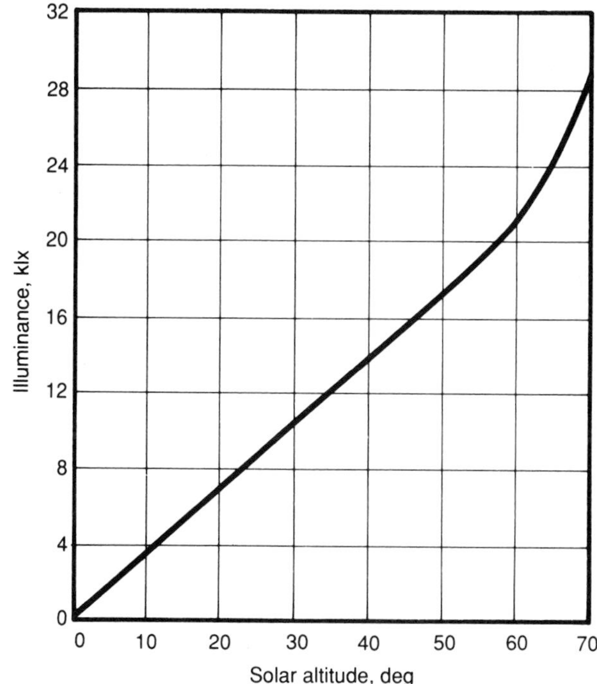

Figure 8.4
Global illuminance on a horizontal surface on an overcast day, $E_{GH,o}$, according to the Kimball and Hand model. Source: Robbins (1986).

however, each was then multiplied by an additional coefficient, such that

$$E_{GH,o} = (x)E_{DH,o} + (y)E_{dH,o} \qquad \text{(klx)} \tag{8.16}$$

and

$$E_{DH,o} = 0.26(23.2 \sin h \, 10^{-0.1/\sin h}) \qquad \text{(klx)}, \tag{8.17}$$

$$E_{dH,o} = 0.54(16.25 \sin^{0.5} h) \qquad \text{(klx)}. \tag{8.18}$$

In the United States, in 1954 Bennett developed footcandle per hour data for the U.S. Weather Bureau (Bennett, 1962). Although this form of data has not been used in daylighting analysis, Bennett was able to show a close correlation between daily total illuminance data and daily total irradiance data for the Washington, DC, area. In addition, because he made measurements at several locations in the Washington, DC, area, he demonstrated the wide variation that can occur in such measurements in relative proximity. Although the equations take on a form similar to other illuminance models, the irradiance in Langley's is the dependent variable rather than solar altitude.

For the U.S. Weather Bureau, Washington, DC, the equation is

$$E_{\mathrm{GH/day}} = 19.8 + 1.05I \qquad (1{,}000\ \mathrm{fc/h}), \tag{8.19}$$

for American University, the equation is

$$E_{\mathrm{GH/day}} = 23.8 + 1.142I \qquad (1{,}000\ \mathrm{fc/h}), \tag{8.20}$$

and for the Silver Hill Maryland Observatory, the equation is

$$E_{\mathrm{GH/day}} = 5.5 + 1.182I \qquad (1{,}000\ \mathrm{fc/h}). \tag{8.21}$$

The total global illuminance for a day, $E_{\mathrm{GH/day}}$, is in thousands of foot-candles, and the total irradiance per day, I, is in langleys.

Chroscicki (1971) published his illuminance algorithms in 1971 at the Barcelona meeting of the CIE. For clear skies the global illuminance on a horizontal surface was defined as

$$E_{\mathrm{GH,c}} = E_{\mathrm{DH,c}} + E_{\mathrm{dH,c}}$$

and

$$E_{\mathrm{GH,c}} = (96 \sin h)/1 + 0.2 \operatorname{cosec} h) + (3.0 + 0.17h) \qquad (\mathrm{klx}). \tag{8.22}$$

Chroscicki did not propose an algorithm for the overcast sky condition.

Krochmann and Seidl (1974) published a model in 1974 based upon work done in Berlin, Federal Republic of Germany. They proposed both a clear and overcast set of equations. For the clear sky the global illuminance on a horizontal surface was

$$E_{\mathrm{GH,c}} = [130(\sin h) \exp -0.2/\sin h] + (1.1 + 15.5 \sin^{0.5} h) \qquad (\mathrm{klx}). \tag{8.23}$$

For the overcast sky the global illuminance on a horizontal surface was

$$E_{\mathrm{GH,o}} = 0.3 + 21 \sin h \qquad (\mathrm{klx}). \tag{8.24}$$

A great deal of work has been done in Japan, India, and Australia in daylighting resource assessment and modeling. Unfortunately, much of this work has not been published in U.S. journals. In India, the work has been performed primarily at the India Institute of Science, and in Japan and Australia, the majority of the work has been conducted at various universities (Matsuura, 1984). In some recent work for the Tokyo area, Nakamura and Oki (1979) proposed the following equations for clear and overcast skies, respectively

$$E_{GH,c} = 1.5 \sin^{1.2h} \qquad \text{(klx)} \tag{8.25}$$

and

$$E_{GH,o} = 0.5 + 42.5 \sin h \qquad \text{(klx).} \tag{8.26}$$

This model is typical of several published in Japan and Australia and follows closely the mathematical concept of those published in Europe and the United States.

Probably the most detailed locale-specific illuminance model was the one developed by Crisp and Lynes (1979) at the BRE in Watford, England, that was first published in 1979. The model differs considerably from the other models in the use of irradiance data and the luminous efficacy term η in addition to the solar altitude variable. The luminous efficacy makes the equation suitable for clear and overcast sky conditions. The equation is

$$E_{GH} = (K_s I_{DN} \sin h + K_a I_{H,c})\eta \qquad \text{(lx),} \tag{8.27}$$

where K_s and K_a are the Ångström monthly constants (Ångström and Drummond, 1972), I_{DN} is the direct normal irradiance (W/m^2), I_{dH} is the diffuse irradiance (W/m^2), and η is the luminous efficacy (lm/W). For the clear sky the luminous efficacy value is 125 and for overcast skies it is 110 lm/W.

The CIE has also published models for the clear and overcast sky conditions. However, unlike the other models, which are based upon local measurements, the CIE models are a theoretical representation of a "stan-- dard" clear or overcast sky and represent sky luminance values rather than illuminances. The clear sky has a luminance distribution that depends upon the position of the sun in the sky and the scattering of light in the atmosphere. The models assume that no clouds are visible over the entire sky dome. In many daylight evaluation techniques, the direct component (that is, sunlight), is not included in the calculations; therefore, the CIE clear-sky equation considers only the luminance of the sky dome without the direct sun. First proposed by Kittler (1962) of Czechoslovakia, the CIE clear sky equation states

$$L_{p,c} = L_{z,c}\left[\frac{(1 - e^{-0.32 \sec \varepsilon})(0.91 + 10_{e}^{-3\delta}) + 0.45 \cos^2 \delta}{0.274(0.91 + 10_{e}^{-3z\theta} + 0.45 \cos^2 z\theta}\right] \qquad (\mathrm{ln/m^2\ st}), \tag{8.28}$$

where ε is the altitude of the point p in the sky, δ is the angle between the sun and point p, and $z\theta$ is the angle of the sun from the zenith.

The CIE overcast sky model was proposed by Moon and Spencer (1942) as a basis for design in those climates where the prevalence of overcast conditions was sufficient to use an overcast sky model as the critical design condition. An interesting feature of the CIE model is the assumption that the zenith luminance L_z is three times as bright as the sky luminance near the horizon. This model also assumes that the entire sky dome is covered with a layer of clouds. The CIE overcast sky model is

$$L_z = (1 + 2 \sin \theta)/3. \tag{8.29}$$

Considerably fewer locale-specific models establish illuminance values for vertical and angled surfaces. Vertical illuminance models are often derivations of the models used to establish horizontal illuminance. Thus, the accuracy of the vertical illuminance model is based on the accuracy of the corresponding horizontal illuminance model.

The *IES Handbook* (IES, 1981) provides graphs of vertical surface illuminance data. The handbook does not describe the model used to establish the graphs; however, IES uses the Kimball (1919) equation to establish vertical illuminances:

$$E_{d,v} = E_{DN} \sin h/r, \tag{8.30}$$

where r was the measured ratio of vertical to horizontal illuminance in Washington, DC, and Chicago, IL.

Vertical illuminance values for clear winter and summer skies are shown in figures 8.5 and 8.6. Note that the values are for diffuse illuminance only and do not include the direct solar component.

The Crisp and Lynes (1979) vertical model was derived from their horizontal illuminance model [equation (8.27)]. Two cases were considered: the vertical surface exposed to the sun (solar azimuth = 0°), and the vertical surface oriented away from the sun (solar azimuth = 180°).

The case with the vertical surface facing the sun is given by the expression

$$E_{GV,e} = E_{dV,c} + (K_s I_{DN} \cos i) \qquad (\mathrm{klx}), \tag{8.31}$$

where $E_{GV,e}$ is the global illuminance on a vertical surface facing the sun,

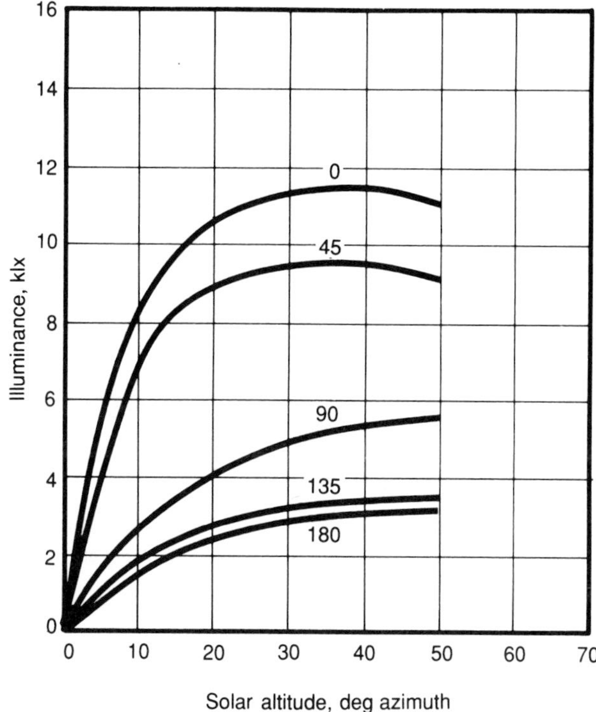

Figure 8.5
Diffuse illuminance on a vertical surface on a clear day in winter, $E_{dV,c}$, according to the Kimball and Hand model. Source: Robbins (1986).

$E_{dV,c}$ is the diffuse illuminance on a vertical surface—clear sky—and i is the angle of incidence of direct solar radiation.

The diffuse illuminance striking a vertical surface $E_{dV,c}$ is given by the expression

$$E_{dV,c} = 0.5E_{dH,c} + 0.1E_{GH,c} \qquad \text{(klx).} \tag{8.32}$$

The case with no direct exposure to the sun is given by the expression

$$E_{GV,op} = E_{dV,c} \qquad \text{(klx),} \tag{8.33}$$

where $E_{GV,op}$ is the total (global) illuminance on a vertical surface opposite the sun.

With the exception of the IES daylight data, none of the locale-specific daylight availability models have been used very widely to generate detailed availability data or as a design aid for the lighting enginner.

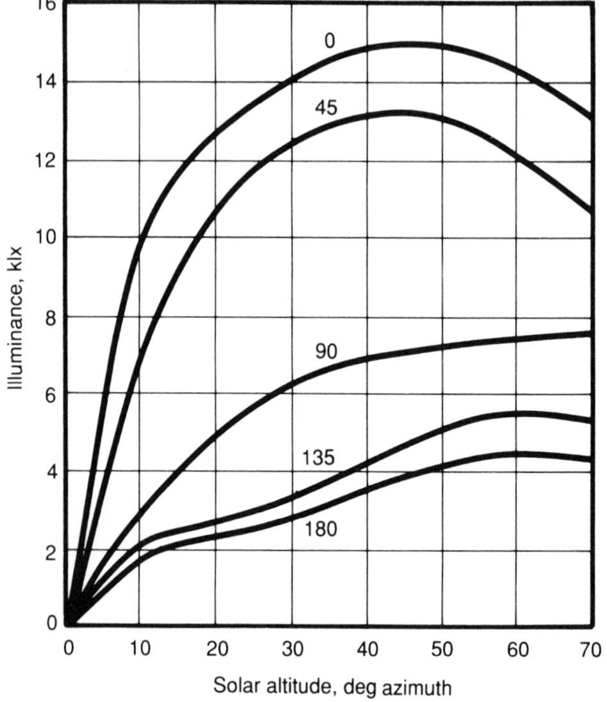

Figure 8.6
Diffuse illuminance on a vertical surface on a clear day in summer, $E_{dV,c}$, according to the Kimball and Hand model. Source: Robbins (1986).

8.2 General Illuminance Models

General illuminance models are much more complex than the locale-specific models and are primarily used to generate hourly daylight availability data as part of a building energy simulation computer code such as DOE 2 or BLAST. General illuminance models have also been used to establish hourly illuminance data in tabular form. Such a prediction of exterior natural illuminance on a surface of any orientation is a function of the astronomical coordinates of the sun (solar altitude, azimuth, and time), geographic location of the place being analyzed (latitude, longitude, and altitude above sea level), and the local atmospheric conditions (sky clearness, atmospheric turbidity, and water vapor). Although all of the models deal with solar and site location in a similar manner, local atmospheric conditions are handled differently.

In all of the general models the total instantaneous natural illuminance E_G on a surface is made up of three components: the direct illuminance E_D, the diffuse illuminance E_d, and the ground-reflected illuminance E_g. These components make up the total illuminance as given by

$$E_G = E_D + E_d + E_g \qquad \text{(lx)}. \tag{8.34}$$

One of the first general illuminance models was developed by Dogniaux et al. (1967) for CIE in the mid-1960s. Other general models to be discussed include the work by Gillette, Pierpoint, and Treado (1982) at NBS, Robbins and Hunter (1983) at SERI, and Littlefair (1982) at BRE in England. The work at SERI was sponsored by DOE. The work at NBS was sponsored by DOE and the National Fenestration Council (NFC). Similiarities exist between the general models, as between many of the locale-specific models.

8.2.1 The Dogniaux Model

The Dogniaux model (Dogniaux et al., 1967) was developed for CIE in 1967 but has never been formally accepted as an agreed CIE recommendation, although it was approved by a majority of the Daylighting Technical Committee (TC 4.2). This model defines the direct component of daylight on a clear day on a horizontal surface $E_{GH,c}$ as

$$E_{GH,c} = E_{DH,c} + E_{dH,c} \qquad \text{(lx)}, \tag{8.35}$$

in which the direct component is defined

$$E_{DH,c} = E_{SCj} e^{-amT} \qquad \text{(lx)}, \tag{8.36}$$

Where E_{SC} is the apparent extraterrestrial solar illuminance constant for a given day j. The value of E_{SCj} was determined by

$$\begin{aligned} E_{SCj} = {} & 126{,}820 + 4{,}248 \cos wJ + 0.0825 \cos 2wJ \\ & - 0.0043 \cos 3wJ + 0.1691 \sin wJ \\ & + 0.00914 \sin 2wJ + 0.01726 \sin 3wJ \qquad \text{(lx)}, \end{aligned} \tag{8.37}$$

where w is 2/365 and J is the Julian day of the year (Jan. 1 = 1, Dec. 31 = 365). The term a represents the atmospheric extinction coefficient, m the optical air mass, and T the turbidity factor.

The solar illuminance constant E_{SC} was defined by Dogniaux to be 126,820 lux. The atmospheric extinction coefficient was defined by the following regression equation:

Table 8.2
Equation (8.38) coefficients for D and F

β	D	F
0.05	0.1512	0.0262
0.10	0.1656	0.0215
0.20	0.2021	0.0193

$$a = D - F(T), \tag{8.38}$$

in which the D and F coefficients vary with the turbidity coefficient β. These values are shown in table 8.2.

The optical air mass was difined as a function of the solar altitude h such that

$$m = 10.01 + \cfrac{h - 5}{-1.217 + \cfrac{h - 11}{-10.034 + \cfrac{h - 24.5}{150.343 - \cfrac{h - 40}{1.821}}}}, \tag{8.39}$$

and the turbidity factor T, used in equations (8.35) and (8.37), was defined as

$$T = \frac{h + 85}{39.5e^{-w} + 47.4} + 0.1 + (16 + 0.22w)\beta, \tag{8.40}$$

where w is the precipitable moisture in the atmosphere.

The clear sky diffuse illuminance $E_{\mathrm{dH,c}}$ was represented by the following regression equation:

$$E_{\mathrm{dH,c}} = a_0 + a_1 h^2 + a_2 h^3. \tag{8.41}$$

Values for a are shown in table 8.3.

For the overcast condition Dogniaux used the CIE standard overcast sky equation [equation (8.29)] as discussed in section 8.1.2. This model has been used in several property daylighting analysis computer codes to estimate exterior illuminance on an hourly basis.

8.2.2 The Robbins-Hunter Model

The Robbins-Hunter Model (Robbins and Hunter, 1983) is a general illuminance model based on multiyear measurements at SERI of the daylight

Table 8.3
Dogniaux model coefficients for a_0, a_1, and a_2

	$\beta = 0.05$			$\beta = 0.10$			$\beta = 0.20$		
w	a_0	a_1	a_2	a_0	a_1	a_2	a_0	a_1	a_2
0.5	0.9361	0.0020	$-2 \cdot 10^{-7}$	1.0239	0.0024	$-2 \cdot 10^{-7}$	1.1399	0.0030	$-3 \cdot 10^{-7}$
1.0	0.9093	0.0018	$-1 \cdot 10^{-7}$	1.0206	0.0022	$-2 \cdot 10^{-7}$	1.1499	0.0028	$-3 \cdot 10^{-7}$
2.0	0.8727	0.0014	$-0 \cdot 10^{-7}$	1.0074	0.0019	$-1 \cdot 10^{-7}$	1.1698	0.0025	$-2 \cdot 10^{-7}$
3.0	0.8547	0.0012	$-0 \cdot 10^{-7}$	1.0019	0.0017	$-1 \cdot 10^{-7}$	1.1833	0.0023	$-2 \cdot 10^{-7}$
4.0	0.8460	0.0011	$-0 \cdot 10^{-7}$	0.999	0.0016	$-1 \cdot 10^{-7}$	1.1899	0.0022	$-2 \cdot 10^{-7}$
5.0	0.8410	0.0011	$-0 \cdot 10^{-7}$	0.998	0.0016	$-1 \cdot 10^{-7}$	1.1968	0.0021	$-2 \cdot 10^{-7}$

Table 8.4
Constants for daylight availability calculations

Month	E_{SC} apparent solar constant (lx)	(fc)	B: atomspheric extinction coefficient	C: E_{dH}/E_{DN} ratio
Jan	13,4783	12,526.3	0.142	0.058
Feb	13,4127	12,465.4	0.144	0.060
Mar	13,2628	12,326.0	0.154	0.068
Apr	13,0472	12,125.6	0.177	0.092
May	12,8316	11,925.3	0.194	0.118
Jun	12,6629	11,768.5	0.206	0.133
Jul	12,6067	11,716.2	0.207	0.137
Aug	12,6535	11,759.8	0.203	0.126
Sep	12,8129	11,907.9	0.182	0.096
Oct	13,0284	12,108.2	0.163	0.076
Nov	13,2534	12,317.3	0.151	0.064
Dec	13,4127	12,465.4	0.143	0.058

resource and on comparisons with similar measurements made at the Lawrence Berkeley Laboratory and NBS.

Illumination at the earth's surface is considered to be a function of illuminance at the outer edge of the atmosphere, known as the extraterrestrial illuminance constant E_{SC}. Direct normal illuminance E_{DN} is calculated from the equation

$$E_{DN} = (E_{SC})e^{-\alpha} \qquad \text{(lx)}, \tag{8.42}$$

where E_{SC} is the apparent extraterrestrial illuminance constant and α is the local atmospheric condition. Table 8.4 lists the apparent extraterrestrial illuminance constant for the 21st day of each month, based on calculations and measurements made at SERI. The atmospheric condition α is given by

$$\alpha = \tau \times m, \tag{8.43}$$

where τ is the total atmospheric optical depth and m is the relative air mass. The depth τ includes atmospheric attenuation of molecular scattering, aerosol/particulate scattering, and atmospheric turbidity. A monthly constant was used for τ (SERI, 1980), and values are listed in table 8.5. Values for m are obtained from (Buhl, et al., 1982):

Table 8.5
Average values of atmospheric optical depth

Month	Optical depth (τ)
Jan	0.142
Feb	0.144
Mar	0.156
Apr	0.180
May	0.196
Jun	0.205
Jul	0.207
Aug	0.201
Sep	0.177
Oct	0.160
Nov	0.149
Dec	0.142

$$m = [\cos\theta_0 + 0.15)93.885 - \theta_0)^{-1.253}]^{-1}, \qquad (8.44)$$

where θ_0 is the solar zenith angle.

Direct illuminance under a clear sky $E_{D,c}$ is determined from E_{DN} as

$$E_{D,c} = E_{DN}\cos\theta \qquad (\mathrm{lx}), \qquad (8.45)$$

where θ is the solar incident angle between surface and sun. For a horizontal surface, the clear-day direct illuminance $E_{DH,c}$ would be

$$E_{DH,c} = E_{DN}\cos\theta \qquad (\mathrm{lx}). \qquad (8.46)$$

Diffuse illuminance on a horizontal surface on a clear day $E_{dH,c}$ is

$$E_{dH,c} = (C)E_{DN}/(\mathrm{CN_d})^2 \qquad (\mathrm{lx}), \qquad (8.47)$$

where C is the ratio of diffuse luminance to direct normal illuminance from table 8.4, and $\mathrm{CN_d}$ is the sky clearness number for daylight. Recent work indicates that $\mathrm{CN_d}$ is approximately equal to the sky clearness number CN commonly used in radiation studies.

Ground-reflected illuminance under a clear sky $E_{g,c}$ is described by

$$E_{g,c} = E_{GH,c}F_g\rho_g \qquad (\mathrm{lx}), \qquad (8.48)$$

where F_g is the angle factor between surface and ground, and ρ is ground reflectivity. The term F_g is given by (ASHRAE, 1977)

Table 8.6
Equation (8.52) coefficients for a

Solar azimuth (deg)	a_0	a_1	a_2	a_3
Winter (Dec 21)				
0	1,797.577	837.968	−25.783	0.267
+/−45	635.005	775.031	−23.281	0.233
+/−90	−114.284	320.616	−7.113	0.058
+/−135	310.715	175.837	−3.606	0.026
180	228.572	116.277	−3.641	0.030
Summer (June 21)				
0	5,042.856	531.310	−8.881	0.042
+/−45	985.713	688.889	−11.827	0.057
+/−90	61.429	366.367	−6.029	0.033
+/−135	1,285.713	68.453	0.547	-8.33×10^{-3}
180	1,057.142	90.912	−0.666	2.77×10^{-3}

$$F_g = (1 - \cos \Sigma)/2, \tag{8.49}$$

where Σ is the angle between surface and ground. When the surface is horizontal, $\cos \Sigma - 1$ and $F_g = 0$. Therefore, total global illuminance on a horizontal surface is

$$E_{\mathrm{GH,c}} = E_{\mathrm{DH,c}} + E_{\mathrm{dH,c}} \qquad (\mathrm{lx}), \tag{8.50}$$

where the direct and diffuse components are described by equations (8.46) and (8.47), respectively.

Total (global) illuminance on a vertical surface under a clear sky is given by

$$E_{\mathrm{GV,c}} = E_{\mathrm{DV,c}} + E_{\mathrm{dV,c}} + E_{\mathrm{g,V,c}} \qquad (\mathrm{lx}), \tag{8.51}$$

where the direct, diffuse, and ground-reflected components are $E_{\mathrm{DV,c}}$, $E_{\mathrm{fV,c}}$, and $E_{\mathrm{gV,c}}$, respectively. Direct illuminance is described by equation (8.45), the ground component is described by equations (8.48) and (8.49) (with $F_g = 0.5$), and the diffuse component is described by

$$E_{\mathrm{dV,c}} = (a_0 + a_1 h + a_2 h + a_3 h^3) \qquad (\mathrm{lx}). \tag{8.52}$$

Values for the a coefficients are given in table 8.6 by solar azimuth and season. Straight-line interpolation is used to determine intermediate values for other months. Thus, global (total) illuminance on a vertical surface

Table 8.7
Equation (8.56) coefficients for C

C_1	C_2	C_3	C_4	C_5
797.8926	−49.0255	1.7956	−2.7548E-2	1.5221E-4

under a clear sky $E_{GV,c}$ is

$$E_{GV,c} = E_{DN}\cos\theta + a_0 + a_1\delta + a_2\delta^2 + a_3\delta^3 + (E_{dH,c} + E_{DN}\cos\theta)0.5\rho_g \quad \text{(lx)}. \tag{8.53}$$

Using a standard ground reflectivity of 0.2 (ASHRAE, 1977), equation (8.53) can be rewritten as

$$E_{GV,c} = E_{DN}\cos\theta + a_0 + a_1\delta + a_2\delta^2 + a_3\delta^3 + 0.1(E_{dH,c} + E_{DN}\cos\theta_0) \quad \text{(lx)}. \tag{8.54}$$

With this technique, calculations for vertical surfaces under clear skies are made for primary (N, E, S, W) and secondary (NE, SE, SW, NW) compass headings.

The general equation for illuminance on a surface of any orientation under an overcast sky $E_{g,o}$ is

$$E_{G,o} = E_{D,o} + E_{d,o} + E_{g,o} \quad \text{(lx)}, \tag{8.55}$$

where E_D, E_d, and E_g are direct, diffuse, and ground-reflected components, respectively. Overcast values for illuminance can be determined as follows:

$$E_{GH,o} = a(C_1\delta + C_2\delta^3 + C_4\delta^4 + C_5\delta^5) \quad \text{(lx)}, \tag{8.56}$$

where $E_{GH,o}$ is global (total) illuminance on a horizontal surface, a is the station variable, and C_0 is set to zero. Values for the C coefficients are given in table 8.7, and the station variable a is computed from

$$a = A_0 + A_1\delta + A_2\delta^2 + A_3\delta^3 + A_4\delta^4 + A_5\delta^5 \quad \text{(lx)}, \tag{8.57}$$

where $A_0 = 1$ for all cases, and the remaining coefficients are listed in table 8.8.

Similarly, total (global) illuminance on a vertical surface for an overcast day $E_{GV,o}$ is obtained from

$$E_{GV,o} = a(E_1\delta + E_2\delta^2 + E_3\delta^3 + E_4\delta^4 + E_5\delta^5) + 0.5\rho_g E_{GH,o} \quad \text{(lx)}, \tag{8.58}$$

Table 8.8
Equation (8.57) coefficients for A

Station elevation (m)	A_1	A_2	A_3	A_4	A_5
0	0	0	0	0	0
450	6.0079E-3	3.5319E-4	−2.1357E-5	3.2753E-7	−1.5548E-9
900	2.2954E-2	−6.5710E-4	4.8760E-6	2.5372E-8	−3.0384E-10
2,250	4.7556E-2	−2.1753E-3	4.1302E-5	−3.5649E-7	1.1598E-9
3,000	6.7569E-2	−3.5961E-3	7.8822E-5	−7.7968E-7	2.8774E-10

Table 8.9
Equation (8.58) coefficients for E

E_1	E_2	E_3	E_4	E_5
152.6658	−2.8152	1.1202E-1	−1.8129E-3	1.0911E-5

where a is the station variable derived in equation (8.57), and E_o is set to zero. Values for the E coefficient are given in table 8.9.

The Robbins-Hunter model was designed to use TMY and ETMY input data to localize the output for a specific city and surrounding area. The TMY control weather stations were used to define the general climate of large regions of the United States. In addition, TMY and ETMY data were used to localize the model for a given city. Table 8.10 lists the TMY control weather stations and their locations. Figure 8.7 illustrates the regions of the country for which each control weather station was used as a reference.

Using the Robbins-Hunter model, daylight availability data have been published for 80 cities throughout the United States. The data (shown in table 8.11) can be used in manual or computer-aided design of daylighting systems and are in a format similar to the American Society of Heating, Refrigeration, and Air Conditioning Engineers (ASHRAE) radiation data found in the *ASHRAE Handbook of Fundamentals* (ASHRAE, 1977). The Robbins-Hunter model is also being used in several microcomputer codes developed for daylighting design.

8.2.3 The Gillette Model

The Gillette model (Gillette, Pierpoint, and Treado, 1982) was developed at NBS with joint support from DOE and NFC. It is based on extensive measurements made in Gaithersburg, MD. The Gillette model uses solar

Table 8.10
SOLMET rehabilitated control stations

Map number	Station number	Station name	Location (Lat.	Long.)
1	23050	Albuquerque, NM	35°3′N	106°37′W
2	12832	Apalachicola, FL	29°44′N	85°2′W
3	24011	Bismarck, ND	46°46′N	100°45′W
4	94701	Boston, MA	42°22′N	71°2′W
5	12919	Brownsville, TX	25°54′N	97°26′W
6	93729	Cape Hatteras, NC	35°16′N	75°33′W
7	14607	Caribou, ME	46°52′N	68°1′W
8	13880	Charleston, SC	32°54′N	80°2′W
9	3945	Columbia, MO	38°49′N	92°13′W
10	13985	Dodge City, KS	37°46′N	99°58′W
11	23044	El Paso, TX	31°48′N	106°24′W
12	23154	Ely, NV	39°17′N	114°51′W
13	3927	Fort Worth, TX	32°50′N	97°3′W
14	93193	Fresno, CA	36°46′N	119°43′W
15	24143	Great Falls, MT	47°29′N	111°22′W
16	3937	Lake Charles, LA	30°7′N	93°13′W
17	14837	Madison, WI	43°8′N	89°20′W
18	24225	Medford, OR	42°22′N	122°52′W
19	12839	Miami, FL	25°28′N	80°16′W
20	13897	Nashville, TN	36°7′N	86°41′W
21	94918	North Omaha, NE	41°22′N	96°31′W
22	23183	Phoenix, AZ	33°26′N	112°1′W
23	23273	Santa Maria, CA	34°54′N	120°27′W
24	24233	Seattle/Tacoma, WA	47°27′N	122°18′W
25	93734	Washington, DC	38°57′N	77°27′W
26	94728	New York City, Central Park, NY	40°47′N	73°58′W

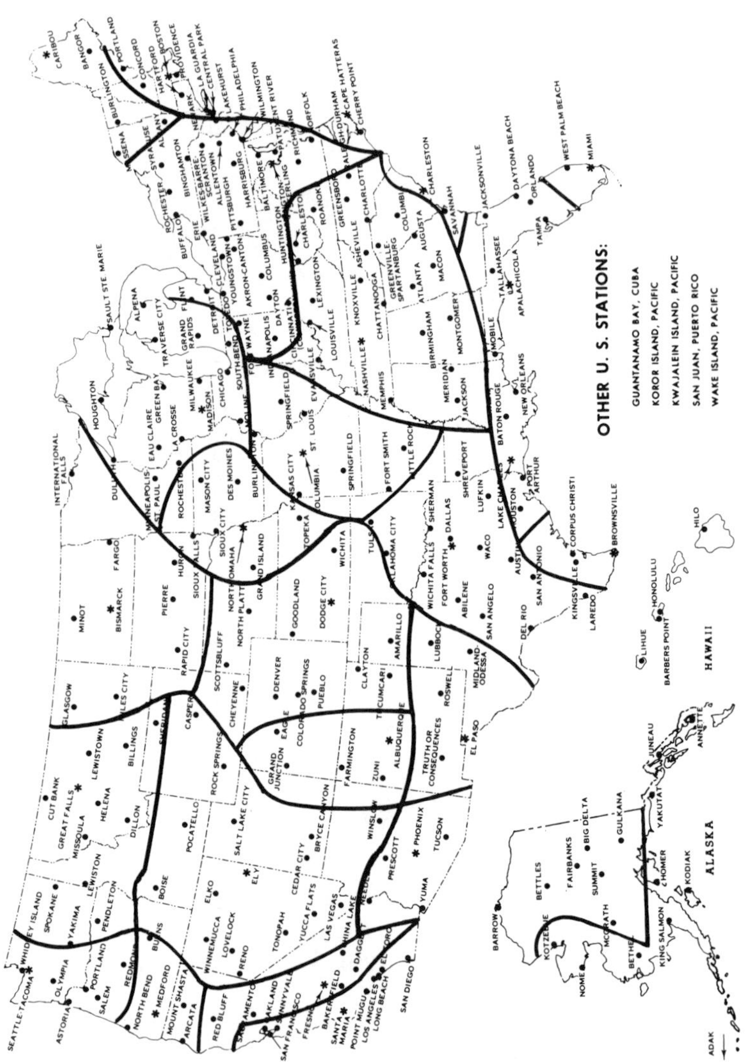

Figure 8.7
Typical Meteorological Year (TMY) regions. Source: Robbins (1986).

Table 8.11
Hourly global illuminance values for horizontal and vertical surfaces in Denver, CO

Time		Direct normal clear (lx)	North vertical global clear (lx)[a]	Northeast vertical global clear (lx)	East vertical global clear (lx)	Southeast vertical global clear (lx)	South vertical global clear (lx)	Vertical global overcast (lx)	Horizontal global clear (lx)	Horizontal diffuse clear (lx)	Horizontal global overcast (lx)	Time	
Jan	8[b]	46,110	2,770[c]	12,330	44,700	52,920	31,760	1,890	10,250	3,530	5,060	16	Jan
	9	72,910	5,130	6,470	59,630	82,570	61,210	3,680	27,130	5,590	8,000	15	
	10	83,130	6,530	8,090	51,120	88,280	78,980	5,070	40,490	6,370	10,240	14	
	11	87,480	7,020	8,640	32,850	82,350	89,310	5,990	48,940	6,700	11,990	13	
	12	88,730	6,760	8,340	10,190	69,030	93,120	6,320	51,820	6,800	12,650	12	
	13	87,480	7,020	7,450	9,150	49,870	89,310	5,990	48,940	6,700	11,990	11	
	14	83,130	6,530	6,040	7,450	28,390	78,980	5,070	40,490	6,370	10,240	10	
	15	72,910	5,130	4,220	5,170	7,610	61,210	3,680	27,130	5,590	8,000	9	
	16	46,110	2,770	2,240	2,510	3,090	31,760	1,890	10,250	3,530	5,060	8	
Feb	7	23,080	1,610	13,570	27,160	25,310	9,640	1,010	3,470	1,710	3,050	17	Feb
	8	69,730	5,100	27,030	70,500	76,150	40,680	3,240	23,200	5,160	7,340	16	
	9	85,490	7,330	16,060	73,670	93,140	63,780	5,130	41,810	6,330	10,350	15	
	10	92,310	8,660	10,610	60,350	94,290	79,840	6,690	56,230	6,840	13,430	14	
	11	95,400	9,130	10,950	38,500	85,560	89,650	7,740	65,320	7,070	15,710	13	
	12	96,310	8,760	10,440	12,580	69,990	93,480	8,110	68,420	7,130	16,530	12	
	13	95,400	9,130	9,400	11,310	48,500	89,650	7,740	65,320	7,070	15,710	11	
	14	92,310	8,660	7,850	9,400	24,890	79,840	6,690	56,230	6,840	13,430	10	
	15	85,490	7,330	6,160	7,010	8,830	63,780	5,130	41,810	6,330	10,350	9	
	16	69,730	5,100	4,170	4,350	5,450	40,680	3,240	23,200	5,160	7,340	8	
	17	23,080	1,610	1,440	1,390	1,590	9,640	1,010	3,470	1,710	3,050	7	

Table 8.11 (continued)

Time		Direct normal clear (lx)	North vertical global clear (lx)[a]	Northeast vertical global clear (lx)	East vertical global clear (lx)	Southeast vertical global clear (lx)	South vertical global clear (lx)	Vertical global overcast (lx)	Horizontal global clear (lx)	Horizontal diffuse clear (lx)	Horizontal global overcast (lx)	Time	
Mar	7	53,690	4,430	38,190	62,180	52,080	14,420	2,530	15,510	4,820	6,230	17	Mar
	8	77,100	7,650	41,180	81,460	79,150	35,530	4,750	36,560	6,930	9,690	16	
	9	86,680	9,600	28,350	78,290	88,780	54,290	6,800	54,910	7,790	13,680	15	
	10	91,330	10,680	12,920	63,550	87,880	68,910	8,550	69,010	8,200	17,490	14	
	11	93,550	11,090	13,030	40,970	78,260	78,040	9,750	77,880	8,400	20,000	13	
	12	94,220	10,580	12,290	14,510	62,360	82,000	10,200	80,910	8,460	20,860	12	
	13	93,550	11,090	11,070	13,010	40,440	78,040	9,750	77,880	8,400	20,000	11	
	14	91,330	10,680	9,400	10,920	16,880	68,910	8,550	69,010	8,200	17,490	10	
	15	86,680	9,600	8,120	8,650	10,640	54,290	6,800	54,910	7,790	13,680	9	
	16	77,100	7,650	6,170	6,090	7,450	35,530	4,750	36,560	6,930	9,690	8	
	17	53,690	4,430	3,470	3,220	3,800	14,420	2,530	15,510	4,820	6,230	7	
Apr	6	31,060	8,580	31,660	38,620	23,620	2,950	1,700	7,750	3,690	4,670	18	Apr
	7	66,360	7,580	56,690	77,240	57,210	7,960	4,030	29,480	7,880	8,540	17	
	8	80,600	10,400	54,660	86,090	74,100	24,840	6,310	50,430	9,580	12,630	16	
	9	87,550	12,140	41,270	80,250	80,170	41,090	8,500	68,430	10,400	17,390	15	
	10	91,240	13,100	22,550	65,310	78,430	54,680	10,450	82,230	10,840	21,320	14	
	11	93,100	13,400	15,320	43,280	69,280	64,080	12,040	90,900	11,060	24,160	13	
	12	93,660	12,780	14,330	16,440	53,240	68,650	12,750	93,850	11,130	25,480	12	
	13	93,100	13,400	12,920	14,660	30,750	64,080	12,040	90,900	11,060	24,160	11	
	14	91,240	13,100	11,560	12,540	14,660	54,680	10,450	82,230	10,840	21,320	10	

	15	87,550	12,140	10,330	10,320	12,130	41,090	8,500	68,430	10,400	17,390	9	
	16	80,600	10,400	8,290	7,840	9,070	24,840	6,310	50,430	9,580	12,630	8	
	17	66,360	7,580	5,600	5,100	5,620	7,960	4,030	29,480	7,880	8,540	7	
	18	31,060	8,580	2,590	2,270	2,420	2,950	a1,700	7,750	3,690	4,670	6	
May	5	2,260	2,540	6,250	5,830	2,210	1,070	450	430	350	1,460	19	May
	6	46,710	19,270	50,180	54,970	30,380	4,910	2,790	17,580	7,280	6,650	18	
	7	69,180	17,920	64,980	79,200	53,130	9,100	5,030	38,960	10,790	10,170	17	
	8	79,760	12,340	61,880	84,620	66,230	15,590	7,320	58,800	12,430	14,810	16	
	9	85,410	13,870	48,970	78,040	70,380	29,880	9,560	75,730	13,320	19,610	15	
	10	88,560	14,750	31,400	63,550	67,870	41,880	11,740	88,680	13,810	23,630	14	
	11	90,180	14,940	16,730	43,120	59,710	50,810	14,020	96,810	14,060	28,120	13	
	12	90,680	14,370	15,530	17,440	43,920	55,610	15,360	99,580	14,140	31,380	12	
	13	90,180	14,940	14,140	15,450	22,420	50,810	14,020	96,810	14,060	28,120	11	
	14	88,560	14,750	13,250	13,560	15,300	41,880	11,740	88,680	13,810	23,630	10	
	15	85,410	13,870	11,860	11,460	12,890	29,880	9,560	75,730	13,320	19,610	9	
	16	79,760	12,340	9,730	9,080	9,920	15,590	7,320	58,800	12,430	14,810	8	
	17	69,180	17,920	7,360	6,600	6,910	9,100	5,030	38,960	10,790	10,170	7	
	18	46,710	19,270	4,470	3,800	3,870	4,910	2,790	17,580	7,280	6,650	6	
	19	2,260	2,540	1,010	1,250	1,230	1,070	450	430	350	1,460	5	
Jun	5	9,200	7,230	14,700	13,370	5,340	1,770	950	2,320	1,650	2,900	19	Jun
	6	49,240	23,300	54,440	57,440	30,340	5,630	3,190	21,390	8,860	7,270	18	
	7	68,280	22,090	66,700	77,930	50,170	9,680	5,400	42,120	12,280	10,850	17	
	8	77,760	14,280	63,720	82,550	62,170	13,120	7,690	61,190	13,980	15,610	16	
	9	82,980	14,640	51,280	75,930	65,600	25,390	9,930	77,430	14,920	20,360	15	
	10	85,950	15,440	34,210	61,400	62,240	36,250	12,240	89,830	15,460	24,520	14	
	11	87,490	15,600	17,150	41,220	53,630	43,590	14,950	97,600	15,730	30,350	13	

Table 8.11 (continued)

Time	Direct normal clear (lx)	North vertical global clear (lx)[a]	Northeast vertical global clear (lx)	East vertical global clear (lx)	Southeast vertical global clear (lx)	South vertical global clear (lx)	Vertical global overcast (lx)	Horizontal global clear (lx)	Horizontal diffuse clear (lx)	Horizontal global overcast (lx)	Time
12	87,970	15,270	15,990	17,550	37,820	47,490	16,760	100,250	15,820	35,360	12
13	87,490	15,600	14,890	15,740	18,880	43,590	14,950	97,600	15,730	30,350	11
14	85,950	15,440	14,090	14,140	15,550	36,250	12,240	89,830	15,460	24,520	10
15	82,980	14,640	12,550	12,130	13,230	25,390	9,930	77,430	14,920	20,350	9
16	77,760	14,280	10,350	9,800	10,280	13,120	7,690	61,190	13,980	15,610	8
17	68,280	22,090	8,110	7,280	7,430	9,680	5,400	42,120	12,280	10,850	7
18	49,240	23,300	5,290	4,450	4,470	5,630	3,190	21,390	8,860	7,270	6
19	9,200	7,230	1,740	1,750	1,710	1,770	950	2,320	1,650	2,900	5
			Northwest vertical (lx)[d]	West vertical (lx)	Southwest vertical (lx)						

a. Illuminance category and orientation.
b. Solar time.
c. Hourly illuminance values in lux.
d. Orientation: use with solar time at right of table.

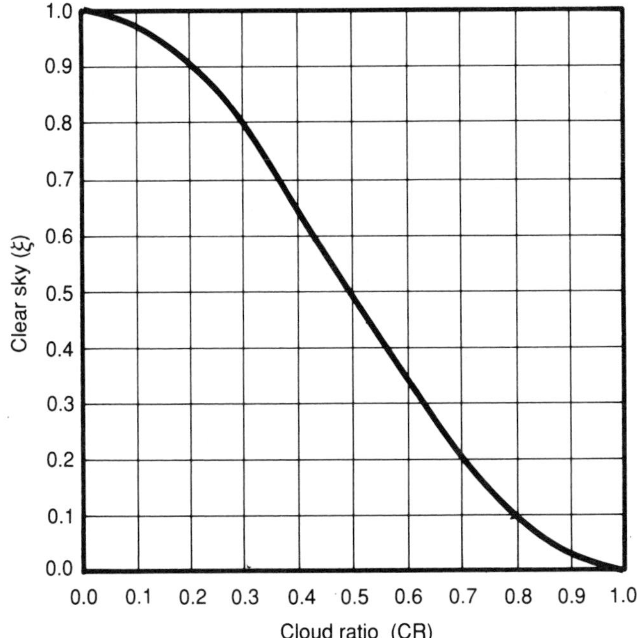

Figure 8.8
Cloud ratio curve established by the Gillette model. Source: Robbins (1986).

location (solar altitude, azimuth, and time), global horizontal radiation, and the ratio of diffuse to total horizontal radiation as the principal inputs. The solar location data are used to determine the zenith clear sky luminance using the CIE clear sky equation [equation (8.28)], the horizontal radiation is used to determine the horizontal illuminance and the overcast sky luminance, and the ratio of diffuse to total radiation (called a cloud ratio CR) is used to establish partly cloudy conditions. The curve used to obtain intermediate sky luminance values as a function of the cloud ratio is shown in figure 8.8 (Gillette, Pierpoint, and Treado, 1982).

The zenith luminance L_z used in the CIE clear sky equation is determined from the Dogniaux model equation (8.41) with the a_3h^3 term excluded, so that the equation is

$$L_{z,c} = a_0 + a_1 h^2. \tag{8.59}$$

The overcast equation is based upon the Moon and Spencer model adapted by CIE as shown in equation (8.29). If this equation is integrated across the entire sky dome to obtain horizontal luminance, the resultant for a luminance at angle θ is

$$L_z = (3/7\pi)E_{GH,c}(1 + 2\sin\theta). \tag{8.60}$$

A luminous efficacy approach is used to derive global and diffuse illuminance on a horizontal surface such that

$$E_{GH,c} = (93 + 18\,CR)I_{GH,c} \qquad (lx) \tag{8.61}$$

and

$$E_{dH,c} = 111 I_{GH,c} \qquad (lx), \tag{8.62}$$

In this model the mean extraterrestrial illuminance was derived for air mass zero by integrating the ASTM standard spectral irradiance curve of the solar constant while correcting for the CIE standard eye response. By using the method of numeric integration used by ASTM (1974), the mean extraterrestrial solar illuminance is found as follows:

$$E_{SC} = 683 \int_{380}^{760} V\lambda I\lambda\, d\lambda, \tag{8.63}$$

where V is the CIE standard photopic spectral eye response (ASTM, 1974), and I is the standard ASTM solar spectral irradiance averaged over the small wavelength band $d\lambda$.

The resultant is

$$E_{SC} = 127{,}500 \text{ lx}.$$

The amount of direct normal illuminance that passes through the atmosphere can be represented by

$$E_{DN} = E_x e^{-am}, \tag{8.64}$$

where a is the optical atmospheric extinction coefficient and n is the optical air mass.

The air mass is defined as (Gillette, Pierpoint, and Treado, 1984)

$$m = \frac{1}{\sin h}, \tag{8.65}$$

where h isthe solar altitude. Furthermore, if a single average value is used for the extinction coefficient, the direct normal illuminance becomes

$$E_{DN} = E_{SC} 1 + 0.033 \cos \frac{360^* - J}{365} \exp - a/\sin h. \qquad (8.66)$$

The sin h term corrects for the flux on the horizontal plane and corrects for the loss in intensity due to partly cloudy conditions.

The Gillette model was developed specifically for use in DOE 2. It is part of a daylighting subroutine added to the DOE model to include daylight system performance as part of the energy analysis capability of the DOE 2 code.

Neither the Robbins-Hunter model nor the Gillette model has been accepted by the IES/US to replace the current IES daylight availability model developed by Kimball and Hand.

In addition to the work being conducted at SERI and NBS, there is a significant body of daylighting research being performed at the Lawrence Berkeley Laboratory (LBL). In the Passive Group, Wayne Place is in charge of the daylighting research effort. His work primarily concerns analyzing the effect of daylighting on building energy performance; however, he has developed a model using luminous efficacy values to establish daylight availability from radiation data for use in BLAST. This daylight prediction model has not yet been published.

In the Windows and Daylighting Group, under the direction of Steve Selkowitz, a major research effort is being conducted in daylight measurements and resource models. Measurements have been taken since 1978 around San Francisco, CA. They have published several papers on this effort but have not yet published a resource model. All of the work being at LBL is under the direction of DOE through the Solar/Hybrid and Conservation offices of DOE.

8.2.4 European Models

There have been several general illuminance models published in Europe. The best known of these works are the models by Aydinli (1983) in West Germany, Petersen (1982) in Denmark, and Littlefair (1982) in England. The Aydinli and Petersen models are quite similar to the work conducted in the United States. The model by Littlefair, developed at BRE, differs in that it describes the global illuminance in terms of a direct component, a diffuse background component, a circumsolar diffuse component, and a ground reflected component such that

$$E_G = E_D + E_{d,bk} + E_{d,sol} + E_g \qquad (\text{lx}). \qquad (8.67)$$

The Littlefair (1982) model used data taken at Kew, England, and cloud cover data from Germany to predict average daylight availability on horizontal and vertical surfaces. The model was used to develop an average sky for use in daylighting design in Great Britain.

8.3 A Sunlight Probability Model for Daylighting Applications

The design and performance of many daylighting systems depend greatly on the duration and frequency of sunshine occurring over the year at the proposed location of the building. Sunlight as percent-sun is recorded at all of the over 600 weather stations throughout the United States. These data are provided by NOAA in the form of monthly and annual percent-sun data for each weather station. The NOAA percent-sun is based on the minutes of sunshine available from sunrise to sunset over the month. To consider the sun to be shining requires an intensity measurement greater than 200 W/m^2.

Architects and building lighting designers need sunlight probability data primarily to analyze the performance of a daylighting system, which varies in penetration characteristics depending on whether the day is clear or overcast. Another major use of sunlight probability data is to study the performance of direct-beam daylighting systems that use the direct component, or direct normal component, of daylight as an illuminant. Typically, such systems can only function to their full potential during the clear portions of a given time period.

Daylighting system performance can also be estimated using sunlight probability data that correspond to the typical operating schedule of the building. The operating schedule of the building for a month, or a year, is called the standard work month and standard work year, respectively. For convenience, a series of standard work months and years have been established for such system performance analysis using sunlight probability data.

As part of the analysis and establishment of rehabilitated hourly solar radiation surface meteorological (SOLMET) data, the Air Resources Laboratory (ARL) established a series of polynomial equations to determine solar radiation based on the solar zenith angle, the opaque cloudiness (sky cover), the minutes of sunshine, and a precipitation counter. Based on measured minutes of sunshine SS and solar radiation from a clear sky

SRCS, solar radiation SR can be expressed by the following regression equation (Cotton, 1979):

$$SR = SRCS\,(b_0 + b_1\,SS + b_2\,RN), \tag{8.68}$$

where RN is the precipitation counter data from the SOLMET stations. If both sunshine and opaque cloudiness observations are available, then

$$SR = SRCS\,(d_0 + d_1\,SS + d_2\,OPQ + d_3\,OPQ^2 + d_4\,OPQ^3 + d_5\,RN), \tag{8.69}$$

where OPQ is the opaque cloudiness value from the SOLMET station data.

Since SR data are available for each SOLMET station and since one can solve for minutes of sunshine by using either equation (8.68) or (8.69), it is possible to estimate sunlight probability for the SOLMET control stations and the derived SOLMET stations. The sunlight probability model presented in this section was developed as part of the DOE daylighting research effort at SERI based upon an analysis of the SOLMET radiation data in this manner (Robbins and Hunters, 1983).

ARL used 25 of the 26 SOLMET control cities to generate SR data for the country. Actual minutes of sunshine data SS were collected at only 15 of the 26 stations. Therefore, b and d coefficients, used in equations (8.68) and (8.69), are available for only 15 stations. For the SERI model, 11 control stations and accompanying regions for which no b values were generated were merged with regions for which these values had been computed.

Two criteria were used to establish which regions should be merged. These were (a) regions with similar patterns of percent-sun using NOAA weather data and (b) regions with similar patterns of cloudiness using the cloudiness index $\bar{K}_t$ data available at SERI (Knapp, Stoffel, and Whitaker, 1980).

A deviation of $\pm 7\frac{1}{2}\%$ for monthly values and $\pm 5\%$ for annual values was allowed in establishing the new regions because this deviation range already existed in the existing regions. The new regions are shown in figure 8.9. These regions have only been merged based on percent-sun and $\bar{K}_t$ patterns and may not be a suitable merging for other climate characteristics.

Using the calculated SS values, the sunlight probability SA for any hour the sun is shining is expressed as $SA_h = SS_h/SM_h$, where SM_h is the maximum number of minutes of sunlight in a given hour h. The SA for 0800 is

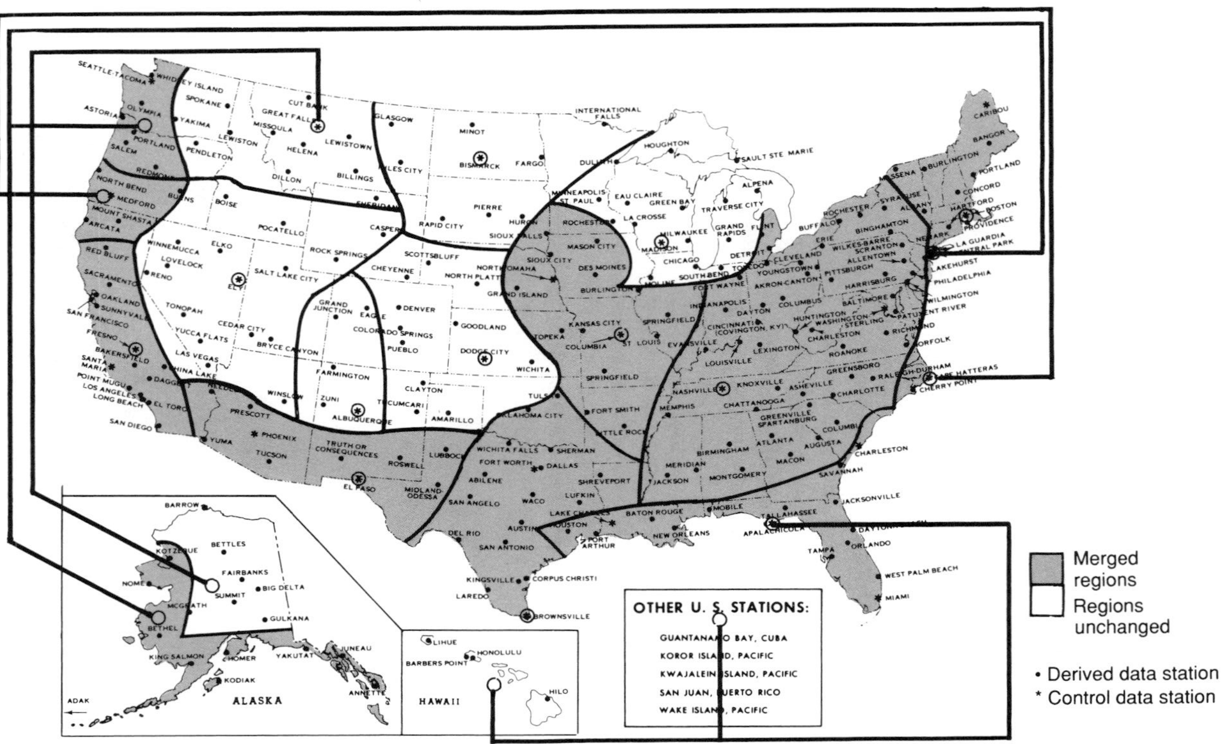

Figure 8.9
Merged TMY regions used in generating the SERI sunlight probability data. Source: Robbins (1986).

based on data for the hour from 0700 to 0800. The sunlight probability for a given month SA_m is expressed as

$$SA_m = \frac{\sum_{i=1}^{n} (SS_h/SM_h)}{n}, \tag{8.70}$$

where n is the number of days in the month the sun is shining during hour h. Two methods were used to apply the equation:

1. Determine the actual minutes of possible sunlight during the hours in which sunrise and sunset occur and use as the SM_h for those hours. This will be called the monthly average sunlight probability.
2. Use 60 min as the SM_h for all hours, including the sunrise and sunset hours. This will be called the monthly fraction of sunlight probability.

The monthly average data most closely matches the actual occurrence. The monthly fraction matches the NOAA methodology for accounting for the sunrise and sunset hours.

The annual fraction is the sunlight probability for a given hour over the entire year. The annual hours data represent the total number of probable hours of sunlight for a given time period for the year. Similarly, the monthly average hours represent the probable average number of hours of sunlight for a day of a given month. Data for Denver, CO, are illustrated in table 8.12.

The operating schedule of a building is used to determine the standard work month and year. For example, a building with an operating schedule from 0800 to 1700 h would use the same time frame to establish the standard work month and standard work year. For daylighting purpose 12 typical operating schedules were established to determine the sunlight probability by standard work month and by standard work year (see table 8.13).

The standard work year is used to establish the operating time period of the building; the actual operating schedule of the building varies with the number of days per week (month or year) the building is actually operating.

The sunlight probability for a given standard work year SA_s is expressed as

$$SA_s = \sum_{i=1}^{p} \frac{\sum_{i=1}^{m} SS_h/SM_h}{m/p}, \tag{8.71}$$

Table 8.12
Hourly sunlight probability data for Denver, CO

Hour, solar	Jan[a]	Feb	Mar	Apr	May	Jun	Jul	Aug	Sep	Oct	Nov	Dec	Annual fraction	Annual hours
0500–0600	.000	.000	.000	.339	.836	.782	.875	.376	.000	.000	.000	.000	.537	98.3
0600–0700	.000	.146	.792	.754	.804	.790	.894	.823	.849	.288	.000	.000	.672	187.5
0700–0800[b]	.621[c]	.627	.805	.761	.790	.810	.911	.824	.822	.853	.729	.752	.777[d]	283.5[e]
0800–0900	.732	.642	.811	.755	.804	.818	.913	.830	.831	.860	.749	.798	.797	290.8
0900–1000	.754	.678	.843	.758	.786	.821	.896	.836	.846	.864	.744	.772	.801	292.3
1000–1100	.775	.750	.872	.764	.792	.823	.877	.852	.871	.851	.764	.784	.815	297.6
1100–1200	.818	.711	.829	.766	.796	.837	.864	.791	.866	.847	.812	.810	.813	296.7
1200–1300	.839	.721	.789	.734	.758	.794	.879	.754	.859	.852	.776	.822	.799	291.5
1300–1400	.854	.739	.745	.707	.717	.769	.814	.752	.861	.860	.763	.845	.786	286.9
1400–1500	.829	.716	.726	.675	.684	.774	.741	.710	.821	.854	.748	.878	.764	278.7
1500–1600	.831	.715	.704	.649	.649	.715	.641	.711	.797	.865	.747	.861	.741	270.3
1600–1700	.736	.731	.693	.623	.627	.694	.662	.724	.748	.855	.760	.816	.723	263.8
1700–1800	.000	.150	.679	.639	.618	.712	.756	.718	.774	.276	.000	.000	.582	162.4
1800–1900	.000	.000	.000	.227	.618	.699	.796	.355	.000	.000	.000	.000	.448	82.0
Monthly average	.779[f]	.610	.774	.654	.734	.774	.823	.717	.829	.760	.759	.814		
Monthly fraction	.779[g]	.618	.735	.654	.669	.677	.720	.713	.746	.760	.721	.814		
Monthly average Hours of sunlight per day (h)	7.8[h]	7.3	9.3	9.2	10.3	10.8	11.5	10.0	9.9	9.1	7.6	8.1		

a. Month.
b. Hour, in solar time, during which sunlight probability was calculated.
c. Average fraction of probable sunlight per given year.
d. Annual fraction of sunlight for a given hour.
e. Total hours of sunlight for a given time period for a year.
f. Monthly average.
g. Monthly fraction.
h. Average hours of sunlight per day, per month.

Table 8.13
Standard work year for sunlight probability

0700–1600	0800–1600	0900–1600
0700–1700	0800–1700	0900–1700
0700–1800	0800–1800	0900–1800
0700–1900	0800–1900	0900–1900

where m is the number of days in the year (365) the sun is shining for each hour p in the standard work year chosen. Similar calculations can be made to determine the sunshine probability on a monthly basis.

An example of sunlight probability data by standard work month and year for Denver, CO, is shown in table 8.14.

An analysis of the sunlight probability data indicates an important difference between these data and the more common percent-sun provided by NOAA. The distinctions lead to the following conclusions:

1. A monthly percent-sun is a poor indicator of the available sunlight when used to design or analyze daylighting systems because it includes data outside the operating schedule of the building.

2. The NOAA methodology for accounting for the sunrise and sunset hours can be very misleading in terms of representing the actual amount of sunshine during that period. For example, using the NOAA technique, if the sun rises at 0730 and the sky is clear until 0800, the percent sun is 0.5 (based upon a 0700–0800 time frame). Using the actual minutes of possible maximum sunlight the sunlight probability would be 1.0 for the same period. Thus, the sunlight probability data summaries by month and standard work year indicate that many cities are sunnier than the NOAA data would indicate.

Using a method developed at SERI (Robbins, 1986), one can determine the fraction F of the standard work year that a given level of daylight can maintain a predetermined design illuminance. The weighted fraction DUF is determined from the following equation:

$$\mathrm{DUF_a} = \mathrm{DUF_c}(\mathrm{SA_s}) + \mathrm{DUF_o}(1 - \mathrm{SA_s}), \tag{8.72}$$

where

$\mathrm{DUF_c}$ = daylight utilization fraction of the specified standard work year that daylight can replace and supplement, and

Table 8.14
Sunlight probability data by standard work month and standard work year for Denver, CO

Standard work year	Jan[a]	Feb	Mar	Apr	May	Jun	Jul	Aug	Sep	Oct	Nov	Dec	Annual (SA_S)
0700–1600[b]	.705[c]	.644	.792	.732	.758	.795	.843	.788	.842	.799	.683	.732	.760[d]
0700–1700	.708	.652	.783	.722	.746	.786	.827	.782	.834	.804	.690	.740	.756
0700–1800	.649	.610	.774	.716	.735	.780	.821	.777	.829	.760	.632	.678	.730
0700–1900	.599	.564	.715	.678	.726	.774	.819	.743	.765	.702	.584	.626	.691
0800–1600	.784	.700	.792	.730	.753	.796	.837	.784	.842	.856	.759	.814	.787
0800–1700	.779	.703	.782	.719	.740	.786	.820	.778	.832	.856	.759	.814	.781
0800–1800	.708	.653	.773	.712	.729	.779	.814	.773	.827	.803	.690	.740	.750
0800–1900	.649	.598	.708	.672	.720	.772	.813	.736	.758	.736	.632	.678	.706
0900–1600	.804	.709	.790	.726	.748	.794	.828	.779	.844	.857	.763	.821	.789
0900–1700	.796	.711	.779	.715	.735	.783	.810	.773	.833	.856	.762	.821	.781
0900–1800	.717	.655	.769	.707	.723	.776	.804	.768	.827	.798	.686	.739	.747
0900–1900	.652	.596	.699	.663	.714	.769	.804	.728	.752	.726	.624	.672	.700

a. Month.
b. Standard work year operating hours.
c. Sunlight for a given month and standard work year.
d. Sunlight probability for a standard work year.

DUF_o = fraction of the specified standard work year that daylight can replace and supplement electric lighting.

The weighting of the clear F_c and overcast F_o values is determined by the sunlight probability data for the given standard work year. An hourly weighted value of F can be determined using a dynamic simulation technique and the hourly sunlight probability data. The weighted value F_T can then be used as part of a building energy thermal analysis based on the changes in electric lighting power budget and cooling load. This method of energy analysis depends on hourly, monthly, or annual sunlight probability data to determine clear and overcast conditions for daylight system performance analysis.

8.4 Conclusions

This chapter demonstrates that daylighting resource research and modeling has had a long, full history. Further, DOE has sponsored a significant body of research conducted in this area, which has led to the development of a general illuminance model as well as the sunlight probability model. These research efforts offer, for the first time, both computer and manual techniques to perform daylighting system design and analysis for a wide range of cities in different climate zones of the United States. These measurements need to be continued at the various locales where multiyear data have already been collected, but the measurement capability needs to be expanded to include more cities in different climates of the United States.

This does not mean there is no longer a need for resource assessment research for daylighting—just the opposite. As additional mathematical models of the daylight resource are proposed, more measured data are needed from a wide range of cities to test and verify the models. As the phenomenon is better understood, other research areas will surface; however, a few of the key issues that still need additional research include turbidity as it applies to the visible spectrum, direct normal illuminance, sky luminance, the partly cloudy sky, and more detailed illuminance models that can be applied to a wider number of cities around the country.

References

American Society for Testing and Materials (ASTM).1974. *Standard Solar Constant and Air Mass Zero Solar Spectral Irradiance Tables.* ASTM E490-73a, Philadelphia, PA: ASTM.

Ångström, A. K., and A. J. Drummond. 1972. "The global illuminance and its dependence on air mass and turbidity." *Appl. Optics* 1:455.

ASHRAE. 1977. *Handbook of Fundamentals.* New York: American Society of Heating, Refrigerating, and Air Conditioning Engineers.

Aydinli, S. Feb. 1983. "Availability of solar radiation and daylight." *Proceedings of the International Daylighting Conference.* Phoenix, AZ, pp. 15–19.

Basquin, O. H. March 1906. "Daylight illumination." *The Illuminating Engineer* 1.

Bennett, I. March 1962. "Natural daylight illumination ... its relation to insolation." *The Illuminating Engineer* 47:145–149.

Blackwell, M. J. 1954. "Five years continuous recording of total and diffuse solar radiation at Kew Observatory." *Meteorological Research Publication 895.* London: Meteorological Office.

Buhl, Marshall, et al. March 1982. *Thermodynamic Limits on Conversion of Solar Energy to Work or Stored Energy: Effects of Temperature, Intensity, and Atmospheric Conditions.* SERI/TP-233-1565. Golden, CO: Solar Energy Research Institute.

Chroscicki, W. 1971. "Calculation methods of determining the value of daylight intensity on the ground of photometrical and actinometrical measurements with unobstructed planes." *CIE: Proceeding of the CIE Meeting*, Barcelona, Spain, Session 71.24.

Commission Internationale de l'Eclairage (CIE). 1957. *Sun and Sky as Sources of Light: Definition 10-010, International Lighting Vocabulary: Vol. 1.* 2nd edition. Paris: CIE.

Commission Internationale de l'Eclairage (CIE). 1970. *Daylight: International Recommendations for the Calculation of Daylight 2.* CIE Publication 16 (E-3.2). Paris: CIE.

Cotton, G. F. 1979. "ARL models of global solar radiation." Appendix VI, *SOLMET User's Manual.* TD-9724. Asheville, NC: Air Resources Laboratory.

Crisp, V. H. C., and J. A. Lynes. Nov. 1979. "A model of daylight availability for daylighting design." Paper PD143/79. CIBS Lighting Division Conf.

Dogniaux, R. 1960. "Donnses meteorologiques concernat l'ensoleillement et l'eclairage natural." *Cah. Cent. Sci. Batim.* No. 44, Cahier 351, 24.

Dogniaux, R. 1968. "Distributions spectrales energetique et lumineuse de la luminare naturelle." *Lux* 366.

Dogniaux, R., et al. 1967. *The Availability of Daylight: Computer Procedure for Calculation of Irradiance and Illuminance as Parameters of Microclimate.* Draft report prepared for CIE TC-4.2.

Elvegard, E., and G. Sjostedt. 1940. "The calculation of illumination from sun and sky." *The Illuminating Engineer* 35:333–342.

Gillette, G., W. Pierpoint, and S. Treado. 1982. *A General Illuminance Model for Daylight Availability.* Washington, DC: National Bureau of Standards.

Gillette, G., W. Pierpoint, and S. Treado. July 1984. "A general illuminance model for daylight availability." *Journal of the Illuminating Engineering Society.*

Hunter, K. C., C. L. Robbins, and K. S. Harr. 1983. "Calibration laboratory for the calibration of photometric sensors." *Proceedings of the International Daylighting Conference.* Phoenix, AZ, pp. 243–244.

Illuminating Engineering Society (IES). 1981 *Illuminating Engineering Society Handbook: Reference Volume.* New York: IES.

Kendrick, J. D. 1984. Personal correspondence. Australian National Committee. CIE TC4.2.

Kimball, H. H. 1919. "Variations in the total and luminous solar radiation with geographical position in the United States." *U.S. Weather Review* 47:769.

Kimball, H. H., and I. F. Hand. 1921. "Sky brightness and daylight illuminance measurement." *U.S. Weather Review* 48:481. Also *Trans. Illum. Eng. Soc.* XVI:255.

Kimball, H. H., et al. 1923. "Daylight illumination on horizontal, vertical and sloped surfaces." *U.S. Weather Review* 50:615. Also *Tran. Illum. Eng. Soc.* XVIII:434.

Kittler, R. 1962. "Rozlonzenie Jasu no Bezoblacnej Oblohe Podl'a Meranf a Teoretickych Vztahov (Measured sky luminance distributions and comparisons with theoretical values)." *Bull. Met. Tchecosl.* 15(34).

Knapp, C., T. Stoffel, and S. Whitaker. 1980. *Insolation Data Manual.* SERI/SP-755-789. Golden, CO: Solar Energy Research Institute.

Krochmann, J., and M. Seidl. 1974. "Quantitative data on daylight for illuminating engineering." *Lighting Research and Technology* 6(3):65–171.

Littlefair, P. April 1982. "Designing for daylight availability using the BRE average sky." *Proceedings of the CIBS National Lighting Conf.*, Univ. of Warwick, pp. 40–62.

Lynes, J. 1969. *Principles of Natural Lighting.* London: Elsevier.

Matsuura, K. 1984. Personal correspondence. Kyoto University.

McCluney, R. 1983. "Sensors for daylight research." *Proceedings of the CIE Daylight Availability Conference.* May 19–20, 1983, Berkeley, CA.

McDermott, L. H., and G. W. Gordon-Smith. 1951. "Daylight illumination recordings at Teddington." *Proceedings of the Building Research Congress.* (London Congress) Division 3, Part III.

Moon, P., and D. E. Spencer, 1942. "Illuminance from a non-uniform sky." *The Illuminating Engineer* 37:707.

Nakamura, H., and M. Oki. 1979. "Study on the statistic estimation of the horizontal illuminance from unobstructed sky." *Journal of the Illuminating Engineering Society*, Japan.

National Building Research Institute (NBRI). 1984. Personal correspondence. Pretoria, South Africa: NBRI.

National Climatic Center (NCC). 1979. *SOLMET, Volume 2—Final Report: Hourly Solar Radiation-Surface Meteorological Observations.* TD-9724, Asheville, NC: NCC.

Petersen, E. Oct. 1982. *Solstraling og Dagslynmalt og Beregnet.* Report No. 34. Lyngby, Denmark: Danish Illuminating Engineering Laboratory.

Robbins, C. L. 1986. *Daylighting Design and Analysis.* New York: Van Nostrand Reinhold Co., Inc.

Robbins, C. L., and K. C. Hunter. May 1983a. *A Model for Illuminance on Horizontal and Vertical Surfaces.* SERI/TR-254-1703. Golden, CO: Solar Energy Research Institute.

Robbins, C. L., and K. C. Hunter. Aug. 1983b. *Sunlight Probability Data for Select Cities in the United States.* SERI/TR-254-1687. Golden, CO: Solar Energy Research Institute.

Robbins, C. L., K. C. Hunter, and T. Cannon. 1984. "Mapping sky and surface luminance distribution using a flux mapper." *Energy and Buildings* 6:247–252.

Roswell, W. A., and R. Loudermilk. July 1968. *An Oceanographic Illuminometer for Light Penetration and Reflection Studies.* Scripps Institute of Oceanography. Ref. 68-11.

Solar Energy Research Institute (SERI). 1980. *SIM (Solar Irradiance Model) Users Manual.* SERI/TR-06022-1. Golden, CO: SERI.

Walsh, J. W. T. 1961. *The Science of Daylight.* London: MacDonald and Co., Ltd.

Contributors

Raymond J. Bahm

Raymond J. Bahm has been involved with solar energy research since 1976. Mr. Bahm has been professionally affiliated with the University of New Mexico Bureau of Engineering Research for much of that time.

Mr. Bahm's major accomplishments in solar energy research include review and summary of calibration methods for solar radiation measuring instruments, review and evaluation of existing historical solar radiation data bases for the United States, and refinement and application of a technique for preparing maps of solar radiation availability from earth satellite photography.

Richard Bird

Richard Bird has been involved with solar energy research since 1978. Dr. Bird was professionally affiliated with The Optical Society of America; The Naval Weapon Center, China Lake; and the Air Force Geophysics Laboratory.

Dr. Bird's major accomplishments include the development of broadband and spectral solar radiation models and internationally recognized spectral solar radiation standards. He has done theoretical and experimental research on effects of various atmospheric constituents on optical systems, particularly the effects of atmospheric aerosols. He has written 40 reports in the field of radiative transfer and light scattering phenomena.

Kirby Hanson

Kirby Hanson has been involved with solar energy research since 1957. Dr. Hanson is professionally affiliated with the National Oceanic and Atmospheric Administration in Miami, Florida.

Dr. Hanson serves on many national and international committees, including the US-USSR Joint Committee on Cooperation in the Field of Environmental Protection, where he chairs the Project on Effects of Pollution of the Atmosphere on Climate.

Roland L. Hulstrom

Roland L. Hulstrom has been involved with solar energy resource assessment since 1971. Mr. Hulstrom has been professionally affiliated with Martin Maretta Aerospace, Denver Division, from 1967 to 1977, and with the Solar Energy Research Institute since 1977.

Mr. Hulstrom's major accomplishments include production of the first comprehensive solar radiation energy resource atlas for the United States and the invention of the first atmospheric optical calibration system (AOCS, U.S. Palent No. 4,779,980) for solar energy applications. He received the National Weather Service's Inaugural Forecast of Solar Energy Distribution over the U.S. Award in 1981 and a certificate of recognition for the creative development of technology from The National Aeronautics and Space Administration in 1980.